FIGHTING FOR SPACE

FIGHTING FOR SPACE

TWO PILOTS AND THEIR HISTORIC BATTLE FOR FEMALE SPACEFLIGHT

AMY SHIRA TEITEL

GRAND CENTRAL PUBLISHING

NEW YORK BOSTON

Cover design by Philip Pascuzzo. Cover photo of red-haired woman by Bettman/Getty Images.
Cover photo of blond woman by Ralph Crane/The LIFE Picture Collection/Getty Images.

Grand Central Publishing
Hachette Book Group
1290 Avenue of the Americas, New York, NY 10104
grandcentralpublishing.com
twitter.com/grandcentralpub

Originally published in hardcover and ebook in February 2020
First trade paperback edition: February 2021

Grand Central Publishing is a division of Hachette Book Group, Inc. The Grand Central Publishing name and logo is a trademark of Hachette Book Group, Inc.

Material from *John Glenn: A Memoir* by John Glenn with Nick Taylor, copyright © 1999 by John Glenn, used by permission of the author's estate.

Print book interior design by Thomas Louie

Library of Congress Cataloging-in-Publication Data

Names: Teitel, Amy Shira, author.
Title: Fighting for space : two pilots and their historic battle for female spaceflight / Amy Shira Teitel.
Description: First edition. | New York : Grand Central Publishing, 2020. | Includes bibliographical references.
Identifiers: LCCN 2019041812 | ISBN 9781538716045 (hardcover) | ISBN 9781538716038 (ebook)
Subjects: LCSH: Women air pilots--United States--Biography--Juvenile literature. | Air pilots--United States--Biography--Juvenile literature. | Cochran, Jacqueline. | Cobb, Jerrie. | Aeronautics--United States--History--20th century.
Classification: LCC TL539 .T35 2020 | DDC 629.13092/520973--dc23
LC record available at https://lccn.loc.gov/2019041812

ISBNs: 978-1-5387-1605-2 (trade paperback), 978-1-5387-1603-8 (ebook)

Printed in the United States of America

LSC-C

Printing 1, 2020

For women everywhere—past, present, and future—who stand their ground when they know they're right. You inspire more people than you realize.

And for Jackie. It's time the world remembers you.

PREFACE

The first half of the twentieth century was an interesting time for America.

Our story starts in 1912, the year the *Titanic* sank while carrying more than 2,200 people from England to New York. Steamships and other boats were the only way people could cover vast oceanic distances. There were no transatlantic flights; airplanes were fodder for the rich and daring to soar moderately high above a gathered crowd for minutes at a time.

The First World War changed that. Aerial officers flew ahead of soldiers on foot to get the lay of the land, then they carried guns as protection, and finally, those guns were fixed to the front of the planes. A pilot could aim his gun by aiming his aircraft, relying on an interrupter gear to ensure the propeller blades never got in the way of automatically firing bullets. Wood and fabric fuselages soon gave way to all-metal vehicles, and by the end of the war pilots routinely engaged in swirling aerial dogfights. For civilians, these stories imbued flying with unparalleled excitement and a feeling the future was right over the horizon.

As these high-performance planes became available to private pilots, the Everyman took a step closer to the sky. Aviators dazzled audiences with aerial displays. Sometimes, lucky onlookers could even take a ride in a plane, either at a county fair or at a small airport that offered rides for a small fee. Flying held allure. It was romantic, thrilling, dangerous, and the pilots who flew were like no one else on Earth.

Coincident with the end of the First World War was the first wave of feminism; suffragettes fought for equality and won women the right to vote in 1920. Women had freedoms they'd never known, a change in status that saw echoes in fashion and culture. Women eschewed restrictive clothing in favor of skirts and trousers that offered physical freedom. In this newly emancipated climate, women took to the sky, though this wasn't without challenges. Few men would consent to teach women to fly, and even fewer would take on African American students. When Bessie Coleman decided she wanted to earn her pilot's license, she had to train in France, where she met no racial barrier in getting a license. When she returned to America in 1921 at the age of twenty-nine, she was greeted with press coverage and enthusiastic audiences. Female aviators held a different appeal; if men flying was exciting, women flying was a novelty like no other. The sky was starting to open to women.

Regardless of gender, flying had the power to turn everyday people into heroes, and no one embodied this phenomenon as completely as Charles Lindbergh. After he made the first solo flight across the Atlantic, he went from unknown air mail pilot to celebrity overnight. When Amelia Earhart became the first woman to cross the Atlantic in a plane, her fame similarly skyrocketed even though she was merely a passenger on that flight. Regardless, women watching these feats were inspired to follow suit, and by 1929 there were ninety-nine licensed women pilots of all backgrounds in the United States. That same year, Amelia was a founding member of an all-female flying society called the Ninety-Nines, which is still active today.

As the United States sank into the Depression in the 1930s, women working became increasingly commonplace. Wives, mothers, and daughters did their part to help their families escape the poverty that touched nearly every corner of the country. For the women who could afford luxury activities in the period, flying became more popular; by the end of the decade, there were more than 500 licensed women pilots in the country. Women's independent streak was mirrored in pop culture. Fictional heroines, even in love stories, were often defined by their jobs and passions as much as their womanhood. Take Rosalind Russell (a friend of our heroine Jackie, whom you'll meet in a moment) in *His Girl Friday* (1940). Rosalind is prepared to quit her job as a reporter to marry Ralph Bellamy and enjoy the domesticity that comes with married life. But her boss and ex-husband Cary Grant (another acquaintance of Jackie's) isn't prepared to let her go. So he woos her, not with the promise of a big house and children but with a story of an escaped accused murderer she can't resist covering. Rosalind isn't just a woman, she's a hard-hitting reporter, and that's what Cary loves most about her.

The Second World War brought new opportunities for women. As men went off to fight, women took their places in factories earning the same pay as their male counterparts. Military programs gave women more immediate ways to serve their country. The Army Air Force had the WASPs (Women's Auxiliary Service Pilots), the Navy had the WAVES (Women Accepted for Volunteer Emergency Service), and the Army had its WACs (Women's Army Corps). For the first time since the Civil War, American women could earn military recognition. They took pride in their work, and almost all hoped their independence would persist after the war's end.

But the same government that made Rosie the Riveter a wartime icon for working women put her in the kitchen in the 1950s. After decades of upheaval from the Depression and Second World War, postwar America focused on family values, and government propaganda campaigns followed suit. God was added to currency and the Pledge of Allegiance. A woman's role was now that of wife and mother with ideals of "womanhood" wrapped up in her biology. The postwar woman married younger and had more children. Her home was her castle, equipped with appliances to free her from the drudgery of cleaning, television for entertainment, and the PTA as a social outlet. Fictional women changed, too. Marilyn Monroe and Jane Russell in *Gentlemen Prefer Blondes* (1953) want nothing more and are consequently defined by their search to find rich husbands who can shower them in diamonds. "Career woman" became a dirty word; it was wrong for a woman to express desires outside her biology.

Postwar America was also defined by a widespread fear of Communism. Political leaders, most notably Republican senator Joe McCarthy of Wisconsin, told the public that they should be fearful of subversive Communist influence in their lives. Communists could be lurking anywhere. They could be teachers, artists, or journalists using their platforms to advance the spread of Communism throughout the world, and the danger lay in its challenge to the American way of life. In this climate, women choosing career over family were sometimes viewed as a similar threat to American family values. To this end, in the late 1950s and early 1960s, a woman declaring herself a feminist was, to some, akin to declaring herself a Communist. By and large, women were suddenly marginalized by society in a way they hadn't been for decades, and minority women faced renewed discrimination on account of both their gender and their race.

This return of women to the home coincided with the postwar technological boon that saw the advent of jet planes, supersonic flight, rocket-powered planes, and the first satellites. Spaceflight was on the horizon at the same time that women were barred from a number of the opportunities they'd had for decades.

It was the women who grew up in the prewar years, the generation who had never known a world where so much was denied them, who took on this new world order and forced the change that became the second wave of feminism. In 1963, the year our story ends, Betty Friedan's *The Feminine Mystique* was published, which shed light on the epidemic of ennui affecting housewives throughout America. Though segregation was already being dismantled, the Civil Rights Act of 1964 guaranteed equal employment and desegregated public facilities. The second wave of feminism characterized by the women's liberation movement of the late 1960s truly began to allow all women to define themselves as more than their biology.

This is the changing world in which our story takes place.

Born in 1906, Jackie Cochran entered her adult years right around the time women started to enjoy post-suffrage freedoms. She learned to fly not long after Amelia Earhart was elected president of the Ninety-Nines and had the fortune to cement her reputation as a pilot in the 1930s when air races made pilots celebrities. It helped, too, that she was married to one of the richest men in the country; Floyd Odlum could buy her the cutting-edge planes that won her races. She was perfectly positioned with both a skill level and celebrity to lead the WASPs in the Second World War. Postwar, nothing was going to send Jackie back into the kitchen. She became the first woman to fly faster than sound in 1953 and ultimately secured more records than any pilot in the twentieth century, male or female, full stop. She also ran a luxury cosmetics company, was personal friends with multiple presidents, and was so determined to hide her true parentage that she invented the story that she was raised an orphan. She is, in short, an incredibly complex woman.

Jerrie Cobb, twenty-five years Jackie's junior, comes into our story and America during the Depression. She never knew a world where women didn't have the vote. When she was a preteen, her father briefly owned an airplane, so flying was accessible. Her formative teenage years were thus spent pursuing flight licenses at a time when it was neither common nor unheard of for a woman to aspire to a flying career. In the 1950s, however, she came face to face with the biases against women flyers. She fought tooth and nail to earn her living as a pilot—it wasn't glamorous or comfortable, but it was the life she loved.

Both Jackie and Jerrie, neither of whom let social norms compromise their love of flight, wanted to move into the space age with America. By the time NASA began recruiting astronauts in 1959, Jackie was too old. Jerrie, however, was the perfect age to take on what she felt was injustice barring women from joining the astronaut corps. She took the same medical tests as NASA's first group of astronauts, albeit through a private program completely unaffiliated with the space agency, and believed this qualified her to fly in space. After a handful of other female pilots followed in her footsteps, she hoped the group might convince NASA to start a female astronaut program so she could get a mission.

When we talk about women wanting to fly in space in the 1960s, it's easy to look at the story from our modern standpoint and cry gender discrimination. But we can't ignore the social and political context in which our story takes place, nor can we overlook the other players who impact what happened. We have Randy Lovelace, a pioneer in aviation medicine and the physician behind the medical testing program, who had more important professional interests he had to protect. We also must get to know Lyndon Johnson, the senator whose lingering depression after being elected vice president impacted the way he dealt with issues crossing his desk. We also have to look at the broader context of what was happening in space. The early 1960s was the time when President Kennedy challenged America to land a man on the Moon within the decade. Managing that Herculean task fell to Jim Webb, NASA's administrator, who didn't want to complicate things by adding women to the space agency.

The history underlying this story and the varying perspectives is vital, and there's a lot of history layered into the narrative because the story needs context. In the interest of exploring every avenue, I spent the better part of four years researching not only these women and the players who shaped their stories, but the media's portrayal of women pilots, domestic and international politics, as well as the evolution of flight, spaceflight, and women's rights. In addition to reading dozens of books, I recovered thousands of pages—letters, memos, records of invitations, photographs, interview transcripts, and even handwritten notes—from the Roosevelt, Eisenhower, Kennedy, and Johnson presidential libraries, the NASA archives, and other collections both public and private throughout the country. I also interviewed a handful of experts, among them Gene Nora Jessen, one of the female pilots whom you'll meet in these pages. Sadly, most of the people in this story have passed.

But I endeavored to bring them to life in these pages. I want you to marvel at Jackie's daring, get swept up in Jerrie's romance, feel the excitement of a record-breaking flight, and experience the excruciating frustration of a dream slipping away. All the women in this story are like the screen heroines of the 1930s and 1940s. They are not defined by "womanhood." They are pilots and identified as such. They are flyers as well as wives, mothers, and businesswomen. They are incredible, multifaceted people trying to eke out their place in aviation and space while navigating love, family, and friendship, always without compromising who they are.

MAIN FIGURES

Jackie Cochran, née Bessie Pittman (1906–1980)

After learning to fly in 1932, Jackie became the foremost pilot of the twentieth century, holding more speed, altitude, and distance records than any other pilot in the world. She also won multiple aviation trophies, owned the Jacqueline Cochran Cosmetics company, and counted notable figures like Dwight Eisenhower, Lyndon Johnson, Jimmy Doolittle, and Amelia Earhart among her closest friends.

Floyd Odlum (1892–1976)

Floyd was a lawyer, industrialist, and businessman described as perhaps the only man to profit off the Great Depression. He and Jackie met in 1932 and married four years later. Throughout their lives together, Floyd facilitated giving Jackie everything she could ever want.

Amelia Earhart (1897–1937)

One of the most notable pilots of her era, Amelia pursued distance flights while Jackie pursued speed records in the 1930s. The two were such close friends that Amelia was the only guest at the Indio Ranch to sleep in Jackie's own bed when she was there alone.

Lyndon Baines Johnson (1908–1973)

A congressman when he and Jackie met in 1940, the two became fast friends and remained close as Lyndon became a senator, then vice president, and finally president.

Randy Lovelace (1907–1965)

Trained at the Harvard Medical School before doing his residencies at New York's Bellevue Hospital and the Mayo Clinic in Rochester, Minnesota, Randy gained an interest in aviation medicine working at the School of Aviation Medicine at Randolph Field, Texas. From there, he became a leading pioneer in aviation medicine. He helped design the oxygen mask Jackie used in the 1938 Bendix for which he won the 1939 Collier Trophy, and in 1959 NASA asked him to run the medical portion of the Mercury astronauts' selection process at his own Lovelace Clinic in New Mexico.

Don Flickinger (1907–1997)

Like his colleague Randy Lovelace, Don became acquainted with aviation medicine at the School of Aviation Medicine at Randolph Field, Texas. The pair worked together for many years, including the Mercury astronauts' medical testing.

Jerrie Cobb (1931–2019)

Jerrie discovered flying before she was a teenager, and knew she wanted to spend her life in the clouds. She earned licenses as quickly as she could and managed to find enough flying jobs to become a career pilot. In 1960, she took the same medical tests given to the Mercury astronauts, launching her on her personal quest to become the first woman in space.

Jim Webb (1906–1992)

Jim was a trained lawyer and worked as a congressional assistant, served with the Marine Corps, and held the position of director of the Bureau of the Budget and undersecretary of state under President Truman before taking the role of NASA administrator under President Kennedy.

Hugh Latimer Dryden (1898–1965)

Hugh spent his career on the cutting edge of aviation and aeronautics. He was among the pioneers of supersonic flight before working on missile development during the Second World War. Post-war, he joined the National Advisory Committee for Aeronautics and eventually became its first ever director. He led the NACA into supersonic and hypersonic flight programs before taking the role of deputy administrator when the NACA morphed into NASA.

Dwight "Ike" Eisenhower (1890–1969)

During the Second World War, Ike was a five-star general in the United States Army (as of 2019 he remains one of only five men to ever hold this rank) and supreme commander of the Allied Expeditionary Force in Europe; in this capacity he played a major role in orchestrating the D-Day landings at Normandy in 1944. Between 1953 and 1961, he served as the thirty-fourth president of the United States.

Robert Pirie (1905–1990)

A naval aviator himself, Robert served as deputy chief of naval air operations from 1958 to 1962, the period in which the women were hoping to take additional astronaut tests at the Naval Air Station in Pensacola.

Henry Harley "Hap" Arnold (1886–1950)

Hap was a general of the army and then chief of the Army Air Force during the Second World War. The air force became a separate service branch in 1947, and in 1949 Hap was named its first general and awarded a five-star rank by the US Congress.

Chuck Yeager (1923–2020)

Chuck was a United States Air Force officer who moved into experimental flight testing after the Second World War. In 1947, he became the first person to fly faster than the speed of sound in level flight. Known as one of the most intuitive pilots, he flew a variety of high-speed aircraft in his career, as well as instructing other test pilots, Jackie being one of his many students.

John Glenn (1921–2016)

After dropping out of college in the wake of the Japanese attack on Pearl Harbor, John became a marine aviator who served in both the Second World War and the Korean War. He eventually moved into flight testing, training at the Naval Air Test Center in Patuxent River, Maryland, before working for the Fighter Design Branch of the Navy Bureau of Aeronautics. He had about 9,000 hours of flying time, 3,000 of which were in jets, when he was selected as a Mercury astronaut in 1959.

Wally Funk (1940–)

Flying professionally since 1957, Wally had a stunning career. In addition to working as a flight instructor, Wally notably served as the first female air safety investigator for the National Transportation Safety Board and served as the first female Federal Aviation Administration inspector. She also applied for NASA's astronaut corps twice once the agency started accepting women.

Jerri Sloan (1929–2013)

Jerri learned to fly as a teenager and enjoyed a lifelong career as a pilot. Notably, she worked as a research pilot flying North American B-25s for Texas Instruments and served as vice president to both Air Freighters International and Air Services, Inc.

Bernice "B" Steadman, née Trimble (1925–2015)

B Steadman was an American aviator, businesswoman, and frequent face in the air race circuit in the mid-century. She founded the Trimble Aviation School before she was married and thus became one of the few women in the country to own and operate her own school. She also served as president of the Ninety-Nines.

Myrtle "K" Cagle (1925–)

When K earned her wings at the age of fourteen, she was the youngest pilot in her home state of North Carolina. This started her on a career path that saw her earning her commercial and multi-engine licenses and working as a flight instructor, flight instrument instructor, and ground instructor. She also ran her own charter plane service near Raleigh.

Gene Nora Jessen, née Stumbough (1937–)

Gene Nora quit her job as a flight instructor at the University of Oklahoma for the Pensacola tests. After the tests were cancelled, she joined a sales demonstration team of Beechcraft aircraft. As one of the Three Musketeers, she flew in formation over the contiguous forty-eight states before setting up her own Beechcraft dealership in Boise, Idaho.

Rhea Woltman, née Hurrle (1927–)

Rhea earned her commercial pilot's license with single- and multi-engine ratings and was even a licensed seaplane pilot and held multiple instructor ratings. She retired from flying midway through the 1960s.

Irene Leverton (1927–2017)

Irene held multiple licenses and taught hopeful pilots to fly gliders, single-engine planes, and multi-engine planes. She even tested pilots for the FAA. In the early 2000s, she was presented with the Civil Air Patrol's Commander's Commendation Award and its Meritorious Service Award, as well as the FAA's Master Pilot Award. She's been inducted into multiple aviation halls of fame.

Jean Hixson (1922–1984)

Jean joined the WASPs late in the war, graduating in Class 44-6. Postwar, she worked as a flight instructor while earning her degree in education, which launched her second career as a schoolteacher. She continued teaching and flying until her retirement a little over a year before her death.

Jan Dietrich (1926–2008)

One of two girls in her aviation class in high school, Jan parlayed her love of flight into a full-time career. She worked as a flight instructor as well as a corporate pilot for a construction company in California, and earned advanced licenses including her airline transport pilot license.

Marion Dietrich (1926–1974)

The other of two girls in her aviation class in high school, Marion's love of flying led to a career in journalism, though she flew charter flights and ferried aircraft in her spare time.

Janey Briggs Hart (1921–2015)

After earning her pilot's license during the Second World War, Janey added helicopter pilot to her résumé, becoming the first licensed woman in her home state of Michigan. Outside of flying, Janey had a keen interest in politics; she was a liberal democrat along with her husband, Senator Philip Hart. She was also an avid sailor and mother of eight.

Sarah Ratley, née Gorelick (1933–)

Sarah learned to fly before she learned to drive and earned a slew of licenses including her commercial pilot's license, multi-engine land and water instrument rating, and helicopter pilot license. She earned a bachelor's of science in mathematics, minoring in physics, chemistry, and aeronautics, before switching tracks completely in the mid-1960s to work as an accountant with the IRS.

CHAPTER 1

Muscogee, Florida, 1912

BESSIE PITTMAN RAN AS FAST AS HER SIX-YEAR-OLD LEGS COULD CARRY HER. She ran over dirt roads blanketed with wood chips and sawdust to stop them from becoming impassible when it rained. But as today was hot and dry, Bessie's bare feet kicked up clouds of dust with every step. The flour sack she wore as a dress was soon covered in filth, but she didn't care. She just wanted to get home before the ice block clutched in her hands melted. She'd packed it in sawdust and wrapped it in newspaper to protect it against the summer heat, but it was a losing battle. Bessie ran past townsfolk. She ran past the school. She ran past a Presbyterian church, a Methodist church, a post office, a barbershop, and a general store. She ran past the shacks residents called home, little one-room dwellings on stilts with windows permanently open to the elements such that there was nothing to do but tack paper over the holes when it rained. It was the same kind of home her family lived in.

Sawdust Road, she privately called the ramshackle little town of Muscogee, Florida. The small community near the east bank of the Perdido River existed because of the sawmill. Like most of the men in town, Bessie's father, Ira, worked for the Southern States Lumber

Company for $1.25[1] a day, but the workers weren't paid in money, they were paid in chips that could only be redeemed at the company commissary. It was almost a form of indentured servitude that meant mill families rarely escaped mill towns. It wasn't a good life, but the Pittmans never thought of themselves as poor. Like all of their neighbors, they got by, growing what food they could and leaning on friends in tougher times. When there was more work to get, Bessie's brothers, Joseph and Henry, could nearly double the family's earning while her sisters, Mary Elbertie and Maybelle Myrtle, were in school. But today, one of Bessie's brothers had the mumps and her mother, Mollie, needed the ice block to cool his fever.

The ice was nearly gone by the time Bessie got home. Taking the soggy package from her daughter, Mollie, headstrong and quick to anger, reached her breaking point. She grabbed a switch from the backyard and turned toward her youngest child, but Bessie was her strong-willed mother in miniature. They locked eyes. Without looking away, the girl leaned down and picked up the first piece of firewood her little hand found and raised it with fire flashing in her eyes. For a moment neither moved; then, sensing that her daughter wouldn't back down, Mollie let the switch fall and backed away. Bessie, who already regarded her mother as slovenly, lazy, and mean, now added cowardly to the list. She threw the piece of wood to the ground and walked away victorious.

The Pittmans weren't in Muscogee long. Once all the trees in the area had been felled, planked, and shipped out as lumber, the mill closed and the workers were left scrambling to find new jobs at other mills where the forest was abundant. Ira packed up his family and its meager possessions, which included Mollie's Bible that doubled as the

---

1 About $31 in 2019.

only record of family births and deaths, and relocated to Bagdad, a port town near the junction of Pond Creek and the Blackwater River, about twelve miles east of Pensacola. Outwardly it was just another mill town—in this case, owned by the Bagdad Land and Lumber Company—but for Bessie it was the most luxurious place she'd ever seen. Routinely skipping lessons after the strict schoolmarm had taken a ruler to her hands on her third day of class, Bessie spent her days wandering through town alone. She marveled at the motorcars driving on paved roads lined with electric lamps that shone as it got dark. She'd find a spot atop a hill and watch the long freight trains lazily carrying lumber from the mill. She practiced her ABCs deciphering the names on the boxcars, imagining the distant places each train was coming from or going to. But the millinery shop was by far Bessie's favorite. She could happily spend hours staring at the window displays, taking in every detail of the beautiful hats trimmed with fur and feathers. Between the foreign boxcars and the exotic fashions, Bessie began dreaming of a life far from any mill town.

But her days of fantasizing came to an end when she was found by a truant officer and forced to attend the one-room schoolhouse. Initially angry over the loss of her freedom, Bessie changed her tune when she met her teacher, Miss Bostwick. To Bessie, Miss Bostwick was elegance incarnate. She had an exotic northern accent, wore pressed dresses, and though she was prone to rapping students on the knuckles with a ruler, Bessie found she was fair in her discipline. Miss Bostwick, in turn, took an immediate shine to Bessie. The teacher offered her student ten cents[2] a week to bring firewood to her house, and before long the girl was making daily trips with armfuls of small logs. More important than the small sum were Miss Bostwick's lessons about table manners and the value of personal hygiene. Uncommon though it was in the Pittman house, Bessie developed the habit of brushing her teeth

---

2 About $2.50 in 2019.

with a gum tree branch and taking daily baths in the tin tub filled with cold well water. Miss Bostwick gave her a comb, a ribbon for her hair, and a proper dress so she could hold her head up high with the other kids in her classroom. She also gave her books. Bessie slogged through *David Copperfield*, underlining words for Miss Bostwick to explain later. This tutelage imbued Bessie with a sense of self-worth she'd never known, but her education came to an abrupt halt after two years. Miss Bostwick traveled back north and, with all the trees around Bagdad gone, Ira moved his family to Columbus, Georgia, where the Bibb City cotton mill held the promise of jobs for the whole family, including a now ten-year-old Bessie.

There was no question of Bessie returning to school when she could earn money for the family. Every night from six o'clock in the evening to six o'clock in the morning, Bessie pushed a cart up and down the aisle of weavers, her bare feet slapping against the concrete floor and her little hands deftly replacing bobbins. At midnight, the weavers ate whatever meal they had brought for lunch while Bessie crawled into her cart for a nap. When Bessie proudly brought home $4.50[3] after her first week—real money, not commissary chips—Mollie confiscated the full sum for the family. The next week Bessie handed three dollars to her mother and kept the remaining $1.50 for herself. Money, she quickly realized, could be her path to independence.

Week by week Bessie added to her hidden fortune until she had enough money to buy a pair of used shoes from a peddler. Her choice was a pair of too-big high heels like the ones she'd seen the grown-up ladies wearing in Bagdad; on her next shift, the sound of her slapping feet was replaced by the telltale clicking sound. Bessie quickly realized they were more fashionable than practical; her second purchase was a pair of tennis shoes. As the weeks went by, she saved enough money to buy a georgette blouse, a colored corset cover, and a black wool skirt

3 About $113 in 2019.

on installment payments from the same peddler. On Sundays, her one day off, she donned the whole outfit and walked up and down Broad Street. Face pressed against shop windows, she looked longingly at the beautiful ladies' clothes that were for sale, knowing full well she looked like she was playing dress-up in her mother's closet.

The cotton mill turned out to reward hard work. In a matter of weeks, Bessie was promoted to the inspection room where she examined rolls of material and rejected any with flaws. Soon after she was made foreman, put in charge of fifteen older children and some young adults in the inspection room. Each promotion brought a small raise, but the overall conditions remained deplorable. The building's facilities consisted of an outhouse on the other side of a nearby river. There was no break room. The foremen were lecherous, pinching Bessie in ways no little girl should be pinched. Healthcare was rudimentary but available. When Bessie suffered an attack of appendicitis, a young doctor removed the organ for a small fee but left talcum powder behind in her abdomen; the talc provoked scar tissue that adhered to her intestine, causing frequent blinding cramps and blockages that often demanded medical intervention. The poor working conditions became too much, and workers formed a union. Bessie paid a dollar for her membership card, and soon after that, her mill along with twenty-six others in the area went on strike.

After three months with no resolution, Bessie the eleven-year-old foreman went to the woman at union headquarters. She wanted to know when her workers could expect to get back to the mill. The woman considered Bessie for a moment. She was struck by the girl's drive and energy, things rarely seen in the oppressed workers and almost never in a child growing up in a mill town.

"Get out of the factory and make a new start," the woman advised Bessie. "Go see Mrs. Ryckley."

Mrs. Mattie Ryckley was a small Jewish woman who co-owned and ran three beauty salons and a hair goods store on Broad Street with her husband, Charles. The couple needed an errand girl in the shops who could double as domestic help with their six children, and Bessie assured Mrs. Ryckley that she could do both. Satisfied with this self-professed experience, Mrs. Ryckley offered Bessie $1.50[4] a week plus room and board. Unwilling to turn down the opportunity to escape the cycle of mill life, Bessie, just eleven years old, moved in with the Ryckleys for a full-time job.

Working for the Ryckleys was hardly better than the mill. Bessie was up at five o'clock every morning to make kosher breakfast for the family before going to work in the salon, where she spent hours in a windowless back room mixing shampoos, transformations, and dyes by the light of a gas flame. But the strong work ethic she'd developed in the mill remained, and she was rewarded. As she learned more about the products she mixed, she got the chance to shampoo clients before a hair treatment, apply color, and even learned how to give a permanent wave. Clients and technicians alike would tip her for her help, and soon she was making close to ten dollars a week.[5] She sent half back to her family and kept half for herself.

Bessie's favorite part about working in the salons was the clientele, especially the well-off women in elaborate dresses. "Fancy ladies," Bessie called them. One of these fancy ladies mentioned that she owned a house, and since the other women referred to her as a "madam," Bessie assumed she was a schoolteacher like Miss Bostwick. She even had a similar northern accent. Bessie loved how glamorous the Madam was and delighted in her stories of distant cities and grown-up parties. Every time the Madam came into the shop, Bessie dawdled as much as she dared, stretching out appointments for a little bit more time with her favorite client. The Madam noticed, and, in spite of the stories she told, warned Bessie of the dangers that came with her profession.

---

4 About $38.50 in 2019.
5 About $201 in 2019.

"It's easy for a pretty girl to make a living without mixing shampoo and waving hair," she told Bessie one day as the girl flitted about the shop. "It's much better to work and maintain self-respect."

As Bessie grew into a teenager, the Madam's lessons reinforced her willingness to work and taught her always to keep one eye looking for her next opportunity.

Two years after she began working for Mrs. Ryckley, nothing had changed. Bessie was still up before dawn serving as a nanny at home and working in salons every day of the week, and though the tips were nice Bessie thought it was time for a raise. It wasn't long before a man came into the shop and gave Bessie the opportunity she needed. He was checking that all employees were of age. Without skipping a beat, Mrs. Ryckley assured the man that Bessie, who had taken to wearing her hair tied up in an attempt to look older than her thirteen years, was sixteen and that she was the girl's legal guardian. Satisfied, the man left, and Bessie promptly maneuvered her employer into a quiet corner.

"I will go on full pay as an operator, or I will leave," she announced.

Mrs. Ryckley was struck by Bessie's ingratitude; she was teaching the girl a trade, she should be grateful. But Bessie stood her ground, confident she could manipulate the situation to her advantage. She calmly told Mrs. Ryckley that the clients who came to see her brought in about two hundred dollars every week, meaning she brought more business to the shop than she was compensated for. These regulars, she continued, would surely follow her to another shop if she left. If Mrs. Ryckley wanted to keep these clients, Bessie would happily continue living at the family house rent-free, but she wouldn't take care of the children or cook. She would keep working in the salons, but she'd have a retainer of thirty-five dollars[6] a week plus commission, and she'd work regular hours, just like every other stylist. Then she laid down her trump card: Bessie told Mrs. Ryckley she'd overheard her conversation with the inspector, and that she wouldn't hesitate to use the falsified information to ruin her reputation if need be. Mrs. Ryckley had to admit defeat, and Bessie realized the power that came with confidence.

Now fourteen, Bessie was already a striking young woman, equal parts feminine and tomboy. She dressed in fashionable, well-made clothes that flattered her budding figure, and her experience in the salons meant she could style her hair in a way that drew attention to her large brown eyes. She was also gaining something of a reputation as a rebel, attending church nearly every Sunday morning, then hanging around behind the local grocery store in the evenings smoking cigarettes with her male friends. Men were inexorably drawn to her, including a tradesman ten years her senior named Robert Harvey Cochran. Their courtship was accelerated when Bessie became pregnant within weeks of their meeting, and the pair married in Blakely, Georgia, on November 14, 1920.

It wasn't long before a traveling salesman stopped by the salon and unwittingly presented Bessie with another opportunity to advance her career. The man came to the shop selling Nestle permanent wave machines, and as he talked to Mrs. Ryckley, he mentioned there was a store in Montgomery, Alabama, to which he could sell one if only he could offer an expert technician in the deal. The Nestle wave process took about six hours and used a dozen two-pound, chemically treated brass hair rollers heated to over 210 degrees Fahrenheit. The rollers were suspended with counterweights so they wouldn't burn the client's scalp, giving the client a look akin to Medusa during the process and polished waves after. It was finicky, but Bessie was adept at the system.

---

6 About $810 in 2019.

Seeing a chance to get even farther from her family and the mill town, she told the man she was the expert he needed. He was somewhat doubtful that this young, pregnant girl was the expert he needed, and Bessie had to admit that she might need to take care of her family first.

The new Mr. and Mrs. Cochran went to Florida, first to Robert's hometown of Noma then on to Bonifay, where Bessie's family had relocated. There, less than four months after their wedding and three months before her fifteenth birthday, Bessie gave birth to a son, Robert H. Cochran Jr., on February 21, 1921. Within months the young family moved clear to the other side of the state and settled in Miami.

The marriage wasn't a happy one. Bessie loved her baby boy with his blond curly hair and sweet disposition, but she still wanted more from her own life. And so, less than a year after they moved to Miami, Bessie traveled alone with Robert Jr. to DeFuniak Springs, Florida, where Ira and her oldest brother, Joseph, were working at the W. B. Harbeson Lumber Mill. Leaving her baby with his grandparents, she took a job giving Nestle permanent waves in the salon in the Nachman & Meertief department store in Montgomery, Alabama. Her son and her marriage remained a secret.

Once again, Bessie quickly accumulated a roster of loyal clients. Even the salon's floor manager, Mrs. Lerton, was taken by Bessie's quick wit and even quicker hands and took the girl under her wing. Picking up where Miss Bostwick and the Madam had left off, Mrs. Lerton taught Bessie about sewing, furthered her cooking lessons, and helped introduce her to the local social scene. By seventeen, Bessie had enough money to buy a Ford Model T. The manual, however, was so technical she could barely understand it, so she studied the engine until she figured out how it worked and became her own mechanic. She also took dance lessons that she put to use when friends or clients invited her out for a social evening. Though on the surface Bessie was beginning to live the life of luxury she'd dreamed of back in Bagdad, she lacked stability. She was bouncing between homes, living off and on with family in DeFuniak Springs for stretches at a time, often with her favorite brother, Joseph, and his wife, Ethel Mae. Sometimes she brought Robert Jr. with her and other times she left him with her parents.

Watching Bessie flit around between Montgomery and DeFuniak Springs, Mrs. Lerton began to push the girl toward a more stable life. She wanted the girl to study nursing, but Bessie protested. She liked the lifestyle that came with working in the salon, and though she was literate, she worried she wouldn't be able to keep up with lessons. She doubted any hospital would accept a grade-school dropout at all. But Mrs. Lerton was adamant and talked the training director at St. Margaret's Hospital into admitting Bessie as a student.

Never satisfied with anything less than an all-out effort, Bessie approached nursing with the same dedication she had everything else. Instead of slogging through complicated medical texts, she memorized the lessons and committed the layout of surgical instruments to memory. She was enthralled by surgery, fascinated with how the procedures worked and never put off by the blood. This, coupled with her continued insistence on absolute cleanliness, made her a favorite assistant among surgeons. On her days off, she gave the patients haircuts and learned how to give a straight razor shave. The only thing she didn't complete was the license exam; she was so self-conscious of her poor writing skills that she never made her new career official. All the while, her husband, Robert, was nowhere to be seen, and it wasn't long before real cracks started showing in the marriage. Locals in DeFuniak Springs often saw Bessie spending time alone with the local grocery store owner, Will Meigs. Robert, meanwhile, sought the companionship of another young woman, Ethel May Mathis. When Bessie learned of the affair, she moved home with four-year-old Robert Jr. to DeFuniak Springs and promptly filed for divorce.

* * *

In the spring of 1925, the days were warm enough that four-year-old Robert Cochran Jr. was left to play by himself out in the backyard of his grandparents' house. Somewhere in the yard on the afternoon of Friday, May 29, he found a matchstick, and in playing with the little wooden piece set fire to some paper. In an instant, his clothes became engulfed in flames. A neighbor spotted the emergency and rushed over, burning his hands as he did his best to rip off the boy's flaming clothes. Bessie was called back to the house, and she arrived to find physicians attending to her son, but it was too late. The boy was so badly burned he couldn't be saved. The only thing to do was keep him comfortable. The family moved him to his bed, where he regained consciousness long enough to tell his mother and grandparents what had happened. Bessie stayed sitting on the edge of his bed all afternoon, but before the day was out, Robert Jr. was gone.

Two days later, Bessie buried her son beneath a small, heart-shaped headstone in the local cemetery. Then she packed some of his things away in a trunk and tried to pack her sadness away with them, but she couldn't. Depression settled in as she tried to return to work, but her failed marriage only made things worse. Her own divorce filing had been dismissed for lack of proof of Robert's infidelity, but Robert succeeded in ending their union. He filed for divorce claiming Bessie had committed adultery on or around November 20 of the previous year, 1924. Bessie didn't fight it. Instead, she presented herself at the Circuit Court of Montgomery County in Alabama and submitted to the judge's interrogation about their relationship. The divorce was finally granted on February 5, 1927, three months before Bessie's twenty-first birthday.

Between Robert Jr.'s death and her divorce, DeFuniak Springs held more pain than happiness for Bessie. It was time to move again, this time to Mobile, Alabama, where her aunt Amanda had a spare room near another salon. But within a year, Bessie was back in DeFuniak Springs. Her whole family had fallen ill with influenza and needed care. Her sisters Mamie and Myrtle survived, but her father didn't; Bessie helped the family bury Ira in the local cemetery close to Robert Jr. Her brother Henry died not long after in a Coast Guard accident, and she buried him, too. When she left town again, she didn't go far, just eight miles west to Pensacola, where there were high-end salons and it was a short drive home.

Pensacola failed to cure Bessie's malaise. She put her own money into the poorly run Le Jeanne Beauté Shoppe, but her co-owner proved to be an unsavvy business partner. Although she enjoyed the social scene and nights spent dancing with the handsome young ensigns from the nearby navy base, when they told her about the marvelous airplanes they were learning to fly she felt even more restless. She needed to get out into the air herself. She took to the road selling dress patterns from town to town, and though she reveled in the romantic freedom that came with the open road, she knew that it was a dead-end job. She tried to further her career by studying the latest techniques at a beauty school in Philadelphia but found she knew more about trends and techniques than her teachers. Nothing was making her truly happy.

In the span of a decade, Bessie had been married and divorced, welcomed a son then buried him. She had buried her father and her brother. Her sisters and beloved brother Joseph were all raising their own families, and the only family Bessie had left was her mother, with whom she continued to clash. DeFuniak Springs held nothing but painful memories for Bessie. She decided the time had come to make a clean break.

One midsummer day in 1929, twenty-three-year-old Bessie arrived at the train station in Pensacola. Her worldly possessions were packed into suitcases, and her life savings was tucked away in her pocketbook, including the money she'd gained from selling her Model T. She bought a ticket and boarded a train heading north.

Watching the countryside stream past the window, she decided to reinvent her past. She would tell people that she was an orphan, that the Pittmans had taken her in but never really cared for her. This would explain her lack of family ties. She would never admit to knowing her biological family and would instead tell people that her foster parents, Mollie and Ira, had been so poor and unloving that she had been forced to leave the house at eleven years old to find work. She also decided never to tell anyone about her marriage or her son, though she couldn't bear to erase Robert Jr. entirely. He was her happiest memory. She needed to keep him with her somehow, so decided to keep the only thing she had left that they had shared: a name. She would remain a Cochran to keep her little boy alive in her heart, though she would tell people she'd picked the surname at random running her finger through a phone book. As the train sped further north, Bessie Pittman faded into obscurity.

When she arrived in New York City days later, she retained Bessie's skill as a hairdresser and nurse, her obsession with cleanliness, and her moxie, but nothing else. No one would ever know Bessie Pittman, she'd decided, but the world would absolutely know Miss Jacqueline Cochran.

CHAPTER 2

New York City, Summer 1929

NEW YORK CITY WAS LIKE NOTHING JACKIE HAD EVER SEEN. SHE MARVELED at the lights that made the city bright even at night. There were more cars on one road than she had seen in her entire life. Skyscrapers stretched high above her, and the skeletons of new ones being built seemed to disappear into the clouds. Walking down Broadway brought back memories of Broad Street in Columbus, but everything was so much bigger and grander. She was surrounded by noise—engines roaring, streetcars rattling, horns honking, dogs barking, and even music. Huge crowds raced along as though propelled by some unseen energy, unafraid to push past cars and dodge around horses in their immaculate clothes. Everyone she saw seemed to exude effortless wealth. Far from daunted, Jackie felt energized.

It didn't take long for her to find a room to rent in the back of a restaurant. For three dollars a week, she had an apartment to herself with a private shower, access to the kitchen that sat outside her door, and a view of Central Park. She cleaned the room from top to bottom before unpacking. As a final touch, she stuck a strip of felt over the crack at the bottom of the door to keep out food odors. When she was done, she looked around, satisfied with her little haven in the big city.

"Start at the top, Jackie."

She repeated this mantra to herself as she walked to Charles of the Ritz. Located inside the Ritz Hotel on the corner of Madison Avenue and 46th Street, the salon was famous in the beauty industry. Beyond styling, the shop had developed its own cosmetics line that it sold to New York's social elite through its multiple locations. It was the world Jackie wanted to be a part of, so it was where she needed to work.

Jackie found the salon as posh as its reputation. The floor was patterned marble. The walls were lined with wood and glass shelves filled with elegant displays of perfumes and makeup. Mirrors made the room feel more spacious. Jackie filled out an application slip and asked to speak with the salon's owner, Charles Jundt. Mr. Charles was a nearly fifty-year-old mustachioed German man from Alsace-Lorraine who, after working as a hairdresser in both Paris and London, had immigrated to the United States nearly twenty years earlier. As the owner of Charles of the Ritz, he was the driving force behind the salon's sterling reputation. When he came out to meet Jackie, he sized up the twenty-three-year-old self-proclaimed expert at permanent waves and transformations.

"Little girl," he began, "what can you do?"

"I can do everything," Jackie replied.

"You don't look old enough to be an expert," he said dismissively.

Jackie stood her ground. She explained she'd been working in salons for more than a decade and insisted that she could, in fact, do everything, and some things probably even better than he could. Then she asked for a job, offering to take a fifty-percent commission in lieu of a salary.

Mr. Charles chuckled at her brash confidence. He considered her for a moment, noticing her long blond hair that was expertly curled.

"You'll have to cut your hair," he said as though testing how badly she wanted to work for him.

"I wouldn't cut my hair for you even if you promised to turn your whole business over to me," she shot back, deeply put off by his arrogance. With that, she turned on her heel and walked out the door.

The next day the phone in Jackie's apartment rang. Mr. Charles had changed his mind. Impressed by her determination, he was willing to offer her fifty-percent commission and would let her keep her hair, but Jackie steadfastly refused to work for a man who even jokingly attempted to impose on her personal rights. She turned down his offer then set out to his rival salon, Antoine's. Founded by Polish hairdresser Antoni Cierplikowski—known simply as Monsieur Antoine—the original Antoine's in Paris was a favorite of stars like Claudette Colbert, Greta Garbo, and Edith Piaf. When Monsieur Antoine started doing women's hair "à la garçonne" he launched the bob to worldwide fame. As soon as he opened a salon in New York's Saks Fifth Avenue department store, his Parisien cachet followed. To Jackie's delight, the manager at Antoine's hired her on the spot with the understanding she would have to prove herself through her work.

Before long, Jackie had a steady roster of clients so wealthy they asked her to follow them to Miami Beach in the winter months so she could work at Antoine's sister salon and style their hair year-round. With nothing keeping her in New York, Jackie bought a little Chevrolet and began splitting her years between the two cities.

By the spring of 1932, both New York and Miami Beach had changed. The whole country was feeling the effects of the stock market crash of 1929 as the economy hit its lowest point. Banks failed, reducing the money supply. The value of the dollar rose as wages dropped and jobs disappeared altogether. As business generated less revenue, the ripple effect forced both business and personal bankruptcies. People sold their most prized possessions in a desperate attempt to feed their families while farmers' crops sat rotting. But the women who frequented Antoine's in New York and spent the winters in Miami Beach experienced far fewer effects from the Depression. They were wives of businessmen and politicians so could still afford their dyes and permanents, which meant Jackie was able to work full-time and maintain the dual-city lifestyle to which she'd grown accustomed. Some clients even became friends. Though Jackie the grade-school dropout turned stylist might have been firmly in a different social class than these wealthy wives, there was no stigma around Jackie being a career girl, especially when that career afforded her the financial freedom to live the life she wanted. Jackie, a lover of parties, went out to lavish supper clubs three or four nights a week, where she could dine and drink cocktails with friends and dance to a live orchestra. More often than not, the nights would end in a casino, where Jackie kept herself to strict rules. She didn't gamble unless a friend loaned her money, which she repaid in full if she won, and she always went home alone at midnight.

This was her intention the night she found herself sipping before-dinner cocktails at the Surf Club in Miami Beach. She was there at the behest of Molly Hemphill, a friend from the salon. Their host for the evening was Stanton Griffis, the American ambassador to Spain and a business partner of Molly's husband, Cliff. As she stood chatting with the Hemphills, Jackie noticed a man walking toward the coat-check room wearing a simple suit rather than formal dinner dress. He was slender, clean-cut, had a fair complexion, and wore horn-rimmed glasses. She wouldn't say he was necessarily handsome, but he wasn't unattractive, either. She couldn't quite put her finger on it, but there was something about him that grabbed her.

"Why can't you ever introduce me to men like that one?" she asked, leaning over to Cliff. "He looks as though he does something with his life besides gambling."

"Oh, Jackie, you're incredible." Cliff laughed as Molly strode across the room to take the man's arm. Guiding him through the crowd, she gave him a whispered, tongue-in-cheek introduction to her friend Jackie. "She's a working girl—not too pretty. She has very unique hair, which she claims is natural. She has a beautiful figure and all the boys like her. Even Cliff, I think, is slightly on the make for her, but I don't think anyone has made any headway—so you might be interested in meeting her. She is just an ignorant little nobody but, strangely enough, we always have her around. Why don't you come and meet her?"

As the man sat on the arm of Jackie's chair, Molly introduced him as Floyd. Up close, Jackie could see he had brilliant blue eyes that were framed by golden hair that fell loosely over his forehead. As the group moved into the dining room, Jackie was pleased to see he sat next to her. There was something about him that put her at ease. He was funny, kind, and utterly fascinating. As the evening wore on she began to suspect that he was interested in her, too; he was so focused on learning all about her. He listened closely as she told him how she'd gotten her start in the beauty industry, how in addition to working at Antoine's she was also part owner in cheaper salons where girls just out of beauty school gave less expensive treatments women on the dole could afford; she'd often visit these salons in the evening after her own workday. She even told Floyd about her stint as a traveling saleswoman and how the open road still held some allure. Floyd listened with such rapt attention she felt safe opening up about a private dream.

"I've been thinking of leaving Antoine's to go on the road as a cosmetics manufacturer," she told him. "The shop can be so confining and the customers so frustrating and what I really love to do is travel. I want to be out in the air." Jackie knew being a stylist took creativity but not a lot of smarts. She craved a challenge. She also knew that women needed more than a hairstyle for a confidence boost. She firmly believed that it was a woman's right to feel the emotional satisfaction that came from enhancing her natural beauty. To her amazement, Floyd didn't scoff at her ambition. Instead, he offered her some sage advice.

"There's a depression on, Jackie," he reminded her. "If you're going to cover the territory you need to cover in order to make money in this kind of economic climate, you'll need wings. Get your pilot's license."

Floyd stayed by Jackie's side as the party moved into the casino after dinner. Though he'd given up gambling years earlier, Floyd pressed a twenty-dollar[7] chip into her palm, which she promptly lost. So he gave her another, which she also lost. Jackie realized she didn't know what he did for work and worried she was losing his hard-earned money. He had mentioned something about a bank, so she assumed he worked in finance, but his comment about the Depression didn't suggest he knew more than the average person. He was so quiet and serious, so un-tycoon-like, that she assumed he was a bank teller, in which case she couldn't have him spending more money on her. She moved to leave, but Floyd pressed another twenty-dollar chip into her hand, imploring her to stay a little longer. They continued this back and forth until Jackie had lost more than three hundred dollars.[8] Now she was seriously concerned about his financial situation.

It turned out that Floyd could afford to lose much more. Jackie's gentle Floyd was Floyd Bostwick Odlum, a lawyer turned titan of the financial world. Seeing the writing on the wall in the summer of 1929, he had not only survived the stock market crash with his millions intact, but he was also profiting off the Depression. The value

---

7 About $367 in 2019.
8 About $5,520 in 2019.

of companies was plummeting, and banks and brokerage houses were keen to be rid of the responsibility of the associated stocks. So Floyd's Atlas Corporation bought them, in each case purchasing just enough shares to be the controlling party for as little as fifty cents on the dollar. He would reorganize a company, liquidate its assets, then use the profits to do the same thing to another company. While people lost their jobs in the process, Floyd emerged as one of the ten wealthiest men in the country. Jackie didn't know the extent of his wealth that night; all she knew was she felt a connection with Floyd. They had similar backgrounds; fourteen years her senior, Floyd had also grown up poor, the youngest son of a Methodist minister. They had both worked hard for their success. The uneducated beautician and the introverted millionaire shared an insatiable ambition and an impatience with anything but an all-out effort.

Two days later, she sat on his right as the guest of honor at his dinner party at the Indian Creek Golf Club. Their courtship was clearly beginning, but there was a complication: Floyd was married.

By Floyd's own admission his marriage to Hortense McQuarrie was over except on paper. That was why he was in Miami alone; he was taking the time to think through whether he wanted a divorce while Hortense stayed in New York with their two sons. Meeting Jackie tipped the balance. He wanted to make her his wife, but the situation was delicate. Floyd didn't want to drag his sons through a messy public divorce or embarrass Hortense by openly dating before their marriage was legally over, so he wanted to end one relationship before starting another. Jackie, sure that destiny had brought her and Floyd together, trusted that he was a good man who would do the right thing.

When Floyd ran into Jackie weeks later back in New York, he immediately asked Jackie to dinner. She gave him the number at Antoine's and told him to call her anytime. It was against the rules to receive personal calls in the salon, but Jackie didn't care. Floyd did call, and the manager put him through to Jackie, who was so excited she picked her next free night for their first proper date, which happened to be her twenty-sixth birthday. When Floyd found out midway through the night, he gave her a twenty-dollar gold piece from his pocket. Jackie slipped it into her purse and quietly vowed never to spend this first gift from him. From that night on, their courtship progressed under a veil of discretion; Floyd still didn't want to be seen openly dating until his divorce was final. Jackie didn't mind. She appreciated the delicacy of the situation and relished every chance she had to get to know him better.

And all the while, his casual mention of flying continued to percolate in the back of her mind. It was like listening to the navy cadets in Pensacola all over again, but this time she had a reason. She was at the top of her industry in one of the most prestigious salons, but she was starting to get bored. The more she thought about it, the more she wanted to begin the next phase of her career, and she couldn't get the idea of flying as a cosmetics saleswoman out of her mind. It wasn't unheard of for a woman to be a pilot. Amelia Earhart had shown women that flying was in their reach, and in 1929 she'd founded the Ninety-Nines, a sorority of female pilots so named because there were ninety-nine licensed women aviators present at the time of its creation. The more Jackie thought about it, the more her goal of joining their number felt feasible. In a matter of weeks, the books on her nightstand and the magazines in her mailbox were about airplanes. All her conversations—with Floyd, with friends, even with herself in her head—were about flying. After two months of informal study, it was time to move from the page to the cockpit.

Jackie hadn't taken a single day off since starting at Antoine's, which meant she had six weeks of vacation time stored up that she could use to get her pilot's license. Talking over her plans with Floyd, she realized she didn't want to spend all her time off working, so she decided to split it: three weeks for relaxation and three weeks for flying lessons. Floyd scoffed at her ambitious timeline, saying there was no way that she, a complete novice, could get her license in six weeks, let alone three. Jackie shot back that she absolutely could and would. Amused, Floyd decided to make it interesting. The man who didn't gamble bet the woman who'd never been inside an airplane $495[9]—the cost of the lessons—that she would need more time. Jackie accepted the bet without hesitation.

On a Saturday morning in the last week of July, Jackie boarded a nearly empty eastbound train at Pennsylvania Station. Speeding toward Long Island, she watched the urban landscape roll by as the conductor called out the stops in a deep, drawling voice. When he announced Mineola Station, she disembarked and walked the short distance to Roosevelt Field.

She'd picked Roosevelt Field for its proximity to the city, but the walls of the little airport held history. Named for President Theodore Roosevelt's son Quentin, who was killed during the First World War, it was one of the most easterly airports in the country. It was thus a favorite starting point for long-distance pilots, including Charles Lindbergh, who had begun his transatlantic flight in the *Spirit of St. Louis* here in 1927. As Jackie walked into the front office, the little airport was empty except for Husky Llewelyn.

"I'm starting on a six-week vacation," she announced. "Do you think I can learn enough in that time to get my pilot's license?"

Husky had seen his fair share of young women come through the airport. They got it into their heads that they wanted to fly after taking a ride somewhere, then they marched into his office intending

9 About $9,105 in 2019.

to be the next Lindbergh, only to never return for a second lesson. Looking at Jackie with her blond curls, petite frame, and made-up face, he couldn't imagine she'd be any different. As Husky sized up Jackie, she did the same. The man was as big as his name, and she seriously questioned whether the little plane on the runway could support his weight.

Neither teacher nor student voiced their private trepidations as Husky explained that today he would be taking her up for a half-hour teaser lesson to see if she wanted to commit to the full course. Without another word, he led her out to the runway. He climbed into the rear cockpit of the Fleet biplane trainer as Jackie took her seat in front. He told her to pay attention to the stick and pedals—they were connected so hers would move like magic as Husky controlled the plane—but under no circumstances was she to touch anything unless he told her otherwise. With that, Husky sped down the runway until, much to Jackie's relief, the plane lifted into the air.

In an instant, Jackie's life changed. *Why have I waited?* she thought to herself. *I can't believe that I have put this off—a reason for living—for so long.* Up in the sky, on that clear day, flying felt natural. She was on par with the birds, sure that there was some higher power watching over her.

Husky flew them in circles around the airport while Jackie paid close attention to how the stick's movements corresponded with the plane's. He showed her how moving the stick side to side controlled aileron flaps on the wing that made the plane turn left and right. Pushing the stick down lowered the elevator in the back and the plane's nose dove; pulling the stick back did the opposite. Pressing the pedals moved the rudder, causing the plane to roll side to side. It not only made sense to Jackie, but it also seemed so intuitive. As their half hour came to a close, Husky yelled over the engine for her to take control and bring them in for a landing. Jackie didn't know whether he had confidence in her or if he was just bored and hoping for the excitement of jumping in to save the day, but she did as she was told. She grabbed the stick and noticed how natural it felt in her hand, how the plane seemed to move as an extension of her body. She followed his instructions to the letter and made a perfect landing on the nearby field.

"How many hours do you have to fly to get a license?" Back on the ground, Jackie was desperate to return to the air and start her formal lessons.

"You've got to fly twenty hours, then pass a test," Husky told her, adding, "It'll take two or three months if you're lucky."

"I have to do it in three weeks because I don't intend to spend my entire vacation out here," she explained. She left out the part about the bet with Floyd.

"That'll be tough," Husky laughed.

"I don't think so." She defiantly put her $495 on the counter.

Husky happily took her money and signed Jackie up, but made it clear that the fee was nonrefundable and he couldn't guarantee that she would get her license or even make a solo flight.

The next day, Jackie arrived at an empty Roosevelt Field at seven o'clock. The morning was still, and as she wandered around the airport, she couldn't stop looking up. The sky had taken on new meaning. It wasn't just the backdrop to a summer day anymore, it was a realm she knew she had to conquer. Walking between the airplanes parked in the nearby hangar, she studied the portraits of great aviators painted on the walls. She read Amelia Earhart's name, Clifford B. Harmon, Henry "Hap" Arnold, Major Alexander Seversky. A wave of inspired determination washed over her as she imagined herself being on that wall next to them. Jackie snapped out of her fantasy when Husky arrived to begin their lessons.

She flew in that same small trainer every day over the next week. She learned what airspeeds she needed for takeoffs and landings and how to guide the plane through basic maneuvers. He taught her how to fly loops, spins, and rolls in dizzying succession. On their eighth day together, after just seven hours and five minutes in the air, Husky turned Jackie loose for her first solo flight. She didn't consider that she was still a novice. All she knew in that moment was that she wanted to be the world's greatest aviatrix, and it was going to start right now. And so she took off down the runway as she had with Husky so many times. The controls felt familiar, as did the sensation of lifting off the ground into the air. The whole flight felt wonderful. Then, all of a sudden, the engine quit. Shielded by the blissful confidence of inexperience, Jackie didn't panic. She assumed Husky had rigged it to stop as a test, so she did as he'd taught her: she managed her speed as best she could without her engine and glided down to a perfect dead-stick landing on the field.

Husky and the handful of other pilots who put their own flights on hold to watch her first solo were astounded. No one had rigged the engine. Jackie had just weathered a real emergency. Husky couldn't deny that after just hours in the air, this petite blond young woman had proved she was one of the smartest and most natural flyers he'd ever seen. So great was her aptitude that Jackie only needed another ten days before she was able to take her final tests. Flying the practical test was easy. The written exam was harder; Jackie lacked confidence in her written abilities so convinced the examiner to let her take this test orally.

On August 17, 1932, just seventeen days after she first stepped into a cockpit, the newly licensed aviatrix collected her money from Floyd.

CHAPTER 3

A Small Airport in Vermont, August 20, 1932

THE CUSTOMS OFFICIAL WAS SURE HE'D MISHEARD HER. "DON'T YOU KNOW the way to Montreal?" he asked, doubtful.

"No, I really don't," Jackie replied. "I wouldn't be asking you for directions if I already knew."

He still didn't understand. "How did you get here?"

"Well, I came up the Hudson River and then followed the lake shoreline until I found the airport. I landed and here I am." Jackie didn't see the problem. It was a clear day, and she had maps showing she could follow the edge of Lake Champlain almost all the way to Montreal. In fact, the flight had been a lot easier than filling out all those customs forms she'd just completed.

The man explained which heading she had to follow for how many miles, where to turn, and how to adjust her timing depending on the winds she met over the Canadian border. But Jackie had no idea what he was talking about. Why couldn't he just point her toward Montreal so she could go?

"I don't know how to read a compass," she finally admitted.

Now he was really confused, shocked that anyone in their right mind would fly alone internationally without knowing how to read a compass. He walked away and returned moments later with a group of men, each of whom grabbed a part of her plane's wings and tail. As one, they turned the plane right there on the field with Jackie still in the cockpit. "Look at the compass!" the customs official shouted.

For the first time, Jackie noticed that the needle on the compass moved with the plane, telling her where her nose was facing. Flying with Husky she'd learned about airspeeds and level flying, but she'd always been in view of the airport. It had never occurred to her she might need to navigate by more than landmarks. An arm appeared by her window, pointing her toward Montreal. The man's directions made sense now, but for the first time in her short flying career, she was nervous. "Suppose I'm not good at reading this compass," she asked the man attached to the arm. "Suppose I wander off. I don't think I fly that straight. What will I see on my way?" All of a sudden, this last-minute trip to Canada for an air meet felt foolhardy, especially with a three-day-old license in her pocketbook.

"Well, there aren't any highways or railroads you can follow. They just don't go that way. You'd better just head in the general direction. When you're about halfway there, you'll see two big silos. If the visibility stays as clear as it is now, you should start seeing some airplanes at that point. Just go where the other airplanes go and that's where the airport will be. If you get lost, or never see the silos, turn around and come back." Jackie thanked the man for his help as well as his patience and took off. She flew a few circles to get the hang of the compass needle before setting off.

Jackie landed in Montreal as something of a heroine. Fellow pilots were amazed that such a newly licensed pilot had managed the trip alone. Even the media took notice; her picture ended up in the *Toronto Star* above a caption calling her a society pilot. The trip to Montreal taught Jackie two crucial lessons. First, if her picture was going to appear in the newspaper, she'd have to make sure it was a flattering one properly identifying her as an aviatrix, not some flippant woman for whom flying was an amusing lark. Second, she needed to learn to read her instruments properly before she got into real trouble.

By fall, there was no question that flying was Jackie's life, but a simple private license wasn't enough. Of the 588 licensed women pilots in the United States by that time, fewer than sixty held advanced licenses. These transport, commercial, and instrument ratings not only qualified pilots to fly larger planes with more powerful engines, it meant they had demonstrated prodigious skill in the cockpit. So Jackie went where the best pilots were trained. She quit Antoine's, and with Floyd's full support packed up her car and drove the six days from New York to San Diego to attend the Ryan School of Aeronautics. Founded by T. Claude Ryan, who had built Lindbergh's *Spirit of St. Louis*, the school was one of the best in the country for advanced flight training. But Jackie, the student who learned by doing, was disappointed by the heavy emphasis on classroom lessons and the teachers' propensity for playing craps in the afternoons. She'd spent most of her savings and uprooted her life to attend a flying school that did no flying. San Diego's one saving grace turned out to be the navy's pilot school at North Island, where Jackie found some familiar faces. The ensigns from her Pensacola days were air officers now, among them a former suitor named Ted Marshall. Pleased to see a familiar face, Jackie vented her frustrations to an understanding Ted, who made her an offer: if she could get herself a plane, he and his friends would teach her to fly the navy way. Jackie promptly bought a well-worn Travel Air trainer, hired a math tutor to take over the classroom lessons, and moved to Long Beach where Ted was stationed.

Flying with the navy boys gave Jackie the impression that she'd crammed ten years of flying lessons into just six months. She learned to fly with one eye on her instruments and the other eye scanning for a stretch of beach or an empty dirt road in case she had to make an emergency landing. It was a valid concern; her used Travel Air trainer's engine had a nasty tendency to die in the air. In her spare time, she explored the California countryside from the coast all the way inland to the Coachella Valley. That's where she fell in love. It was like nothing she'd seen before. Sitting below sea level, the Salton Sea had been recreated in 1905 when the Colorado River flooded and breached a dam; in the eighteen months before engineers could stem the flowing water, it had grown into a sizable body of water. The surrounding desert of reddish-brown earth was broken by the grays, greens, and purples of the sage, desert holly, yucca, and ocotillo trees. The area was home to bobcats, pack rats, lizards, and rattlesnakes. Temperatures could reach as high as 130 degrees Fahrenheit in the summer. But to Jackie, it felt calm. It was the only place where the buzzing energy that continually propelled her forward was quieted. She bought twenty acres of land and began planning the dream house she would build for herself and Floyd. When she told him about it, he purchased close to 900 adjacent acres.

The Grand Central Air Terminal in Los Angeles was unique. Located in the middle of the city and surrounded by homes and businesses, it was a favorite hangout for locals to spend lazy days watching the airplanes landing and taking off. Jackie hadn't been planning on an audience when she arrived that day to take her commercial license exam, but the watchers weren't leaving. Before the gathered crowd, she took off with an instructor crammed into her Travel Air trainer.

She hadn't reached seventy-five feet when her plane's finicky engine stopped dead. Too low to turn back to the runway, Jackie's scanning eye found a vacant lot. She glided down slowly to a landing, then forced her nose down to try to slow her speed, but the lot didn't have as much space as a wide-open empty field. Cursing and cringing, Jackie banged her way through a metal fence, across the highway, and finally came to a stop against an illegally parked car on the opposite side of the street.

A man got out, hardly thrilled that his day of plane watching had been interrupted by such a close look at one of them. Jackie didn't see what the fuss was about. The car wasn't exactly in great shape, and she'd barely hit it hard enough to do any damage, but the man turned out to be a traffic judge angry enough to slap her with a twenty-five dollar[10] fine for illegal parking. As Jackie, the judge, and the examiner stood on the street arguing, a small crowd gathered around the wreckage to gawk. Two men pushed their way to the front, offering to tow the plane into a hangar to make the necessary repairs so she could get on her way. They introduced themselves as Al Menasco and Jack Northrop, and Jackie couldn't contain her excitement. She knew Jack Northrop by reputation. He was revered throughout the aviation world as the designer behind some of the most beautifully streamlined planes in the air. As the two men assessed the damage, Jackie couldn't hold back her excitement.

"Someday I'll be flying one of your planes!" she gushed to Jack.

"Of course you will," he replied absently, barely taking his focus from the plane. Then he suggested she contact one of his test pilots for some extra lessons. Jackie couldn't deny it looked like she needed all the help she could get.

This embarrassing episode behind her, Jackie eventually earned her commercial license, but it wasn't enough. She wanted some objective

---

10 About $482 in 2019.

measure of success to prove how good she was, and there was no better way than through air races.

Air races had the power to turn pilots into celebrities. Victors won trophies and accolades, but more importantly, they often secured records, and record holders were household names. No one personified this phenomenon better than Charles Lindbergh, the unknown pilot who became an instant hero after crossing the Atlantic. For "Jacqueline Cochran" to become a household name, Jackie needed to fly with the best and win. Women's air races were one option, so-called powder puff derbies that drew a strong crowd excited by the novelty of women flyers. But Jackie had no interest in being the best among women. Jackie wanted to be the best pilot, full stop. She wanted to fly with men in the Bendix, the Cleveland National Air Races, the MacRobertson International Air Race. She wanted to fly across oceans and whole continents, staying aloft for hours at a time before landing amid tens if not hundreds of thousands of cheering spectators at the finish line. These were the races that created stars, so these were the races that she needed to fly. There were just two problems.

The first was easy enough to solve by upgrading her commercial license to the required transport license. That meant learning to fly by instruments alone. The big air races happened regardless of weather, which meant potentially flying through a storm or dense fog. Start times were also staggered, so it was possible she would have to fly at night. Without visible landmarks, it was easy for a pilot not to realize she was flying in a bank with her wings tilted to one side, or with her nose pointed in the wrong direction. She could unknowingly fly off course and slam into an unseen mountain or end up over water with no land in sight; these kinds of fatalities weren't uncommon. To avoid getting lost, pilots had what was called "the beam." The beam was a series of Morse code beeps and dashes sent over a known course between landmarks, almost like a virtual wire strung between beacons.

With the plane's radio tuned to the correct frequency, the pilot would hear a pattern in the signal—"right" if she was north of the wire, "left" if she was south of it, and "on course" if she was right along that line. For the most part, each signal extended far enough that leaving one beacon, she could pick up the next right away, flying along the beam. But there were places where the beam couldn't go, like along some stretches of sky and over airports. In those instances, a pilot had to rely on instruments, trusting that the needle on the compass, the airspeed indicator, and the artificial horizon gave her a true indication of where her plane was in the sky. "Flying" and "blind" were not two words Jackie wanted to put together, but it was the only way to learn to trust instruments. Jackie studied maps, learned to fly on the beam, and flew with her windows covered until she could comfortably fly across the country eating lunch and smoking cigarettes without ever looking out the window.

The second problem was harder to solve. None of these major air races allowed women to fly.

By the summer of 1935, Jackie and Floyd's life together was becoming permanent. Hortense had been the one to initiate their divorce filings citing cruelty, and Floyd, working on settling out of court, was living between the large Manhattan apartment he and Jackie shared and their Indio, California, compound where the main house and surrounding guest houses were still taking shape. Among the construction workers was Jackie's brother Joseph. She brought his whole family out west in a private Pullman car and set them up in a small nearby residence, and though they didn't stay long, she relished the time with her young nieces and nephews during her trips to California.

Jackie sold her little Travel Air trainer in favor of a bigger, higher performance plane in which she traveled between New York and California overseeing the house's construction. She was a more proficient pilot for all the practice, too, adding a better understanding of weather patterns and flight conditions to her repertoire. With Floyd's financial help, she even finagled her way into a few traditionally male air races. She entered the MacRobertson England-to-Australia international race but was forced out of contention in Bucharest when her Gee Bee Q.E.D. succumbed to mechanical problems. Jackie didn't spend time worrying about the loss. Instead, she focused on her next big goal.

At the same time, Jackie had begun conquering the business world; she hadn't lost interest in starting her own cosmetics company. Whether out on the town or up in the air above it, she still valued her femininity. With Floyd's continued support, she hired a cosmetic chemist and a perfume consultant with the goal of creating better products than anything on the market. She wanted a non-greasy hydrating face cream, something she knew had to be possible from working with lubricants on her planes. The result was Flowing Velvet, a moisturizer that promised to give women a natural-looking soft, dewy complexion. For beauty on the go, her team developed the Perk-Up cylinder, a three-and-a-half-inch stick with six compartments containing a cleansing cream, foundation, eye shadow, clear red rouge, solid perfume, and a sifter to fill with the face powder of the user's choice. It was a staple Jackie touted as something every woman should have in her handbag or flight bag, as the case may be. Jackie had the products tested at a salon in Chicago whose customers' feedback helped shape the line. To run the business, she hired Antoine's former office manager and leased office space on the thirty-fifth floor at 630 Fifth Avenue in Manhattan near Rockefeller Center, which offered a stunning view of the city. Next, she ordered letterhead for the Jacqueline Cochran Cosmetics Company featuring the slogan "Wings to Beauty."

Jackie personally took samples out to department stores and found that flying actually helped her in business. She might not have been a household name, but distributors knew the novelty of a female flyer's cosmetics line would attract customers. Pogue's in Cincinnati and Halle Brothers in Cleveland became her first two accounts. B. Foreman's in Rochester followed, then J. W. Robinson's in California, where she was sold alongside Elizabeth Arden; the two brands were billed as the only luxury lines in the upscale department store. Jackie's cosmetics empire was born.

Between continued flying time and running a burgeoning business, Jackie still maintained an active social life. She was in her New York apartment one day when the phone rang.

"I've got quite a treat for you." Her friend Paul Hammond was on the line inviting her to dinner.

"Having dinner with you is always a treat, Paul."

"But this time I have a real treat, Jackie. There is someone I want you to meet who will be here, too. Amelia Earhart."

Jackie walked into the Hammonds' dining room that evening to meet the woman whose fame she so desperately envied. Amelia had become the unquestioned queen of the air after flying across the Atlantic Ocean in 1928, though by her own admission she had been a passenger on that particular flight and about as useful as a sack of potatoes. Nevertheless, fame followed thanks to the publisher George Palmer Putnam. His aggressive promotion turned Amelia into a business, establishing her as a "Lady Lindbergh." He promoted her flights, organized her mid-route checkpoints, booked her speaking tours and appearances, published her books, and in 1931 became her husband.

Tall, slim, good-looking, and formally educated, Amelia was a stark contrast to Jackie, who was nine years younger, stood just five foot three inches, had a slightly fuller figure, and was still self-conscious that her schooling had stopped after the second grade. Naturally quiet Amelia preferred addressing crowds from behind a podium while Jackie craved being at the center of that kind of attention. Differences aside, both Jackie and Amelia quickly realized that they were very like-minded women. They were the emancipated daughters of the suffragettes, women for whom having a career and having a husband weren't mutually exclusive. They had freedoms their mothers had never known, and with that came a feeling that the whole world was open to them. Both were involved with successful businessmen who, far from resenting their partners' success, loved them for their independent streaks and strong wills. Both were successful in the male-dominated world of flying, often eschewing dresses for slacks while each still embodied her femininity in her unique ways. The two pilots immediately became fast friends.

Weeks later, the pair flew Amelia's brand-new Lockheed Electra from New York to California together. The trip was beset by weather delays that more than doubled the time to cross the country, but far from breeding tensions, Jackie and Amelia's time together revealed a magical rapport. Their conversation ran the gamut from politics to religion to science to personal matters. Their flying strengths were different, with Amelia favoring distance records and Jackie pursuing speed flights, so any competition between them was a friendly one. They even discovered a shared fascination with extrasensory perception, something Jackie believed she possessed. They would read about air crashes and Jackie, focusing on the plane and its last reported position, would try to "see" with her mind's eye where the wreckage might be. Whatever she saw—telephone lines or a mountain pass by a highway—she and Amelia compared those visions with maps, then called local rescue services with tips.

Their cross-country adventure didn't end when they reached California. Instead, Jackie brought Amelia out to see the house in Indio, which was just about finished. The flight cemented what both women knew would be a very close, lifelong friendship, and Jackie now had an ally with whom to take on the all-male air races.

A thick fog rolled over the Burbank airport just after midnight on the second of September 1935, but the thousands of spectators standing along the runway didn't leave. They would stay out all night just to watch the planes taking off for the start of the Bendix Transcontinental Speed Race. Established by automotive and aviation pioneer Vincent Bendix as a chance for manufacturers to test new developments, it was akin to the Kentucky Derby for pilots. The starting line was in Burbank, California, and the finish line in Cleveland, Ohio. Between those two points, pilots could fly whatever race plane they chose along any route; the winner was the pilot with the shortest time measured from wheels up to wheels down. Because speed was of the utmost importance, the planes in the Bendix were often the newest, cutting-edge designs and their pilots the most skilled at pushing those planes to their limits in the hope of saving even a few minutes in the air. Between planes, training, and everything in between, the $30,000[11] prize money that came with the first place trophy barely offset the winner's cost of competing, but no one flew the Bendix for the money. Pilots flew the Bendix to prove they were the best of the best, and tonight, Jackie was finally going to be flying right alongside them.

Jackie and Amelia had worked together to change the rules governing the Bendix. Two years earlier, the race had been open to women with the understanding that they were competing against each other, not against their male peers. Angry over this continued inequality, women who might have flown the Bendix the following year boycotted in protest. Jackie had picked up the fight this year to secure her own entrance into the coveted race, lobbying for women's inclusion in the Bendix without restrictions. Race officials had finally consented.

---

11 About $552,000 in 2019.

Jackie and Amelia, the two women interested in the 1935 race, were permitted to fly on the condition that they procure signed waivers from all the male entrants confirming they had no problems with female competitors. With that, the Bendix became a mixed race.

Securing her participation had been the easy part for Jackie. Preparing her plane and herself for the demanding race had been the bigger challenge; her goal, after all, was to win, not just fly. With Floyd's financial support she bought a Northrop Gamma—she'd made good on her promise to Jack Northrop after that embarrassing crash and was finally flying one of his planes—and fitted it with an experimental air-cooled Pratt & Whitney engine that could push the plane to winning speeds. But the plane had gone through some growing pains. She'd had it for a year, and in that time it had had four different engines and weathered one crash that had demanded significant repairs. For whatever reason, the rebuilt plane had developed a bad vibration that engineers hadn't been able to correct. The engine, too, had a tendency to overheat when pushed to its limit, sending vibrations through the whole plane that risked causing structural damage. If she flew conservatively, she would be fine, but this was the Bendix, and conservative flying wasn't a path to victory. Knowing she'd push the plane and fearing the worst, both Northrop and Pratt & Whitney had begged her to withdraw from the race, but she'd refused. If they weren't going to publicly admit it was their technical faults that were forcing her to withdraw, she wasn't going to give the impression that she wasn't up for the challenge of the race. Their stalemate kept her in contention for the Bendix.

Against the backdrop of these looming technical problems, Jackie had proceeded with her own preparations with Floyd by her side living vicariously through her flying adventures. Together they'd mapped her route and decided that three o'clock in the morning was her ideal takeoff time so she could take advantage of daylight flying hours. For weeks leading up to the race, she'd gone to bed a little earlier every night to adjust her body clock for the early takeoff. She'd even adjusted her diet to add a little more meat to her upper body; she would need all the strength she could get to muscle the controls of her Gamma. Jackie was as ready for the Bendix as she could be.

Jackie and her troublesome Gamma were on the Burbank runway as the fog got so thick officials were forced to put the race on hold. Half the competitors had already taken off, including Amelia, while the rest waited for the weather to clear. Jackie silently cursed her decision to leave at three o'clock; the added daylight hours wouldn't be much help if she couldn't get off the ground.

When the fog finally started to break around two o'clock and the race resumed, the on-site field representative for Pratt & Whitney begged Jackie one last time not to fly.

"Are you willing to say to the public that that motor is malfunctioning?" she challenged.

"I can't afford to say that," he admitted.

"Then I can't afford not to take off. If I don't take off, probably no woman will be allowed to fly in the race again." She'd had enough. She'd spent months training for this race, and no one was going to keep her from flying. She went through the same thing with the on-site representative from Northrop. "Go away and hush," she told him.

There was just one pilot to go before Jackie. Focused though she was on her own preparations, she paused to watch Cecil Allen race down the runway in his Gee Bee, a notoriously troublesome plane. She watched him gain speed and start to lift off the runway, then watched in horror as the plane struggled under its heavy fuel load and slammed into the fence at the end of the runway where it disappeared inside a fireball.

Jackie joined the fray that rushed to the end of the runway where Cecil's headless body lay among the wreckage. The sight gave Jackie her first pangs of uneasiness, as though seeing the remains of a crash made the risk she was taking suddenly real, but she knew she couldn't let it show. Maintaining a strong front, Jackie went into the airport restaurant and ordered a bowl of chili while officials cleared the wreckage from the runway. One of the Civil Aeronautics Authority officials followed her in, and as she took defiant bites of her meaty stew to prove her nerves weren't rattled, he implored her not to race.

She hit her breaking point. Jackie had had it with men telling her not to fly. She was qualified, her plane was ready, and she had every intention of leaving California that night. She told the official that if she died, it was her problem, not his, but that she would be sure to haunt him for pestering her. With that, she left the restaurant. The bowl of chili was left largely untouched; the rest she promptly vomited up on the side of the building.

It was finally Jackie's turn. As she went through her final preparations with photographers snapping endless pictures and destroying her night vision with their flashbulbs, her emotional strain became too much. Jackie trusted herself in the air, but if she was being honest, she was worried about taking off in bad weather. She climbed out of the cockpit, found a long-distance telephone, and called Floyd in Albuquerque. Over the phone, she poured out her fears. Ever the source of strength and support, Floyd told her the decision to fly was hers alone and that only she could say why she flew. As she listened to the comforting voice of the one she loved, she knew she had to satisfy her emotional urge; she had to fly. Secretly sure she'd never make it to the finish line but bolstered by Floyd's words, Jackie decided to at least take off. That much she could do, she could prove she wasn't scared of starting the race.

Newly determined, she reentered the Gamma's cockpit. She started down the runway with fire trucks trailing behind her just in case. The plane was so loaded with fuel that even her cutting-edge engine couldn't give her the power she needed. It was touch and go whether she'd be able to take off, and all the while the end of the runway was drawing ever closer, though she could hardly see it through the fog. She hit her point of no return, the point where she wouldn't have enough room to stop safely without hitting the fence and turning into a ball of fiery wreckage herself. She pulled back on the yoke and felt the wheels leaving the ground. She grazed the fence, snagging her antenna and tearing it clean off. Eyes trained on her instruments, she flew out over the Pacific Ocean, getting clear of the mountains before turning toward Cleveland. Rising above the fog, Jackie saw stars appear. At that moment, they were the brightest and friendliest things she'd ever seen.

It was nearly an hour before Jackie's engine started overheating, sending vibrations through the plane that left her bouncing around in her seat. Nevertheless, she flew. As she crossed the border between California and Arizona, the Sun started to rise. Now she could see the Grand Canyon, the majestic but foreboding landscape stretching for hundreds of miles, and a violent electric storm brewing on the horizon. Caution finally replaced pride. Alone in her Gamma, she didn't care if she withdrew from the race. She'd made an instrument takeoff through heavy fog, proving to herself and everyone else that she was worth her salt as a pilot. Now, it was safer to land than battle the storm in an unsteady plane.

She dumped fuel through untested valves to lighten her load for a safe landing, but the gasoline shot back toward the cockpit, drenching her in the flammable liquid. But she had trained for emergency situations. Like a seasoned pilot, she registered that the biggest danger was the fumes, so she ripped the canopy open to breathe fresh air but now had wind whipping her face. Struggling to keep her eyes open between the gasoline and the wind, she spotted a small landing strip. With all her focus on that spot beneath her, Jackie brought the Gamma down to a safe landing. The moment she touched down, she leaped out of the plane, ran to the washroom, stripped, and began rinsing herself as best she could. The last thing she needed was to survive an emergency landing only to have some careless onlooker flick cigarette ash near her gas-soaked flight suit.

Her port in a storm turned out to be Kingman, Arizona. When she emerged from the bathroom, a kind attendant loaned her an overcoat while another ran into town to buy her a shirt and trousers. Sitting in the airport soaking wet in a borrowed coat with her race dreams dashed, she refused to admit defeat. She wasn't going to win, but she had managed an instrument takeoff at night and survived an inflight emergency. And more important, there would be another Bendix.

CHAPTER 4

The Cochran-Odlum Ranch in Indio, California, Late March 1937

AMELIA HAD BECOME A FREQUENT GUEST AT JACKIE'S INDIO RANCH. MORE often than not she arrived alone, staying in one of the many guesthouses, spending her days swimming or riding horses before joining the ever-rotating cast of Jackie and Floyd's notable friends for dinner in the main house. An inscribed copy of Amelia's book, *The Fun of It*, sat on Jackie's bookshelf "in memory of an Electra jaunt across the continent." On the rare occasion Amelia arrived when her hosts were out of town, she would stay in Jackie's own bedroom, an intimacy never afforded any other guest.

Such was their friendship that if Jackie and Floyd had had a formal wedding, Amelia would have been present. Instead, the couple had opted for a private civil ceremony in Kingman, Arizona, in 1936—on Jackie's thirtieth birthday—with just two friends as their witnesses. As a wedding present, Jackie had given Floyd an envelope sealed with a quarter set in wax. "This is for you Floyd," she'd written on the outside in an untidy scrawl. "I have never read the contents. You can burn it or read as you wish. I love you very much." She told him the envelope contained her true history, including everything known about her birth parents and blood relatives. To preserve her story of being raised an orphan, she told him the information had come from a private detective. However touched he was by this moment of vulnerability from his new bride, Floyd said he didn't care about her past. He only cared about her present and future with him. He handed the wedding present back to her unopened. Jackie promptly hid it away in her lockbox.

Jackie never shared anything about her true past with Amelia, either, though the women shared just about everything else. They talked about flying, poring over maps to plan Amelia's distance flights or Jackie's speed runs. They continued their forays into extrasensory perception, Jackie concentrating hard on survivors of some air crash they read about in the newspaper before the pair would fly out to help the rescue effort. More than once, Jackie's visions coincided with the accident site. Even Floyd became a close friend to Amelia after all the time she spent at the Ranch. The only thing Jackie didn't like about Amelia was her husband, George Putnam.

To Jackie, George was nothing more than a mean and manipulative man who knew his wife was his meal ticket. All of Amelia's routes were dictated by media stops George set up and refused to change. He openly berated her manners in front of their friends. But more than anything, Jackie hated that George valued Amelia's celebrity over her safety and pushed her into dangerous flights. George's latest idea was an east-to-west circumnavigation around the Earth's equator. A half dozen men had made the flight already, but Amelia would be the first woman to do it. The whole thing worried Jackie. She knew Amelia was a good flyer, but she also knew her friend's limitations. Circumnavigation was a weeks-long endeavor that demanded complicated and precise stellar navigation while flying over huge bodies of water, and navigation wasn't Amelia's strong suit. She would be wholly dependent on a hired guide, and Jackie shuddered to think about what would happen to her friend if that guide was wrong. The circumnavigation got off to a poor start with a crash aborting the flight in Hawaii. Amelia, shaken but unhurt, went right to Jackie's Indio Ranch when she got back from the island.

Days after that crash in Hawaii, Amelia sat on the floor in Jackie's living room in front of the cavernous fireplace. It was a quiet night; in addition to her host, the only guests that night were the women's close friends Mike and Benny Howard, a pair of married pilots. Amelia gave the group a play-by-play of the failed flight that left her audience silent.

"What," she teased, smiling at them each in turn. "Isn't anyone going to ask me the big question. Don't you want to know whether or not I'm going to try it again?"

Still, no one spoke. Jackie had one house rule: no matter how curious, no pilot ever asked another pilot something that would put them on the spot. They wanted to give Amelia time to recover from the flight and decide free from social pressure whether she would try again, but the conversation nevertheless turned to the circumnavigation plans. By the end of the night, no one, not even Amelia, was sure whether she would make another attempt.

The question was still on Amelia's mind days later when she and Floyd, out for a drive around the property, got stuck in the desert sand in an old car. As they waited for help to come dig them out, Amelia turned to Floyd.

"Do you think I should do it?"

"Do what?" he asked.

"Fly around the world."

The question hung in the still desert air as Floyd considered his answer. He knew Jackie and Amelia were extremely similar, so he had a pretty good idea of what was going through his friend's mind.

"Amelia," he told her, "if you are doing this to keep your place at the top among women in aviation, you're wasting your time and taking a big risk for nothing. No one can topple you from your pinnacle. But if you are doing it for the adventure and because you simply want to do it, then no one else ought to advise you."

Amelia did crave the adventure, but decided it would be her last; she would retire from daring flights after completing this last big journey. She returned to the Ranch weeks later with new plans. The flight path had been reversed: she would now be flying from west to east, a route no pilot had taken before. She and Jackie pored over every detail of the route. At Jackie's urging, Amelia hired a new navigator, Fred Noonan, after her original pick had proved unable to navigate by the stars. But Jackie remained uneasy, particularly about the final leg. The proposed route had Amelia leaving Lae Airfield in New Guinea for Howland Island, a spot nearly 2,000 miles southwest of Honolulu, where the Department of the Interior had built her a landing strip. It was her last planned stop before traversing the rest of the Pacific Ocean and landing in California, but it was a tiny island to find. If Amelia was off course at all, she'd miss her only safe landing spot for hundreds of miles. The more they talked about it, the more Jackie's unease grew.

"You're not going to see that damned island," Jackie told Amelia one night. "I wish you wouldn't go off and commit suicide because that's exactly what you're going to do."

Together, they decided to see if Jackie's ESP could help, if she could track Amelia in flight and "see" where she was in case something happened. Jackie's intuition had always been particularly strong with Amelia. One night while driving Floyd into Palm Springs from the Ranch, she was hit with a flashing vision of a fire in one of the engines in Amelia's plane, though she sensed that it wasn't serious and ground crews were already dousing the blaze. Later that night, she and Floyd heard the news of Amelia's fire on the radio; the next morning's newspapers held the full account, and it lined up with Jackie's vision.

To test this connection, Jackie kept a detailed log as Amelia flew across the country with George and sent them to her friend every night. She "saw" that everything was fine, though Amelia landed in

Blackwell, Oklahoma, one night, fifty miles away from her planned stopover. The next morning she "felt" Amelia take off at nine o'clock for Los Angeles.

When the women compared notes, only George tried to poke holes in Jackie's account. Amelia had left Oklahoma at seven o'clock, he goaded. Jackie calmly pointed out that she, unlike him, understood the concept of time zones.

On June 1, 1937, Amelia took off from Oakland, California. Jackie followed the incoming reports throughout the month, tracking Amelia across the United States, to the northern tip of South America, across Africa, India, to Australia, and finally to New Guinea. When news reached America that Amelia hadn't reported in from Howland Island, Jackie had a powerful intuition about her friend's fate. It was just like when she and Amelia had focused on finding crash sites. Concentrating hard, Jackie could see Amelia in her mind's eye floating along in the ocean on the wreckage, saw that she was fine but that Fred Noonan had hit his head and was in bad shape. Jackie's vision of Amelia persisted for three days before the feeling faded. When the image of Amelia vanished, Jackie went to church and lit a candle for her friend's soul, which she knew had gone off on a journey of its own. The loss of such a close friend hit Jackie hard, but death was a reality. Every flier knew the risk they bore.

Questions about Amelia's disappearance were fast cementing her status as an American hero, leaving the way clear for Jackie to become a living legend in the sky.

Within weeks of Amelia's disappearance, Jackie, still mourning the loss of her dear friend, was in hot pursuit of her own fame as an aviatrix. She was by now a staple of the race circuit, known to fellow pilots and manufacturers as a woman to watch in pursuit of speed records, the most demanding records held by pilots skilled enough to push their planes to their design limits in the quest to save a few seconds in the air. There were smaller records, closed course or straight-line flights Jackie was practicing for, but her white whale remained the cross-country Bendix. She still felt the allure of this prestigious race, and so she started planning.

The first thing Jackie needed was a new, faster plane. She'd lent the Gamma that had forced her out of the Bendix in 1935 to her friend Howard Hughes, which meant she didn't have anything fast to fly. One day back at her old stomping ground of Roosevelt Field, she saw her next Bendix plane—a sleek, silver Seversky Pursuit P-35 monoplane she just had to have. As she stood ogling the magnificent design, Alexander P. de Seversky himself sauntered over.

"I just wrote you a letter to ask you if you want to fly my P-35," he said.

"Why don't you let me fly it in the Bendix?" she answered.

The P-35 was groundbreaking. In addition to the standard fuel tank in its belly, the P-35 featured extra tanks in the wings, which meant it could hold enough fuel to fly as far as 3,000 miles without stopping. It was also aerodynamically advanced. Where most planes had curved metal cowlings covering the fixed landing gear, the P-35's gear folded right up into the fuselage. This decreased drag on the plane and increased its cruising speed to 300 miles per hour. Flying this engineering marvel in the Bendix was Jackie's absolute dream. She knew it would be her winning ticket, but first, she had to test it.

Alexander offered her a way to get a feel for his plane that would benefit them both. He was trying to sell the P-35 to the US Army Air Force. It loved the design but had some trepidation on account of its complexity; the engineering developments that excited Jackie demanded pilots have special training, and the AAF preferred simplicity.

Hoping to salvage the sale, Alexander asked Jackie to make the demonstration flight at Wright Field in Dayton, Ohio. He knew that if a woman with no pursuit training could fly his plane, the army brass would accept that it wasn't overly complicated. The demonstration would double as Jackie's proof she could handle it in an air race.

When the pair got to Wright Field, medical officer Captain Harry Armstrong expressed his very different opinion. He told Jackie that a woman had no business flying a plane that was considered complicated for male pilots, which only strengthened her determination. She committed Alexander's lightning-fast introduction to the controls to memory, then repeated his cardinal rule: "Don't play around up there. Just take the plane up, fly it level, and bring it back down." As the plane was new to her, he wanted to protect her life as much as his reputation.

Thus Jackie found herself sitting in the cramped cockpit of a plane she had never flown taxiing along a makeshift grass runway lined with military personnel. The sky overhead was filled with low clouds, meaning visibility would be bad to boot. In her heart of hearts, she wanted nothing more than to postpone the demonstration flight until after she'd had a chance to test it without an audience, but she knew that wasn't an option, so she took off. In the air, she followed Alexander's advice to the letter, flying a modest, level profile for just twenty-five minutes before lowering the gear to prepare for her landing. The moment she did, the rudder pedals started to vibrate under her feet. Alexander had warned her it could mean the gear wasn't locked, in which case she'd make a crash landing right on the plane's belly. Keen to avoid such a disaster, especially with an audience, she pulled the plane up to gain altitude and lined up for a second attempt. But again, the pedals shook as soon as she lowered the gear. Low on fuel, she had no choice but to land. She braced for the impact as she brought the P-35 down as slowly and gently as she dared, but no impact came. Instead, she was amazed to feel the wheels hitting the grass.

Alexander was thrilled. Jackie's flight had preserved his good standing with the Army Air Force, and he was more than willing to have her fly one of his planes in the Bendix. Harry Armstrong, too, had to retract his earlier objections. He was impressed by Jackie's command of the airplane, so much so that he offered to introduce her to the work he and his colleagues did that went hand in hand with flying. While pilots like Jackie proved new airplanes were safe from an engineering point of view, doctors like Harry made sure those pilots stayed healthy.

The medical side of aviation was inextricably linked to pilots pushing their planes in pursuit of new records. While Jackie was focused on flying fast, other pilots were looking to fly higher than anyone else, so high the very environment was a hazard. Below 10,000 feet, a pilot can breathe just fine, but any higher and the air becomes so thin she can't get enough oxygen with each breath. Without sufficient oxygen, the body shuts down. First, the vision blurs, then darkens around the edges as everything takes on a gray tone. Then fatigue sets in, followed in short order by impaired motor functions. A pilot could succumb to hypoxia without realizing anything was wrong, almost always with fatal consequences. Giving pilots a way to take oxygen with them to these higher altitudes would keep them alive, and this was what Harry's colleague William Randolph "Randy" Lovelace II was working on. The thirty-year-old physician just three years out of Harvard Medical School was a licensed pilot himself, now holding the rank of first lieutenant in the Army Medical Reserve Corps. Working at the Aero Medical Laboratory at Wright Field with doctors Walter M. Boothby and Arthur H. Bulbulian, he was tackling the problem of hypoxia by developing an in-flight oxygen mask pilots could wear while flying. Jackie was pleased to meet Randy and learn about his research. She wasn't looking to fly above the clouds just yet but knew the day would come when she would need exactly the kind of system he was working on.

That day came in the summer of 1938 as Jackie began her preparations for that year's Bendix in earnest. Her star was rising; she'd come in third in the Bendix the previous year, and weeks afterward set her first international speed record on a closed three-kilometer course at an air meet in Detroit. Then she set another speed record flying from New York to Miami. These flights, combined with the positive media attention she was bringing to the world of aviation, earned her the prestigious Clifford B. Harmon trophy in 1937, recognizing her as the country's most outstanding aviatrix. Officially holding court with the best pilots in the world, Jackie was sure this was her year to add a Bendix win to her name. She doubled down on planning every single detail of the flight.

First, as per race rules, she renewed her transport license. Then she worked out her route. She plotted detailed maps of four possible courses that took advantage of every beacon, radio station, and beam between Burbank and Cleveland in case she had to rely on instruments; she would pick her course at the last minute after a final weather check. Her challenge was that none of the routes gave her any wiggle room if she came up against bad weather. Going through a storm was dangerous, but going around one would kill her time. Her best choice was to try to fly above the weather—or at least the worst of it—and stay on the same heading, which meant flying above 10,000 feet and risking hypoxia. She needed that high-altitude oxygen system. She had her own doctor write a note to Harry Armstrong and Randy Lovelace, and to her delight learned that their mask was ready. Randy, keen to get some in-flight data, got Trans World Airlines to lend her a completed version of his system.

On August 16, with the race two weeks away, she sent all her information to the race officials. Her accompanying cover letter outlined all of her records, her awards, and an advisory that her new cosmetics line would be on sale in the fall. The letter ended with a stern warning forbidding anyone associated with the race to release her official image before the event. If she won, she would use the image for paid endorsements.

The last piece of her Bendix puzzle was the plane. She'd bought a brand-new one for the race, but it was still under construction. In its absence, she practiced on a similar model at the Seversky factory on Long Island. She also practiced flying in and out of the Burbank airport at night without floodlights or landing lights to get used to the runway. The way the gear folded up demanded she turn off her lights first, so if she left without them on she could fold the gear up that much faster, maximizing her time with the plane in its most aerodynamic configuration.

With a week to go before the race, Jackie's P-35 came out of the factory with a leak in the fuel system. While mechanics repaired the problem, she sat in the cockpit memorizing the exact location of every switch and dial until she knew the panel blindfolded. She knew her plan inside and out, too. She would fly at 16,000 feet with a slightly less than fully opened throttle; according to designers, this was ideal for optimal performance of her 1,200-horsepower engine. She chose to take off around three o'clock in the morning again, risking fog to take advantage of the crisp night air that would give her the most power out of her engine to get her fuel-loaded plane off the ground. If everything went perfectly, she would land in Cleveland with twenty gallons of fuel left over. Those twenty gallons were her only safety buffer. Floyd knew her plan, too. He knew her rate of fuel consumption and what checkpoints she would cross and could use both as a way to track her progress since she wouldn't be able to talk to him en route.

Alexander himself flew her plane from Long Island to Burbank for the race, setting a male cross-country speed record in the process. With everything fixed, Jackie added the finishing touch. She painted her race number, 13, on the side of the plane two and a half feet tall. She was ready to fly the Bendix.

There were fewer racers this year, and Jackie was the lone female entrant, but the crowd gathered to watch the Bendix planes take off was as large as ever. Jackie sat in her P-35, fueled and ready at the end of the runway. The weather bureau said storms were traveling northward from the Gulf of Mexico so she was planning to follow the route that would keep her far enough south to miss the worst of it on her way to Cleveland. Packed into the cockpit she had a thermos of hot coffee, her makeup bag, and Alexander's leather coat; he'd insisted she take it as an extra defense against the cold she'd meet at her cruising altitude. Alone in the dark, she focused on the electric lights at the far end of the runway for a moment before opening the throttle. She felt her engine's power and responsiveness as she gained speed racing down the runway, and at 3:13 in the morning, her wheels lifted off the ground. She quickly folded up the landing gear as the cheering crowd watched her plane disappear into the night sky. Her time had officially started.

Almost as soon as she reached her cruising altitude, Jackie felt something pulling her to one side. It had to be the wing tanks. Hoping to balance her fuel load, she switched from the belly tank to those in the wings in an attempt to drain them first, then put the issue from her mind. Crossing into Arizona, the predicted storm appeared, and her planned altitude of 16,000 feet put her right in the thick of it. "The soup," she called it. It was so thick she couldn't get radio reception. Clamping the pipe stem from the oxygen tank between her teeth, she focused on breathing through her mouth as she settled in for an instrument flight. Eyes darting between compass, altimeter, and artificial horizon, checking her attitude and engine performance, she gradually climbed to 23,000 feet to try to rise above the storm, but the lower temperature only added a layer of ice to her windshield. Between two bad options of flying blinded by weather or by ice, she opted for the former; if the clouds broke, at least she'd be able to see. She brought the Seversky back down into the storm.

Two exhausting hours later, Jackie realized she didn't quite have her balance problem licked. The wing tanks should have been draining at the same rate to keep her fuel load level and her center of gravity balanced, but her instruments showed the right wing was still heavy and pulling her into a bank. Fighting it took all the strength in her hands and arms. She tried switching to the belly tank again, then back to the wing tanks, but it didn't work. Without warning, the plane gave in to the pull of the overloaded right wing. It fell into a spiraling dive sending Jackie, blinded by the storm, straight toward the ground.

Luckily, she knew her way around the cockpit by feel. She got her bearings on the controls, then muscled the plane into a shallow dive and pulled out of the spin. Recovering her lost altitude, she was now certain that the right wing wasn't draining and quickly devised a solution. If she flew at an angle with the empty left wing below the full right one, she could force it to drain. Flying in a bank like that set her off course, but there was nothing else she could do. And so she entered into a sort of acrobatic flight, dipping her left wing to drain the right fuel tank then switching to the belly tank to feed the engines while she leveled out and recovered her heading. It was a demanding routine on a slightly less straightforward flight path, but it worked.

Though this curving flight path was physically demanding, flying through the storm left the cockpit so cold Jackie's feet felt like blocks of ice on the rudder pedals, and a leak in her oxygen tube had left it painfully frozen to her face. Without taking her eyes from the instrument panel, she reached down for her thermos of coffee only to find that the top was missing. Flying as high as she had been, the change in atmospheric pressure had blown the lid right off, leaving her hot drink as cold as she was. She put it aside in favor of some lozenges to keep her throat from getting dry.

Hours later, the clouds finally cleared. Jackie looked down and recognized the Mississippi River by the jetties and landmarks she knew on the banks. She was within five miles of dead center on her course, only slightly north of her ideal flight path as she began to descend for landing. Passing over Grand Lake St. Marys in Ohio, Cleveland was in view.

At the airport, official timers heard the roar of her powerful engine over the dull buzzing of the smaller planes flying nearby. They rushed to the whitewashed line on the tarmac, clocking the exact moment her wheels flew over it as the moment her race ended. One judge waved a checkered flag as a signal. The 120,000-strong crowd in the stands roared as Jackie's silver plane rolled past them to a stop, but that didn't tell her anything. She had been the third to leave Burbank, which meant there were still six pilots in the air who could be making better time. But when she saw the official timer driving out to meet her with hordes of photographers and journalists in his wake, she knew it could only mean one thing. The other pilots weren't flying fast enough to beat her time, no matter how close they were to the airport. She had won the Bendix.

Officials congregated around her plane. Photographers followed, their cameras poised to capture the victor the moment she stepped out of the plane. Jackie wasn't going to let it be a bad one. With the crowd around her plane growing, Jackie pulled her makeup bag into her lap and, ignoring disgruntled looks, combed her hair and touched up her lipstick. *I don't care what they think,* she said to herself. *No matter how hard the trip, I've got to look my best. And there's nothing wrong with that.*

The moment she opened the canopy, the furor was just as she had expected. She posed for photographs. She shook hands with Vincent Bendix as cameramen captured newsreel footage of the victorious landing. It was the fame she dreamed of, but there was only one person she wanted to see. She had landed late and knew Floyd would be worried.

"Where's Floyd?" She spoke into the crowd gathered around her. "I want some cigarettes... I've been smoking an oxygen pipe all the way from California and I need a cigarette for a change!" When she finally found her husband, they didn't know who was happier to see the other.

"I'm not afraid to fly across the country alone," she told him in a private whisper, "but with a crowd like this I'm afraid without you." Floyd beamed at her.

Jackie's official time was eight hours and ten minutes, but she wasn't done flying for the day. After forty minutes of celebration in Cleveland, she was back in the air, dipping her wing in a salute as she sped off toward New York. Two hours and two minutes later, she landed at Floyd Bennett Field. Her combined flying time of ten hours and twelve minutes allowed her to add a new women's cross-country speed record to her accomplishments that day. Jackie left her plane in New York and boarded a commercial flight back to Cleveland, where she had more than enough time to change out of her flight suit and into an evening gown for the night's Aviation Ball where she danced with Floyd late into the night.

Jackie never found out what had caused her fuel line issues. A post-flight inspection found a wad of paper in the right wing tank was blocking fuel flow, but no one at the Seversky factory could say how it got there. Whether it was mistakenly left in the wing and was shaken loose on the flight or whether someone at the Burbank airport had placed it there was anyone's guess.

After her Bendix win, the media crowned thirty-two-year-old Jackie the First Lady of the Air Lanes, and a boost in publicity followed. Jackie was careful to manage her image, ensuring the press used pre-approved flattering pictures of her and referred to her as Miss Cochran, never Mrs. Odlum, though they were free to mention her marriage to Floyd. She also made sure every outlet knew she was an orphan; her invented origin story proved the perfect rags-to-riches narrative to capture America's heart. Paid sponsorships came in spades. Advertisements for Kendall Oil and General Tire and Rubber boasted that the famous aviatrix used both their products in the plane that she flew in connection with her cosmetics company. She even leveraged the win into her business. "Jacqueline Cochran, the aviatrix, puts a special eye bath which she uses herself," read the copy accompanying her ads. "It must be good because as a flyer she unquestionably realizes the importance of healthy, rested eyes." Her line was touted as marrying functional products and stylish packaging the way she combined her vocation and avocation. "Her cosmetics have been subjected to a severe test—the test of professional use by Miss Cochran in her years of flying." They flew off shelves. Her story was everywhere, including the September 12, 1938, issue of *Newsweek* that featured Adolf Hitler on the cover.

Soldiers at military bases across the United States were starting to feel the effects of the growing unrest in Europe. As Hitler's increasingly hostile activities became known, American soldiers were ordered to relocate for training, bringing their families in tow.

Such was the case for Lieutenant Colonel William Harvey Cobb. Harvey had been a member of the National Guard since college, and though he kept his family afloat selling cars during the Depression, he was still with the service. Every year, he packed his family—wife Helena and daughters Carolyn and Geraldyn—into the car and drove from Norman, Oklahoma, to Camp Perry in Ohio for the National Rifle Matches, a training exercise that doubled as a public demonstration and competition with other units. The older Carolyn, who preferred playing house and dressing up her Shirley Temple doll, hated this forced week of camping just to watch her father play at being a soldier. Geraldyn, on the other hand, two years younger, who preferred to be called Jerrie, loved it. Not only did she love that her father's unit almost always won the state championship, but she also loved spending an entire week outside under the open sky with endless stars as company.

Jerrie delighted in running through the Oklahoma fields, sometimes holding an imaginary rifle like she'd seen her father doing at the Rifle Matches, flopping on her belly and rolling around in the grass. She also loved riding horses. Jerrie's preference was for solitary pursuits that imbued her with a sense of freedom she couldn't find with friends. Though she wasn't shy, she had been born tongue-tied; the too-short frenulum binding her tongue to the bottom of her mouth meant words formed clearly in her mind but came out twisted, blurred, and thick when she spoke. By the time a doctor was able to fix her lingual problem, Jerrie had grown into a quiet child who firmly believed the best fun could be had alone.

In 1938, Harvey's unit, like so many others around the country, was called up to active duty. He was off to Camp Barkeley, which meant the Cobbs were relocating to nearby Abilene, Texas. To ease the transition on seven-year-old Jerrie, her parents promised that Texas had as many open fields, stables for horses, and as much open sky as she could want.

CHAPTER 5

Washington, DC, Fall 1940

JACKIE SAT AT A CONFERENCE TABLE WITH SOME OF THE BIGGEST NAMES IN aviation, the lone woman on a fifteen-person committee. The esteemed group included Chief of the US Army Air Corps Major General Henry "Hap" Arnold. Donald Douglas was there, too; his Douglas Aircraft Company was a leader in the burgeoning business of commercial flight. First World War flying ace turned Eastern Air Lines captain Eddie Rickenbacker was at the table as well, along with Secretary of the National Aeronautic Association William Enyart. Chairing the meeting was Dr. George Lewis, head of America's leading aeronautical research organization, the National Advisory Committee for Aeronautics. The group was charged with determining who had made the greatest contribution to aviation the preceding year and deserved to win the prestigious Collier Trophy. To be invited to serve on this committee cemented Jackie's position as one of the nation's leading voices in aviation—though she hadn't yet said much.

"You certainly are quiet," Bill Enyart remarked to her softly. "You must have something up your sleeve."

"Just wait," Jackie teased. "I've got a cute bombshell."

When a break came in the conversation, she took advantage of the silence to address the men.

"Mr. Douglas here," she began, motioning to the commercial airline manufacturer in the room, "is trying to build an airplane to go to forty thousand feet." The issue, she continued, was weather. She knew from her 1938 Bendix flight that storms could be as tall as they were wide. It was one thing for a pilot to battle through bad weather and fly by instruments while sucking oxygen through a pipe stem in an air race as she had done, but it was out of the question to expect paying customers to do the same. If commercial air travel was going to succeed as a business, pilots would have to fly at 40,000 feet where there wasn't any weather, which meant the cabin would have to be pressurized and heated for passengers' safety and comfort. Air travel was meant to offer luxury as well as convenience. Two planes—the Lockheed XC-35 and the Boeing 307—had tested pressurized cabins at 20,000 feet. But most commercial planes—Douglas's own DC-3 being among the more popular passenger planes in the country—still flew unpressurized at 10,000 feet. Solving the pressurization problem wasn't an engineering challenge, it was aeromedical, so Jackie nominated the doctors who had developed the oxygen system she'd used on her Bendix flight: Harry Armstrong, Walter Boothby, and Randy Lovelace. She was determined to get the doctors the recognition they deserved.

"Well, gentlemen," Dr. Lewis resumed speaking when Jackie had finished her statement. "I take a vote to postpone this meeting for at least a month to six weeks, and appoint Miss Cochran as a committee of one—at least she is interested enough; she's going to do a thorough job of it—and reconvene." The meeting was adjourned, and Jackie was appointed the task of convincing the rest of the committee that her friends deserved the award.

Jackie got straight to work. She called the Mayo Clinic and got hold of the doctors who told her the mask they had been working on was complete but needed some further testing, so she volunteered to help. First, she tested it in a portable low-pressure tank that Randy Lovelace and Walter Boothby brought to Washington. She climbed inside and, with the mask securely in place, the doctors lowered the pressure to the equivalent of 40,000 feet. It worked; Jackie didn't feel any early signs of hypoxia. Next, she took the mask up for some flight testing. The doctors needed some data points, so in an unpressurized plane with her mask firmly in place, she flew animals to high altitudes to see what low atmospheric pressure would do to living things. Flying as high as she could manage, Jackie watched in horror as chickens exploded around the 25,000-foot mark. The chickens she didn't mind—she'd killed chickens as a girl for food—but she drew the line at taking up goats and snakes, especially snakes. She was terrified of snakes, and no medical testing program was going to get her to fly with a snake on board, even if it was securely in a box.

When she brought her results to the Collier committee, the men were impressed, both by the doctors' work and Jackie's enthusiasm. The committee unanimously voted to give the award to US airlines for their safety record with special recognition to the three doctors. It was the first time in the award's history that it was given to medical doctors, and the first time the aviation industry truly recognized the role that medicine would play in the future of flying. Randy was particularly moved by Jackie's passion for his work and her willingness to participate in his research. Jackie considered it the most appropriate way to thank him for helping her win the Bendix. Through it all, they both recognized that they'd made a friend for life.

On December 17, 1940, Jackie hung back behind Walter Boothby as President Franklin Roosevelt presented the trophy to the three doctors in a small ceremony at the White House.

"I've been giving this trophy for a great many years," Roosevelt said once the proceedings were finished. "Are these medical or scientific doctors?"

"We are medical doctors, Mr. President," Walter answered.

"Well, what are medical doctors getting this thing for?"

George Lewis took over, knowing that just because the president awarded the trophy didn't mean he fully understood the science behind the doctors' achievement. "Mr. President, Miss Cochran did this singlehandedly, to bring our attention to this great piece of work that's been done. It's going to change the face of aviation."

"Mr. President," Jackie piped up, "it would take at least thirty minutes for you to hear the story of where it started, what they're doing in Europe..." She trailed off mid-sentence. Ignoring the realization that she was speaking rather informally to the president, she seized the unique opportunity that had presented itself. "I think Dr. Boothby is the person," she suggested, knowing he was the biggest Roosevelt fan in the room. "You might as well either make another appointment for these gentlemen to come back, or you better pull up some chairs and let everyone sit down."

"Pa," the president said to Edwin Martin "Pa" Watson, the United States Army major general and senior aide to the president, who seemed to have appeared out of thin air. "Have chairs brought in."

By the end of Walter's presentation, the president was so convinced of the value of aviation medicine that he promised an influx of funding to both the army and navy to further the research. Jackie objected. She hadn't worked so hard to get the doctors their due recognition to have funding split between laboratories that would doubtlessly duplicate each other's work. She decided to look into allocating that funding herself, but that was a matter for another day. Right now, she had a lunch date with Hap Arnold and Clayton Knight, the acting head of an American recruiting committee for the British Ferry Command.

It wasn't long before the lunch conversation turned to serious matters. Europe was at war, and while the United States remained committed to isolationism, it seemed increasingly likely the country would be forced to join the conflict eventually. As such, America had recently opened the first peacetime draft in history, forcing millions of young men to join the service. But Britain was in bad shape, even with its Commonwealth allies—Canada, Australia, New Zealand, and a host of other countries—fighting by its side. The nearly four-months-long Battle of Britain earlier that year had brought heavy bombing as the Germans attempted to destroy the Royal Air Force and weaken Britain's defenses with the goal of gaining this final stronghold before expanding across the Atlantic. Hitler's army hadn't managed to invade the island, but Britain had suffered devastating losses. It was now buying planes and munitions from America under President Roosevelt's Cash-and-Carry program and flying planes out of Canada since the Royal Canadian Air Force was closely linked to the Royal Air Force. But Britain was running out of gold and couldn't wait for the United States to join the war and come to its aid. It needed more supplies now. To this end, President Roosevelt had that very day announced the Lend-Lease program, under which America would supply Britain what it needed with the understanding that repayment would come later. The United States had already agreed to send some bombers overseas, and Clayton and Hap wanted Jackie to pilot one of them across the Atlantic. It was a practical mission, and having Jackie, the award-winning, record-setting, beautiful aviatrix at the controls could also bring positive publicity to the need for more pilots.

Jackie could barely believe her ears. She would happily be the AAF's poster girl for patriotism, and as a pilot, she was thrilled at the idea of flying a bomber across the Atlantic. It was a dream she'd confided in Floyd nearly a year earlier. They didn't have to ask twice. Jackie accepted without hesitation and Floyd, though slightly nervous about his love flying into a war zone, nonetheless threw his full support behind her. But it wasn't just the flight that excited Jackie. She knew Britain had recruited a small number of female pilots to help with the war effort; they ferried planes for the Royal Air Force, freeing male pilots for combat in the process. She thought it might be interesting to visit this Air Transport Auxiliary and see firsthand how women were contributing to the war. America might need a similar program in the future.

While the bureaucratic wheels began turning to clear a female civilian pilot to fly a US bomber from Canada to Britain, Jackie returned to the question of Roosevelt's promised funding for aviation medicine. She wanted the money to go to Randolph Field in Texas, the lab that had produced Randy Lovelace. She also knew that the funding would be allocated by Congress. To help direct the funds, she decided to meet with some Texas congressmen who would, she hoped, take her advice. The first name on her list was Lyndon Baines Johnson, a fairly new congressman whose schedule was open. Lyndon, who took inspiration from FDR to go by his initials of LBJ, had earned a reputation as a shrewd politician whose trademarks included a Stetson hat and the habit of dictating memos from the toilet.

Jackie traveled to Washington and met with LBJ in his small attic office in the old House Office Building. She explained the issue, he asked pointed questions, and before long, Jackie was blurting out everything she knew about the national need to prioritize aviation medicine lest America lose its technological preeminence.

"I will give it every single thing I have," Lyndon promised. "And we'll see if we can't get your aviation school of medicine funded and done properly." Then he paused and considered her for a moment. "You know, this is refreshing, to have a person come in on a thing of this magnitude, and you are very young!"

"I look younger than I am, probably," she replied. "I've been around a long time, working. And this needs very badly to be done."

"You fly too, don't you?"

"Oh, yes, sir, I fly."

Jackie and Lyndon, both notorious for valuing beneficial friendships, each recognized they'd made a new ally that day.

On March 8, 1941, FDR's Lend-Lease program was signed into law, so while America was still not directly involved in the conflict, it was participating nonetheless. That meant there was nothing stopping Jackie from making her bomber flight. Her biggest roadblock was pushback from Royal Air Force officials who questioned whether a woman should—or even could—fly an overseas ferrying mission. Her résumé spoke volumes with more than two thousand hours in the air, multiple ratings, a handful of records, and three Harmon Trophy wins. What she didn't have was experience with a twin-motor heavy plane like the one she'd be flying to Britain. And so she practiced. She took a captain's course with Northeast Airlines, gaining experience flying bigger planes at night, on instruments, with one engine, and through adverse weather. With Floyd's help, she got her hands on a Lockheed Lodestar for additional practice. Experience gained, Jackie put pressure on the British Ferry Command to formalize her appointment. As luck would have it, Britain's minister for wartime aircraft production and procurement, Lord Beaverbrook, was an old friend of Floyd's. Once he stepped in, Jackie's appointment was approved. She would work with the British Ferry Command flying out of Montreal, Canada. She left for her wartime assignment on June 16.

Jackie arrived at the Royal Canadian Air Force station at St. Hubert outside Montreal to find a Lockheed Hudson bomber waiting for her. She also found a score of angry male pilots. She knew there would be resistance to her on account of her gender, but she wasn't prepared for their outright hostility. Pilots openly challenged her very presence at the airbase. They taunted her, cut the seat belts from her cockpits, called her a publicity stunt, and gleefully warned her she'd be shot out of the sky as soon as the Germans found out she was a female pilot. They even went so far as to say that her flying a bomber was akin to taking bread out of their mouths. Jackie knew she had every right to be there and was as competent a pilot as any of the men, but she was nevertheless bothered enough to need an escape. She rented a car and sat out in the parking lot alone between flights, her own little haven free from sidelong glances and muttered insults.

It fell to an RCAF pilot named Captain Cipher to check her out in the Hudson and clear her for the overseas flight, and he was visibly unhappy about the assignment. He grudgingly took her up in the air, showing her the ins and outs of the bomber with as little explanation as possible. His lack of instruction forced Jackie to pay close attention to his movements, committing everything he did to memory so she could duplicate it herself. Then he asked her to land from the back seat, a wholly unconventional arrangement since, from that position, the handbrake was so low on the floor it was nearly out of reach. She had to lean so far over to engage it during landing that she couldn't see where she was going, not to mention it took considerable muscle to control the plane in the process. Regardless, she managed, and after nearly a dozen flights decided they were done for the day. She began taxiing the plane toward the hangar, but Captain Cipher apparently disagreed. He demanded she return to the runway and keep flying.

"I'm going to go in and tell you what I think of you!" Jackie shouted, her temper boiling over. "You are going to give me a ticket right now on this airplane, or I am going to turn you in and when I get through with you, you'll wish you never heard my name."

"Who do you think you are?" he demanded, matching her anger.

"I mean a hell of a lot more to aviation than you do, boy!" She challenged him to land the plane from where she'd been sitting. He did, three times, nearly crashing on the final landing.

Captain Cipher ultimately wrote Jackie's ticket clearing her for the flight, though he made the insulting recommendation she only pilot the plane in the air; she shouldn't take off or land. At least she was approved to fly, but finding her a crew was another matter. Hardly anyone wanted to serve on a woman's crew. Captain Grafton Carlisle and a radio operator named Coats, however, didn't mind at all, even when the other pilots threatened to quit in protest and conspired to bar them from flying professionally after the war. To them, a good pilot was a good pilot, regardless of sex.

Just days after her arrival in Canada, it was time for Jackie to fly the bomber to Britain. But the journey got off to a challenging start. When she arrived at the airport around midnight the day of her transatlantic flight, she found the emergency tank of antifreeze for her propellers was empty, her life raft was gone, and her tools had been stolen. She didn't make a scene. Instead, she just went out, bought a new all-purpose wrench from a nearby mechanic, and resumed her preflight checks. Captain Carlisle made an incident-free takeoff before handing the controls to Jackie for their first leg to Gander, Newfoundland, where they spent the night. The next morning, Jackie woke up to find one of her windows was broken and her wrench was missing. She wanted to scream, but instead, she taped it up and bought another wrench from a secondhand store, which she suspected was the same one that had just been stolen. Once again, Captain Carlisle got them into the air.

Jackie maintained a steady 135 miles per hour across the Atlantic. It was a dull flight made mildly more interesting by clouds. Then, right before daybreak, tracer bullets shot up in front of and around them. Jackie froze while her crew sprung into action. Someone grabbed a Very pistol and fired a color-coded signal, hoping that the fire was friendly and would stop upon seeing the flare. After a moment, the bullets stopped. Through a break in the clouds, Jackie could see smoke rising up from a ship on the water, obviously in distress. She couldn't tell if it was a German or English ship, or if it was even a military ship at all. It was her first view of the devastation of war, and she felt helpless. Finally, the coast of Ireland came into view, and Jackie felt her whole body relax. She had never been happier to see land. After a brief stop, they made their final leg to land at an airbase in Prestwick, Scotland, the endpoint for the North Atlantic ferry routes.

From Prestwick, Jackie traveled to London, where congratulatory letters poured in. Women wrote that they felt a surge of pride for her every time they used a product from her cosmetics line. A telegram from the Royal Air Force thanked her for the bomber and extended a standing invitation for her to return anytime with another one. Floyd telegrammed, simply, that his heart was filled with pride. Though British officials had tried to impose their rule of silence on her military flight, the media got wind of America's flying glamour girl. Journalists clamored to get into her hotel room for an interview, but she made them wait as she changed into a dress and refreshed her makeup. When she did speak to the media, she downplayed the drama of her training, called the Lockheed Hudson a "grand" plane, and shaved three years off her age, telling the world she was thirty-two. Amid the hubbub, Jackie made sure to take a moment and realize what she had just accomplished.

Jacqueline Cochran Odlum -
Savoy Hotel
London England -
great relief many congratulations much ~~pride~~ love stop call when you can stop
dont sign any stories or articles because there
are many here after such things and Harry
says they may try even to use my name or his
~~stop~~ as authority stop heart filled with
pride

Floyd

Eisenhower Presidential Library

For ten days Jackie met with Royal Air Force officials about her flight and took occasional calls from local cosmetic manufacturers. She also met with Pauline Gower, head of the female ferry pilots flying for the Air Transport Auxiliary. The ATA recruited pilots who were otherwise unfit to fly with the Royal Air Force because of age, gender, or some physical handicap. If a pilot could fly with one leg or a missing eye, he was fit for ATA duty; the pilots themselves joked that ATA stood for Ancient and Tattered Airmen. These men and women flew planes between factories and air bases so combat pilots could focus on combat duty. Jackie's flight had cleared the way for the ATA girls to ferry bombers as well as smaller planes, and she was amazed to find it wasn't a small group, either. There were dozens of female ATA pilots doing a vital service, and their number was growing. It was a model she thought was worth bringing back to the United States.

After her brief tour in England, Jackie caught a ride home on a B-24 with a dozen other pilots. Packed in like sardines, they took turns lying on the makeshift floor of wooden planks to look out the window and smoked a dwindling supply of cigarettes. After landing in Montreal, Jackie went right back to New York, where she changed into a proper dress at her Manhattan home before an impromptu press conference. Leveraging her publicity, she told the gathered media of her intention to write a report recommending that the US Army Air Force set up a female ferrying program. There were surely hundreds of female pilots who, with a little training, could do for the United States military what the British women were doing for England. It would be a worthwhile preparation in the event of war. Then, exhausted, she told her staff she was going to bed and that she was not to be disturbed by anyone except the president of the United States. Kidding aside, she intended to sleep in her own bed until noon.

The next morning at nine o'clock, the phone in Jackie's apartment rang. It was someone from the White House on the line asking if Jackie would please join President Roosevelt, his wife, Eleanor, his mother, Sarah, and Princess Martha of Norway for lunch in Hyde Park. One mad dash through her closet and a police escort later, she sat down with her esteemed company.

She told the president how the RAF was utilizing women pilots in the war effort. The American War Department, she knew, was only recruiting pilots who could be trained for combat, which excluded women by default. Nevertheless, she thought it prudent for America to consider some equivalent program, a ferrying or support role from women pilots. Following a presidential order to determine what that role might be, Jackie ended up in the office of Lieutenant Colonel Robert Olds, head of the Air Transport Command. Olds hired her as a dollar-a-year volunteer tasked with finding him 100 female pilots each with 500 hours in the air. But Jackie had something bigger in mind.

There were some 2,000 licensed women pilots in the country. Off the top of her head, Jackie could think of a handful who would happily join a women's corps. If she could recruit a small group, they could start changing attitudes about women flyers and lay the foundation for a larger program. She knew they would face prejudice as she had from the RCAF, the so-called "woman driver" problem. Men who held the view that women were inherently weak, unable to take criticism, and prone to tears wouldn't want them flying. Most Army Air Force brass assumed a menstrual period made women unstable and therefore unsafe in the air. But flight instructors knew otherwise. The few men who consented to train women found them more measured and balanced than hotshot male pilots who refused to admit weakness. Social bias also forced women to work that much harder. The result was often a better pilot. Jackie figured she could find enough women among her own contacts to form a core group within a month. This first group would be trained to lead the next recruits as the women's corps expanded. Eventually, the women would be split by skill and type of plane, giving them a presence throughout the country.

For the moment, though, Jackie knew she had to be careful in how the program was initiated. So other women weren't accused of "taking bread out of pilots' mouths" as she had been in Montreal, Jackie wanted her pilots organized on a temporary basis with the understanding that they would not interfere with the male pilots' combat roles; the women would step in when men were unavailable or needed elsewhere. It would be an experiment aimed to change the minds of the men in charge. Though this women's corps would be subsidiary to but affiliated with the Army Air Force, Jackie wanted them held to the same disciplinary and training standards. They should have uniforms, ranks, promotions, and even sleep in barracks just like their male counterparts. And of course, they would have to be compensated fairly. Once the women had proven themselves as capable as men, which she knew they would, she could fight for their militarization not as women but as pilots. They would thus work on a civilian basis until they could be properly militarized.

Jackie recommended all this to Lieutenant Colonel Robert Olds, who formalized the recommendation to the US Army Air Force, adding that Jackie be the leader of the eventual program. No one blinked at the suggestion. The top AAF brass considered her perfectly qualified, and some noted that her name alone promised the publicity needed to get the program started. With this first bit of momentum, both Jackie and Robert began reaching out to various women's groups. The War Department, meanwhile, put out a call for women pilots to submit their flying history and credentials.

By the fall of 1941, Jackie realized Robert was moving ahead without her. Though they'd agreed on a methodical, organized recruitment program to ensure positive publicity and good results, Robert was hiring women without structure and assigning them to jobs on an as-needed basis. Frustrated, Jackie went to see her friend Hap Arnold—now chief of the Army Air Forces—and was dismayed to find that he didn't see the need for a women's program at all. Hap argued that, for the moment, there was no pressing national need to free up male pilots for combat by recruiting women. Jackie countered that it was worth being prepared. She urged him to start training women so they would be ready, but the fact remained that Hap had neither the need nor the planes to begin training a group of women. Fortunately, he knew someone who had both. Air Marshal Sir Arthur T. Harris, head of the Royal Air Force delegation in Washington, was in desperate need of pilots for the British ferrying division. Hap sent Jackie to see him, and she agreed to bring a group of American women to England to work with the Air Transport Auxiliary on the condition that she be released from her contract the very instant the US Army Air Force needed her to set up a training school for female pilots at home.

On the first Sunday in December 1941, the seventh, Hap Arnold was taking a break. He'd just returned from Hamilton Field in California and was spending a pleasant morning quail shooting with airplane manufacturer Donald Douglas in Bakersfield. As the men returned to the cabin, they found Donald's father sitting by the car listening to the radio.

"The Japanese have struck Pearl Harbor!"

Details were scarce. The earliest reports said that the Japanese had sunk American battleships, destroyed airplanes, and generally decimated the area around Pearl Harbor and Honolulu. As the day wore on, the full scale of the attack became clear. Hundreds of Japanese fighter planes had damaged or destroyed nearly twenty American naval vessels and more than 300 airplanes. More than 2,400 people had died, and an additional thousand were wounded.

The next afternoon, President Roosevelt addressed a Joint Session of Congress. December 7, 1941, he said, was a "date which will live in infamy." Japan had attacked several American outposts, making it clear that American people, territories, and interests were in danger. There was no other recourse. From the moment of the attack on Pearl Harbor, he said, "a state of war has existed between the United States and the Japanese Empire."

The president received a standing ovation as the room erupted into cheers, and less than three hours later, he signed the declaration. America was at war, and its military was about to need a lot more pilots.

CHAPTER 6

England, May 1942

THREE MONTHS INTO HER SIX-MONTH CONTRACT WITH THE ROYAL AIR Force, Jackie had adjusted to life in wartime London. On any given night from her little house behind Harrods department store, she could hear the roar of a bomber flying overhead, antiaircraft guns firing from Hyde Park, and the eerie silence that said something was happening too close for comfort. Some nights she'd sit out on the front step and watch the fighter planes high above the city before going off to some dinner party or a social night playing games of backgammon. With every night potentially her last, there was no reason not to enjoy herself.

More important, her hand-selected group of twenty-five women pilots flying with the Air Transport Auxiliary was performing brilliantly. Holding the rank of captain, she fought their administrative battles while her girls—she thought of the women under her command as "her girls"—worked alongside British civilian pilots. They learned their way around the foreign countryside, flying planes from factory to field, between air bases, and then back to the factories when they needed repairs. They became socially close-knit, too, carrying on with life as normal against the backdrop of uncertainty and air raid sirens. When one of her girls, Mary Zerbel, wed her pilot fiancé, Roy Hooper, in a small ceremony, Jackie attended in uniform.

Jackie was halfway through her time with the ATA Girls when

Hap Arnold arrived in London toward the end of May for a series of meetings, among which was one with Jackie.

The issue of women pilots in America had gotten complicated since Jackie had left for England. In the wake of the Japanese attack on Pearl Harbor, Hap had given Robert Olds permission to hire fifty more women to fill roles as needed. Jackie had been terribly confused by this since, just weeks earlier, Hap had told her he wasn't going to need women pilots for months. She couldn't understand why he'd suddenly decided there was a pilot shortage but was still ignoring her recommendation of a dedicated women's corps. She'd been so mad she'd hand-delivered a memo to Hap marked "Urgent and Important." Jackie's insistence changed Hap's tune, and he told Robert to make "no plans or open negotiations for hiring women pilots until Miss Jacqueline Cochran has completed her present agreement with the British authorities and has returned to the US."

Now, months later and with her term in Britain coming to a close, the Army Air Force wanted Jackie with her years of experience to spearhead development and training of a women's air corps in the United States just as soon as she finished her remaining three months with the RAF. Hap, firmly on her side, would make sure she was the woman in charge.

When Jackie arrived at the RAF airfield for her flight home in the first week of September, she was given an urgent message. General S. H. Frank wanted to see her, so would she please return to the RAF base. She did and waited a full three days for a meeting that turned out to be little more than a polite dinner that had nothing to do with the war. Irritated by the needless delay, she finally made it home late on Thursday, September 10.

The next morning, luxuriating with a restful breakfast while flipping through the *New York Times*, Jackie stopped at a picture of a woman shaking hands with a two-star general. The headline above the image said, "She Will Direct the Women Ferry Pilots." The caption read "Mrs. Nancy Harkness Love being congratulated by Maj. Gen. Harold L. George." The article announced the establishment of the Women's Auxiliary Ferrying command. Jackie's shock quickly turned to anger. This was the top job Hap had promised to her, the job she had been preparing for over the past year, the job her experience in England was meant to inform. And this wasn't just a memo, it was an announcement in the press. Now everything was confused. Still in a rage, she called Hap, who turned out to be as angry as she was and asked her to come by in the morning. For the rest of the day, memos flew around the upper echelons of the Army Air Force suggesting her months-old proposal for a comprehensive female pilot program be instituted.

Jackie arrived in Hap's office first thing the following morning, the offending press clipping clasped tightly in her hand.

"I understand you already have women pilots?" she said by way of a greeting.

"I don't have any," he insisted.

Jackie demanded to know what had happened. She also wanted Hap to announce that they had been discussing a broad women's program for more than a year and that the article described an experimental trainer plane ferry group that was a small part of her larger program. She was determined to salvage the situation in her favor.

Hap explained that he had been blindsided. The Ferry Command had gone over his head directly to Secretary of War Henry Stimson with a project he hadn't approved. It was an act of insubordination as far as he was concerned, and it also explained Jackie's three-day delay leaving England. General S. H. Frank had called her back from the airfield as a way of delaying her return so the Ferry Command could set up its own program.

Neither knew whether the motivation was ill feeling toward Jackie personally or doubts about her abilities as a pilot, but Hap was determined to get to the bottom of it. With Jackie still present, he called the head of the Air Transport Command, General Harold L. George, and his deputy, General Cyrus Rowlett Smith, into his office. He told them they were to meet with their staffs and Jackie and settle on a program that satisfied *her*. He also insisted that *hers* be the only female flying unit, but as much as Jackie wanted full control, she had to say no. If the WAF program was cancelled a day after its announcement, it would only bring negative publicity. They agreed on a compromise: retain the WAFs but limit its activity to ferrying airplanes. Jackie's own program announced through a War Department press release would be the larger Women's Flying Training Detachment. The WFTD would be an experiment of sorts, a formal training unit that would put no restrictions on non-combat flying with the goal of showing women were as capable as men in the cockpit. It was a minor distinction, but it made Jackie happy that her girls would be separate from the WAFs who, as far as she was concerned, were little more than a bunch of "society dames."

Beginning with the names she had earmarked as potential recruits for Robert Olds a year earlier, Jackie scoured through records of pilots looking to find 500 women between the ages of twenty-one and thirty-five who could complete their training and be on active flying duty by the end of the following year. Applications streamed in. Jackie tried to interview each woman personally but soon had to farm out the job to assistants who traveled the country in her stead. Only Nancy Harkness Love remained a point of frustration. Jackie didn't know why it had been Nancy rather than herself who was appointed chief woman pilot, but she was determined that no woman would challenge her command of this program ever again.

The final piece of the puzzle was addressing Jacqueline Cochran Cosmetics. Jackie hadn't given much thought to the business when she'd left for England. She'd figured six months was a short tour, but she'd returned to find the office in disarray. If she was going to devote herself to the WFTD full-time, she would need someone else to manage her cosmetics business. She found her answer in a Mr. Vaughn, a quiet man who implemented her company's first real management structure. Her company was taken care of, and her female flying unit was taking shape. All Jackie needed now was a place to train her girls.

CHAPTER 7

Washington, DC, December 1942

"HOW DID YOU GET HERE SO FAST?" IT WAS HALF PAST SIX O'CLOCK IN THE morning, and Hap was amazed to see Jackie walking into his Washington office when he knew she was currently stationed in Texas.

"This had better be important," she replied, unwilling to waste time on pleasantries.

"Have some coffee," he offered.

Jackie knew he would only call her for a meeting to deliver bad news, so she'd flown through the night to get whatever it was dealt with sooner rather than later, and his stalling made her uneasy.

Jackie's Women's Flying Training Detachment had been formally activated a month earlier on November 15; Floyd had marked the date by sending Jackie his favorite professional headshot of her in her ATA uniform inscribed with "This is a photograph of the one whom I adore." For the past month, the girls under Jackie's command had been flying out of Howard Hughes Field near Houston, Texas, but the site left much to be desired. A former civilian training center on lease to the US Army Air Force, it had a grass landing strip, three hangars, and a few small buildings. There were no barracks, no restaurant, no place to even make a quick cup of coffee, and only two toilets.

The equipment was no better, a hodgepodge of whatever primary and basic training planes Jackie had been able to get her hands on.

From her office at the Flying Training Command offices in Fort Worth, Jackie had delegated onsite management to Dedie Deaton, a swimming instructor whose cousin worked at Jacqueline Cochran Cosmetics. Dedie's job, as per Jackie's instructions, was to find a place for the first group of twenty-eight girls to eat and sleep. She was also tasked with overseeing their evening study periods and making sure they visited the beauty parlor once a week; Jackie insisted her girls strike the same balance between professionalism and femininity that she valued for herself. The spartan facilities forced Dedie to get creative. She found the girls spare rooms in nearby homes and arranged for army transport vehicles to bus them to the airport every morning. In spite of their unconventional living situation, the first class of girls to join the WFTD maintained high morale as they sunk their teeth into training.

But Jackie faced continued resistance from the same men who had gone over Hap's head to establish the Women's Auxiliary Ferrying command. Harold L. George, Cyrus Rowlett Smith, and commander of the Military Air Transport Service Lieutenant General William Tunner all agreed that women couldn't withstand the rigors of a military flying program, and in the unlikely event they succeeded, saw no reason they shouldn't be managed by the army. Their solution was to fold WFTD into the Women's Army Auxiliary Corps (WAAC) led by Colonel Oveta Culp Hobby.

Jackie wanted nothing to do with the WAAC. She wanted military benefits for her girls, but her girls were pilots, not soldiers. The Army Air Force was under the umbrella of the army so it was barely a distinction, but she still wanted to stay close to the AAF. Long term, she knew the WFTD could help the AAF's campaign to become a standalone service branch. She also didn't think too highly of Oveta Hobby as a leader.

The real issue was that the WAAC and the WFTD had very different physical and background requirements. Pilots needed far different training than did non-flyers, so managing the two groups under one umbrella would be a logistical nightmare. There was also a question of rank. According to the AAF, all pilots were officers, but the number of women in the WFTD exceeded the number of officers allowed in the WAAC, meaning all Jackie's girls would be demoted. Jackie and Oveta would hold the same rank of colonel, but Oveta would still be Jackie's supervisor. To avoid these problems, Jackie wanted her girls militarized directly under the AAF through carefully worded legislation that would circumvent existing language stipulating that all pilots were men.

It had become a battle. George, Smith, and Tunner were as determined to militarize the WFTD into the WAAC as she was to keep them on civilian status until they could be their own group. The men went so far as to send a mole to Jackie for training to get inside information and chip away at her credibility. The WAAC issue, Jackie was sure, was why Hap had asked to see her.

"How would you like your girls to become part of the WAACs?" he asked, confirming her suspicions.

"Those girls will become part of the Women's Army Corps over my dead body." Jackie was livid. "Hobby has bitched up her program and she's not going to bitch up mine."

"Well, it's under consideration," Hap told her, far more calmly. As Jackie sat seething on the other side of his desk, Hap called Oveta and set a meeting.

The following day, Jackie sat boiling in Oveta's office, as much from anger as from the mink coat she wore since no one had offered to take it. Jackie bristled as Oveta gushed that her seven-year-old son was "just crazy about airplanes!" before admitting she didn't know one end of the machine from another. Oveta was ambivalent on the issue of who should command the female pilots. Since the decision wouldn't be made at her level, she didn't feel this discussion was worth her time. Jackie agreed, though on the grounds that no woman who thought airplanes were simply decoration should be in charge of her girls. Livid, Jackie remained firm on the point that a pilot needed to be in command of the WFTD, someone who truly understood the type of work they did. The meeting ended without any resolution, which meant for the moment the WFTD would remain an independent, civilian group separate from Oveta and the WAAC.

As months of training passed, the WFTD thrived. In early 1943, the Army Air Force began expanding the program with the goal of increasing the annual graduation rate. The program relocated to Avenger Field in nearby Sweetwater, Texas, a proper military site where the Royal Air Force was just finishing a training program for Canadian air cadets.

When the second batch of women pilot trainees arrived at the end of February, they joined the Army Air Force Flying Training Command as members of the 319th AAFFTD—the Women's Flying Training Detachment. "You are part of an experiment," they were told at their orientation, "which will do more to advance equality for women than anything that has been done so far."

Life at Avenger Field offered a stark contrast to the common conception of female pilots. These women weren't society flyers training for powder puff derbies. They were women from all walks of life, united by their love of flying and love of country. They were up at six a.m. to clean their barracks and eat breakfast before a full day of training. They flew seven days a week, learning the intricacies of every kind of plane they would be ferrying or testing for the Army Air Force. They did drills, trained to build upper body strength, took ground school classes, and studied every evening before lights out. Social status didn't matter, nor did appearance. They braided their long hair or tucked it under a turban to keep it away from engines and propeller blades and wore oversize coveralls they jokingly called "zoot suits" that were so bulky most girls had to shower in them just to get them clean.

However unglamorous, men seemed to feel an inexorable pull toward Avenger Field. Within days of the WFTD's arrival, more than a hundred male pilots made "forced landings" at the base just to get a glimpse of the all-female flying unit. And Jackie was having none of it. Remembering the handsy foremen from her childhood days in the cotton mill, she barred men from landing at her base unless it was a genuine emergency. She insisted only married officers be assigned to Avenger Field in an effort to protect her girls on their own base. She forbade her girls from fraternizing with male pilots and banned them from drinking and going to nightclubs to maintain a chaste and professional image, though Jackie herself kept whiskey in her desk drawer. They had to wear skirts, not slacks, to officers' clubs. She wanted them to command respect as pilots and as women. When Jackie found out that local prostitutes were passing themselves off as Avenger Field trainees to cash in on the allure of female pilots, Jackie barred the girls from leaving the base at night.

The strict rules earned Avenger Field the nickname Cochran's Convent, which Jackie used proudly. Likewise, the girls weren't bothered. In their unconventional living quarters, camaraderie flourished as they snuck moonshine onto the base for nighttime revelry and wrote a verse to the tune of "The Man on the Flying Trapeze" about the zoot suits. "Once we wore scanties, but now we're in zoots, they are our issued GI flying suits! They come in all sizes, large, large, and large. We look like a great big barge!" Eventually, they bought khaki pants, white shirts, and khaki flight hats to give them a unified look.

On April 24, 1943, Jackie watched with pride as the first group of forty-five girls graduated in a ceremony at Ellington Air Force Base. Hap was there to give the girls their wings, courtesy of Floyd, who'd bought and modified pins to read "319th." Watching from the sidelines, Jackie had to step in when she saw Hap falter over the women's lack of lapels; she walked over and discreetly suggested he simply hand them to each girl. None of the pilots cared that it was unorthodox; they were all thrilled to start flying real missions.

When the second class graduated a month later, Jackie felt far less celebratory. These girls were being sent to Nancy Harkness Love's Women's Auxiliary Ferrying Squadron, the WAFS in the Ferry Command. Jackie had spent months training these pilots in a program designed to prove that women were as capable as men, and now they were being lumped into the catchall ferry command. What Jackie needed was full military control over her program; otherwise, her pilots would be funneled into the program she viewed as below her girls' skill level. She needed the WFTD militarized under the AAF. Floyd tried to help by tapping into his previous life as a lawyer. He researched laws that allowed women to serve as doctors with the army and navy, looking for verbiage Jackie could adapt to apply to pilots.

Things started to fall into place for Jackie that summer. At the end of June, Hap named her head of all women flying for the AAF, a new role that allowed her to expand the program and send her girls to air bases around the country to test planes that had been repaired. Two months later, Jackie's new position was formalized when the WFTD and the WAFS were brought together and renamed the Women's Auxiliary Service Pilots under her full control. Nancy Harkness Love, the woman who had threatened her leadership from the start, was now her subordinate and director of the ferry division. As director of the WASPs, Jackie managed all training and operations in conjunction with the assistant chiefs of air staff and the air surgeon. The only thing missing was military status. Her program was still a civilian one, and Jackie could only hope that its continued success would earn the WASPs the recognition and benefits that would come with militarization.

In Wichita Falls, Texas, just a few hours from Avenger Field, Lieutenant Colonel Harvey Cobb came home to have lunch with his wife, Helena, and their two daughters, both of whom had come home from school for their midday meal. After America joined the war, Harvey transferred to the Army Air Corps hoping to fly overseas, but he was over the age limit to begin flight training. There was, however, a loophole he had finally been able to exploit, which meant he had big news to share over lunch. As the chattering family gathered around the table, Harvey silenced the room with the simple act of holding an envelope over his plate. "Who can guess what this is?" he asked with a smile.

"Orders?" Helena guessed.

"A war bond?" Jerrie asked.

"A commendation?" her older sister, Carolyn, said.

They were all wrong.

"It's nothing very important, I guess." He shrugged. "Just my—are you sure you can't guess?"

"William Harvey Cobb! Enough of this!" Helena cried. "Just your what?"

"Just my license as a private pilot!"

There was a moment of silence as the girls took in the news, then the room erupted in excited squeals. Harvey had been learning to fly in secret. This private license was the first step before his commercial license, which in turn would qualify him for military flight training regardless of age. But that wasn't it. He had one more surprise.

"Just so that I could practice whenever there's enough time—sometimes you can't just walk into the field and get the plane you want that very minute, you know—well, the Cobbs now own their own airplane!"

There was nothing to discuss. The girls were excused from school for the afternoon, and the family piled into the car and drove out to the small air force base at Sheppard Field where a Taylorcraft sat waiting. Carolyn went up first. Then it was Jerrie's turn to climb into the rear cockpit. At twelve she was still short enough she could barely see over the fuselage, but it hardly mattered. As soon as they were airborne, the wind whipping against her face, she fell in love. The Sun somehow felt hotter, even though she was only a few hundred feet off the ground. In that moment she recognized that the sky was her home. She needed to learn to fly, and not when she was grown—she wanted to learn right now.

"Daddy, will you teach me to fly?" she asked just days after that first flight. No, he said. He assumed that her excitement about riding in a plane would pass before long. But she tried again the next day, and the day after that, and every time he gave her another reason why the answer was no. So she changed her tactic.

"Dad, I don't mean that I'll fly *alone*. Just with you. All I want to do is to learn *how*. That's all. Won't you even show me how?"

"Ask your mother," Harvey replied, caving in to his youngest daughter.

Helena was far from thrilled by the prospect. "Don't be ridiculous!"

"Why?" Jerrie asked.

"Because..." she paused, reaching for an excuse, "your school grades are so bad, that's why!" In an attempt to keep her daughter on the ground, Helena had inadvertently given Jerrie the perfect bargaining chip.

Weeks later, her grades acceptably high, Jerrie sat in the Taylorcraft's rear cockpit on a stack of pillows. She could see over the fuselage and reach the controls now, but her feet dangled a foot above the rudder pedals. For the moment, it didn't matter. She listened as her father explained the instruments on the control panel and taught her hand signals. Then he gave her the cardinal rule: "Just sit there and don't touch anything and don't do anything until I tell you." With that, he hopped in the forward cockpit, taxied down the runway, and took the little plane into the air.

Jerrie watched the Texas landscape stretch out below her until she saw her father lift his hands. It was a signal. Carefully, she reached forward and grabbed the stick in front of her. When she saw her father point down, she pushed the stick forward and felt the Taylorcraft's nose dip as it went toward the ground. Next, she saw the signal for *up*, so Jerrie pulled the stick back and felt the plane's nose tilt skyward as it gained altitude. For an hour they flew in concert, Harvey giving hand signals and working the rudder pedals while Jerrie followed his instructions on the engine's throttle and stick. Too soon, Jerrie saw the *hands-off* signal and knew their lesson was over.

Back on the ground, they met a lieutenant from Harvey's squadron in the hangar. "Good flight?" he asked them.

"Training flight," Harvey explained. "Jerrie just had her first lesson."

"Better log it," the young lieutenant said, producing a small blue memo book from his pocket. "Here, Jerrie, let me give you your first log." Jerrie watched as he divided the first page into columns, then filled in the time, date, and place. As her instructor, Harvey signed off on the hour. The men chuckled over the girl's informal logbook, but to Jerrie, its empty pages held immeasurable potential.

Father and daughter spent hours in the little Taylorcraft, often taking off and landing with Jerrie scrunched out of sight; technically, Harvey was accumulating solo hours and shouldn't have had a passenger. But it was his private plane, so he could bend the rules. The real problem for Jerrie was that the days were fast becoming too short for everything she wanted to do.

Jackie felt like time was disappearing as little issues commanded her attention. She, of course, wanted her girls to be the best possible flyers, first and foremost, but she also wanted them to be perceived as virtuous and well-groomed women. They needed real uniforms, not surplus nurses' uniforms or ones made from excess Women's Army Corps uniform fabric that she herself wouldn't be caught dead wearing. She had one designed in New York City at her own expense: a Santiago blue wool jacket, skirt, and beret with a white cotton shirt and a black tie topped with a beige trench coat and a black calfskin utility bag that was a near match to the black alligator Fabrikoid case she sold with her weekend cosmetic kits. For flying, they would wear slacks and a waist-length battle jacket. Now she just needed money to have hundreds of them made, which meant getting approval from Army Chief of Staff General George C. Marshall. Jackie went to his office with three women in tow: a professional model in the perfectly tailored designer uniform, a woman from the quartermaster general's office in an ill-fitting surplus nurse's uniform, and a somewhat rotund woman in a similarly unflattering uniform made of the excess WAC fabric.

"Well," General Marshall said as he considered the women before him, "I like the one you're wearing, Miss Cochran." She was dressed in a hand-tailored gray suit she'd bought in Paris before the war. It wasn't an option, she told him, but she knew where his eye would go next. "The blue one is the best." He predictably picked the closest thing to couture, what the professional model was wearing, then agreed to pay to outfit every one of her WASPs in the designer uniforms. It was a small victory, but it meant a lot to Jackie. Her girls finally looked like the respected pilots they were.

As the months wore on, the WASPs' role expanded. Soon, at 120 Air Force bases around the country, Jackie's girls were assigned to every kind of wartime stateside flying imaginable. They towed targets for male pilots to practice aerial shooting. They flew simulated strafing operations, passing low over ground troops and gun positions to train soldiers how to defend against aerial attacks. Searchlight flights saw the WASPs flying without navigation or lights so searchlight crews could practice finding planes. Tracking missions helped anti-aircraft gunnery crews learn to aim at moving targets. On low-altitude night flights, they dropped flares and smoke screens. When a plane was repaired, it was a WASP who flew it to ensure it was safe. They flew everything, muscling the controls of big planes like the B-17s, B-24s, B-25s, B-29s, C-47s, C-6s, and DC-3s bombers as well as smaller pursuit planes like the P-39, P-40, P-47, and P-61. They delivered flying cargo, top-secret weapons, and personnel around the country. One WASP tested a YP-59A twin-jet pursuit plane. While the army's WAC and the navy's WAVES—Women Accepted for Volunteer Emergency Service—ran recruitment campaigns encouraging women to join the war effort, the WASP program never had to advertise. Press coverage made the program so popular that Jackie had to turn down hopeful WASPs for lack of space.

Jerrie was racking up hours in her father's Taylorcraft when her flying career came to a grinding halt. Harvey was transferred to Denver, Colorado, forcing the family to sell the plane and a devastated Jerrie to give up her horse. To alleviate their daughter's despair over this unexpected move, Harvey and Helena allowed Jerrie to work in a stable so she could board a horse for free and earn some money on the side. On weekends, she was allowed to go to the local private airport to watch the planes landing and taking off, though her only hope of flying was to be invited along for a ride. Though she developed a love of literature, it was these extracurricular pursuits that became the focus of Jerrie's life, to the detriment of her education.

Jerrie was playing hooky one day when her father caught her in a neighborhood park. Rather than lose his temper, Harvey drove his daughter to Rocky Mountain National Park. They found a quiet spot near a stream and sat for a moment before Harvey broke the silence.

"What's the problem?" he asked.

"Well, there's no one problem, really," Jerrie replied. "I just don't like sitting in classes when there's so much going on outside. I can't seem to get interested in most of the stuff in school. Why can't I get a full-time job?"

"Doing what?"

The pony farm was willing to hire her, Jerrie explained, providing she had her parents' consent. Then she admitted that this truancy wasn't her first.

"Jerrie," Harvey sighed as he loosened his tie and unbuttoned the top button on his shirt. "I think it's good that you've had the opportunity to work with ponies and keep your own horse well fed and well sheltered. But I don't think you plan to spend your life taking care of horses."

"It's not a bad life," she said.

"Except that it's not what you want."

Jerrie paused. Her father was right. "No," she said simply.

"I wish you could see the cockpits of some of the new planes on the base, Jerrie. Hundreds of gauges, switches, controls, dials. The pilots and engineers practically have to be mathematicians. People can't just fly by the seat of their pants these days."

"Not much chance of my ever flying that kind of plane." Jerrie remained despondent.

"Now how can you tell, Jerrie? Just look at what's happened to aviation since you were born. From barnstorming little grass-hoppers to B-24s. Just look at how many pilots we train here. Who knows what will happen when the war's over?" But Jerrie still couldn't see how English, history, or gym class was going to get her closer to a cockpit. "You learn a lot of other things, too, Jerrie," her father continued.

"Don't kid yourself. Look, honey, this isn't new to me—I went through the same thing myself." He'd piqued her attention. "I ran away from school because there was so much going on outside that I just couldn't stay in a high school classroom. But by the time I got out of the navy, I knew that I'd been wrong. Education isn't junk. I had to go back, and I was still going to college to get my degree when your mother and I were married. For a while, it put me years behind. And I'm probably not through with school yet."

Jerrie paused. She had, of course, lived through her father's struggle to qualify for flight status with the army all because he hadn't finished school. Sitting there in the park, she realized he was right. She promised to try harder in school, but she wouldn't give up her weekends at the airport. To finagle her way into the air, she joined the Flying Minutemen, an 80,000-person strong volunteer corps of civilian pilots who flew sentry missions on the borders, hunted lost planes in the mountains, and spotted forest fires. She couldn't do any flying, but tagging along on flights and watching the controls was better than nothing.

As 1944 dawned, Jackie decided it was time to revisit the issue of militarization. The WASPs had proven their worth, and she hoped it would be enough to get them militarized the right way, as pilots under the Army Air Force rather than lumping them under the Women's Army Corps. She had Hap's support, which meant she just needed Congress to approve the idea. Representative John M. Costello of California introduced the bill, House Resolution 4219, on March 22. It stated that women would be militarized as part of the US Army but assigned to the US Army Air Force. The so-called Costello Bill went before Congress, falling to the House Committee on the Civil Service, led by Robert C. Ramspeck.

Hap testified that America was experiencing a manpower shortage, the effect of which was considerably lessened by the WASPs. They were professional, experienced, and deserved the benefits of militarization, foremost among which was monetary. As civil servants, the WASPs' pay was lower than their male counterparts' and they still had to cover their travel expenses between bases, their food, their lodging, and some of their clothing. They didn't have the six-figure insurance policies. The military didn't arrange to get the bodies of WASPs killed in action home to their families, nor were funeral expenses covered. The total cost of the program by mid-1944 was not the $20 million[12] some claimed but closer to $6 million.[13] The WASPs cost the government far less than their male counterparts.

Congress disagreed. Most politicians saw the WASP program as little more than a costly way to take jobs away from male pilots. Rumors swirled that the Army Air Force was so desperate for pilots that it was carrying out a large and expensive recruitment program, which Hap adamantly denied. If that was the case, he told the Ramspeck Committee, "then this Nation, insofar as manpower is concerned, is in worse position than any of our allies, and apparently any of our enemies."

While the Ramspeck Committee was debating the Costello Bill in Congress, government-funded civilian pilot training programs were cancelled. These programs that trained hopeful aviators for combat in exchange for their enrollment in the military meant that now, suddenly, thousands of men hoping to fly were subject to the draft as soldiers. Many complained to their congressmen, who put pressure on Hap to find roles for the newly minted aviators, but the commanding general of Army Air Forces didn't change his position; Hap still argued that the WASPs were necessary to free men for combat roles even though America was suddenly flush with male pilots. Discussion over

12 About $291,557,955 in 2019.
13 About $86,000,000 in 2019.

the bill thus turned from an issue of women pilots getting the benefits they'd earned to a discussion of women unfairly taking jobs away from men. Many congressmen argued that the WASPs' militarization meant male pilots with upward of 2,000 flying hours would be stuck cleaning windshields for glamorous lady flyers with less than thirty hours in the air. The opposing voices and displaced pilots alike directed their outrage toward Jackie, suggesting she was using her vivacious wiles to sway Hap to giving her a vanity program. Toward the end of June, the Costello Bill was defeated by twenty votes.

The bill's failure forced the Army Air Force to review the entire WASP program, and by fall Hap had to admit its time was coming to an end. The reality was that the war was nearly won. America needed fewer pilots and certainly didn't need women to free men up for combat missions. Though crestfallen, Jackie asked that the deactivation date be December 20 so the girls would have time to make it home for Christmas. She also asked that they be given certificates and cards outlining their military experience to help them get better jobs in the civilian market. AAF legislation still stipulated that all pilots be men, so there was no chance for any WASP to be hired by the service. Jackie was the exception, hired on as Hap's consultant.

In early October, nearly a thousand WASPs serving all around the country learned their fate. "Each of you has made an important contribution to your country at war and has aided immeasurably in establishing women's place in aviation," their form letter from Jackie read. "It has been a great honor to have been the Director of this program and a pleasure to have known most of you individually. My very best wishes for your future success and happiness."

Both Jackie and Hap were at the final WASP class graduation on December 7. "I am glad to be here today and talk with you young women who have been making aviation history," Hap told the graduates. "You, and more than nine hundred of your sisters, have shown that you can fly wingtip to wingtip with your brothers. Frankly, I didn't know in

1941 whether a slip of a young girl could fight the controls of a B-17 in the heavy weather they would naturally encounter in operational flying. It is on the record that women can fly as well as men. We will not again look upon a women's flying organization as experimental. We will know that they can handle our fastest fighters, our heaviest bombers. On this last graduation day, I salute you and all WASPs. We of the AAF are proud of you; we will never forget our debt to you."

In her heart, Jackie knew her failure to have the WASPs militarized was the cause of the program's cancellation. If she had swallowed her pride and let her girls fall under the WAC, they would be going home with military benefits, but instead, they were going home with nothing. The thought depressed her, made worse by public attacks on her character as she was blamed for the program's cancellation. Publicly, Jackie put on a brave face and maintained the WASP program was cancelled because it had met its goal of leveraging women pilots in a time of war; after all, she had pitched the program as an experiment, knowing it was the only way to get women to fly with the military. As her final act as their leader, Jackie compiled a report for Hap and let the numbers speak for themselves. Of the 25,000 women who had applied, 1,830 had been accepted, 1,074 had graduated, and only 142 had resigned or been let go. When the program was cancelled, just over 900 WASPs, including the sixteen women who had joined the original Women's Auxiliary Ferrying command, were still serving. As a group, the women flew some 60 million miles. Thirty-eight women had died in the line of duty, which equated to the loss of one pilot for every 16,000 hours flown, a number comparable to male cadets.

Jackie was presented with a Distinguished Service Medal for her work as head of the WASPs from her good friend Hap, but the dark cloud surrounding the deactivation lingered. For the first time since striking out in life as Jacqueline Cochran, she had failed. As the war in Europe drew to a close, she retreated to the Ranch with Floyd.

CHAPTER 8

After the War, 1945

JACKIE FELT RUDDERLESS. SHE WANTED TO GET BACK TO FLYING, GET BACK to her business, get back to *something* that could give her distance from the WASP deactivation issue. She settled on travel. She sought permission from the Civil Aeronautics Administration to fly her private Lodestar to France and England under the pretense of expanding her cosmetics business, but her application was denied. Floyd, knowing all this was a thinly veiled excuse to see postwar Europe with her own eyes, facilitated her trip. He bought the failing *Liberty* magazine and made his wife a war correspondent with an overseas assignment just to give her the journey she craved. Jackie knew she was an absolute phony, but figured other people were doing the exact same thing so she might as well enjoy herself.

With an array of clothes, jewelry, and a large stock of whiskey in tow, Jackie visited her friends General Carl "Tooey" Spaatz and General Jimmy Doolittle on Guam before traveling on to Okinawa, Calcutta, New Delhi, Agra, and Karachi. She witnessed the surrender of Japanese general Tomoyuki Yamashita aboard the USS *Missouri* in Tokyo Bay. She watched the Nazi war trials in Nuremberg and visited Hitler's Berlin bunker where he had taken his own life. The articles she wrote were deemed unpublishable by her editor, little more than clumsy rehashings of press releases with no added perspective, but Floyd wasn't about to fire his wife.

When she returned from her European adventure in the spring of 1946, Jackie threw her energy into her cosmetics business. Floyd helped her buy the building at 10 West 56th Street in Manhattan as a new office space where she could also set up a retail store and build corporate culture with a large employee lounge.

She also reconnected with her friend Randy Lovelace. Promoted to lieutenant colonel in the US Army, he was now chief of the aeromedical laboratory at Wright Field, where his staff of 225 doctors explored all known stresses affecting flyers. Like Jackie, Randy had been presented a Distinguished Flying Cross for his wartime work by Hap Arnold. Now thirty-eight and energized by his professional successes, he was considering moving his wife, Mary, and their children to Albuquerque so he could join his uncle Randy Lovelace I—whom everyone called Uncle Doc—at the family clinic. A family tragedy made Randy's decision for him. His son Randy III developed polio and died on July 7. When the Lovelaces' other son, Chuckie, contracted the same disease and was paralyzed, Jackie flew the whole family to the Warm Springs Foundation in Georgia, but the specialized clinic couldn't help. After they lost their second son on August 13, Randy sought solace in his family. In Albuquerque, Randy became a partner in Uncle Doc's clinic alongside a colleague, Dr. Edgar T. Lassetter. Before long, the trio decided to open a new clinic and research department, one Randy wanted to model on the Mayo Clinic, which had played such an influential role in his career.

Their family hardships also brought the Lovelaces closer to Jackie and Floyd. Floyd, in particular, was struck by Randy's kindness, thoughtfulness, and insatiable, almost infectious, curiosity. He saw Randy as someone who could catalyze change and inspire people into action by thinking critically about a problem and acting on what needed to be done. Floyd's confidence in Randy was so complete that when he, Uncle Doc, and Edgar Lassetter launched the Lovelace Foundation as a medical research and development center, Floyd signed on as president and chairman of the board of trustees.

As Jackie settled back into her peacetime life, demobilized troops were similarly returning home to find their postwar lives were very different. Politically, America was experiencing turbulence. President Roosevelt's sudden death in April of 1945 thrust an unprepared Harry Truman into the most powerful office in a country poised to become a global leader as it negotiated its shifting relationship with the Soviet Union. Though allied against the spread of Nazism during the war, the United States and the Soviet Union were ideologically opposed. Soviet communism was a threat to American democracy, making the Soviet Union America's next international adversary. All told, there was more than enough political activity to keep Jackie's friend Lyndon Johnson busy in Washington. Newly reelected to Congress, his politics were defined by working toward American-Soviet cooperation.

While politicians navigated the country's future, Americans nationwide were keen to return to peacetime living and enjoy the financial boon brought by wartime production. After decades of disruption from the Depression and the war, family values took center stage in America's collective psyche. Government advertising campaigns that had encouraged women to join the workforce during the war told women their new job was to take care of the men home from combat. Images of women wearing uniforms were replaced by images of women in dresses and heels taking care of their children while their husbands went off to work. Rosie the Riveter, the symbol of everything women could do, was redefined by her biology and told her priorities should be having and raising children. But the social landscape had changed, and this return to traditional roles didn't resonate with newly empowered women. Some were unwilling to give up their work to return to the kitchen. Others, younger women who had never known an America where their gender didn't have the vote, had grown up in a country that hadn't placed limitations on what women could do. To some, this new model of womanhood defined by marriage and children was baffling.

✈

The Cobb family moved back to Oklahoma after the war, this time to Oklahoma City, and right in time for Jerrie to begin her senior year of high school in the fall of 1946. Classen High, she was thrilled to discover, had a pilot on staff. Coach Conger was a licensed flight instructor who believed there was no reason that a responsible, intelligent, and alert teenager shouldn't learn to fly. And so he consented to teach Jerrie in his Aeronca, a stubby-nosed "knocker" with two seats crammed side by side into its tiny fabric-covered fuselage.

Lessons took place rain or shine. Though taller now, Jerrie still had to sit bolt upright to get a clear view out the little windshield. For a while, it was like flying with her father all over again. Coach Conger would take off and land, allowing Jerrie to fly while they were airborne. Then, one cold March day not long after her sixteenth birthday, Coach Conger turned to her after a flight. "Take her," he said, "she's all yours." Without another word, he jumped out of the plane.

Alone in a cockpit for the first time, Jerrie carefully taxied the plane into position at the end of the runway. The Aeronca's 65-horsepower engine felt like the most powerful thing in the world, and the grass runway seemed to stretch on forever. She checked the wind's heading with a nearby windsock and checked her compass, altimeter, engine speed gauge, and airspeed indicator on the instrument panel. Then she pushed the throttle forward and started to roll. She felt like her heart was beating in time with the engine as she bumped along the runway on three wheels, then two as the nose lifted up. At forty miles per hour, she eased the control stick back and felt the plane lift off the ground in a slow, steady climb. The engine's roar settled into a hum as she leveled out. She circled the field once, then brought the Aeronca down for landing. Repeating the simple pattern two more times completed the requirement for a pilot's first solo flight.

"You did just fine," Coach Conger told her, smiling, as she taxied toward the barn at the end of the run. "I want you to keep right on doing what you're doing."

In the spring of 1947, Dwight David "Ike" Eisenhower sat before the Senate Military Affairs Committee. The issue in question was a new bill, one that called for a national military establishment consisting of an independent army, navy, and air force managed by a secretary of national defense. The five-star army general who'd orchestrated the allied forces' invasion of Europe with the Normandy landings—what Americans had come to know simply as D-Day—knew the future of America's military power hinged on a coequal air force. In his testimony, Ike emphasized the importance and necessity of air power in America's peacetime landscape. Hap Arnold also testified to the importance of aviation in future wars.

On July 26, President Truman finally approved the National Security Act of 1947. Jackie was thrilled that the air force had finally become its own service branch, and was even more pleased when her friend General Carl A. "Tooey" Spaatz was named chief of staff and another friend, William Stuart Symington Jr., was appointed air force secretary.

When Jackie paid a visit to Stuart Symington soon after in the fall of 1947, she found he wasn't alone in his office. Chuck Yeager was there, too, the handsome, twenty-four-year-old pilot with a shock of dark hair whom Jackie knew by reputation. Chuck's flight hadn't been publicly announced yet, but aviation insiders had heard that weeks earlier, on October 14, 1947, Chuck had flown the rocket-powered Bell X-1 experimental aircraft faster than sound in level flight. He was the first person in history to break the so-called sound barrier. Jackie wasted no time introducing herself. "I'm Jackie Cochran," she said, pumping his hand vigorously. "Great job, Captain Yeager. We're all proud of you." Keen to hear more about his flight, she immediately took him to lunch.

Chuck was struck by Jackie. He hadn't expected the restaurant owner to bow to her as they walked in, nor had he anticipated that she'd send back every course of their meal before finally going into the kitchen herself to give the chef hell. She peppered Chuck with questions about his X-1 flight but somehow still managed to work into the conversation the awards she'd won and committees she served on. She bragged that Hap Arnold loved her scrambled eggs and Tooey Spaatz was a drinking buddy but didn't hide the chip on her shoulder. "If I were a man," she told him as they ate, "I would've been a war ace like you. I'm a damned good pilot. All these generals would be pounding on my door instead of the other way around. Being a woman, I need all the clout I can get." Chuck doubted that between her career and Floyd's connections she was lacking in clout. He could hardly even imagine she faced prejudice on account of her sex. Talking to Jackie was like talking to any other pilot, and it was clear she not only knew her planes, she was as crazy about flying as he was.

As they parted ways that afternoon, Jackie urged Chuck to keep in touch. She liked him off the bat and suspected he was both a professional and personal friend worth maintaining.

Still in high school, Jerrie took any job she could find to earn enough money for flying time in pursuit of her private pilot's license. She picked berries, typed for a publisher, ran the cash register at a movie house, and made deliveries for a local drug store. She drove around the city on a three-wheeled motor scooter, dressed in white coveralls with blond pigtails sticking straight out behind her, collecting scavenged automobile parts for a local garage.

The odd jobs paid off when the weekend came. Jerrie took a short bus ride to the town of Moore, stopped at a gas station to buy five gallons of gasoline, then walked the last three-quarters of a mile to the grassy field where Coach Conger's Aeronca was tethered. She filled the tank with the gas she'd bought, then flew until she had nothing left but fumes. She reversed the route to go home, dropping the empty container off at the gas station on her way. Hour by hour, she filled out her logbook.

On March 5, 1948, Jerrie handed her logbook to the towering young man who would be taking her up for her exam. She could tell from his demeanor that he was what one might call a "by the book" examiner. He seemed impressed that she had 200 flight hours in her log, far more than the thirty-five she needed, but he didn't say anything as he folded his long body into the little Luscombe airplane. The all-metal aircraft had a more powerful engine than anything Jerrie had ever flown, but she wasn't nervous.

What should have been a two-hour flying exam lasted only forty-five minutes. Confident in her abilities, Jerrie was sure the examiner was just eager to stretch his legs. He had, in fact, seen enough of her flying to know she was good. He awarded her a private pilot's license. It was the best seventeenth birthday present she could have asked for.

On the night of Saturday, May 22, 1948, Lyndon Johnson stood before a small crowd at a rally in Wooldridge Park near downtown Austin, Texas. The congressman was ready to take his political career to the next level. A wartime tour with the US Navy had raised his profile, and he had cemented his reputation as a politician focused on modernity and equal opportunity. He also knew his constituents wanted to be left alone to drink beer and make money while staving off the growing concern about the spread of communism. The Soviet Union had indeed become America's main adversary. Both countries were developing nuclear weapons, and growing tensions were making the fear of a new war increasingly real. This changing stance had forced Lyndon to reverse his earlier position of pursuing Soviet-American cooperation in favor of a policy of confrontation, and he was determined that Texas wouldn't be treated as some backwater state like Kansas or New Mexico. He would make sure Texans would have their say in the budding Cold War. As he declared his candidacy for an open Senate seat that night, Lyndon's every move was planned. His kidney stone, however, wasn't.

Lyndon had both a fever and chills, but he refused to call off the rally. Dressed in a smart suit and in excruciating pain, he managed to run to the stage after his introduction and rile the crowd up with a speech advocating preparedness, praising the Truman Doctrine and the Marshall Plan before condemning the abuse of big business and big labor. He earned a standing ovation from the crowd, many of whom lingered afterward for a chance to shake his hand.

The next morning, Lyndon boarded a flight for Amarillo with speechwriter Paul Bolton and campaign worker Warren Woodward in tow. The mini campaign crew stopped in San Angelo, Abilene, and Lubbock along the way to call local journalists and local politicians. It wasn't a particularly warm day, but Lyndon was sweating so profusely Warren called the Johnsons' home to ask their maid to send more shirts.

On Tuesday afternoon, May 25, the trio boarded an overnight train bound for Dallas. Sleeping in the bottom berth of an open Pullman car, Lyndon alternated between having bone-rattling chills, during which he'd call for more blankets or for one of his companions to get into the berth to warm him with body heat, and feverish sweats, when he'd have all the car's windows opened. Delirium eventually set in, but he remained convinced that the stone would pass on its own. The next morning in Dallas, Lyndon dressed and took a car to the Baker Hotel. He met with Stuart Symington and General Robert J. Smith about making Sheppard Air Force Base a permanent installation, but the meeting took all his energy. By the end of the day, Lyndon couldn't make it down the hallway in his hotel without throwing up. Despondent over the campaigning days he would lose while recovering, he finally had to admit the stone wouldn't pass on its own. That evening, he consented to go to the hospital.

The next afternoon, following a last-minute invitation from Stuart Symington, Jackie arrived in Dallas for an Air Force Association luncheon. When she arrived fashionably late for the pre-lunch cocktail hour, she spotted Stuart; he was so tall she could easily pick him out above the crowd.

"I thought you'd never get here," he said by way of a greeting as she approached him.

"I told you I'd be here in time for lunch!"

"Well now, Jackie, I've turned out to be your mailman." He handed her an envelope. She slipped it into her purse before ducking into the hallway to see what was so secret. "Jackie," the letter began, written in Stuart's handwriting. "Lyndon is in hospital. Go to the back of the hospital where you'll find some steps. Walk up, and on the second floor . . . you'll find him. Please don't announce yourself. Just go and find him. I don't want you to be recognized."[14] It hadn't escaped her notice

---

14 Recounting this story in her memoir, Jackie writes that the note read "Lyndon

that Lyndon was absent from the luncheon, and now she knew why. She also thought she understood why her invitation to Dallas had been so last-minute: Stuart knew she always traveled in her own plane, and also knew it would be a discreet way to get Lyndon to a specialized clinic.

Jackie's first thought was discretion. If she left right now, so soon after her arrival, it might arouse suspicion, and that was the last thing she wanted. So she reentered the hall just as lunch was served. She sat through the whole meal before excusing herself with a headache just as the speeches were beginning. Leaving the hall, Jackie realized for the first time that she was anxious. She hired a car, looked up the hospital address, grabbed a map, and set off. The promised steps at the back of the hospital were some kind of off-limits emergency stairwell, but the door had been left ajar. She pushed her way inside and followed Stuart's directions to Lyndon's room.

"Stuart Symington sent me," Jackie told Warren Woodward, the campaign aide, as he opened the door to reveal Lady Bird Johnson sitting in a small room. Jackie registered how exhausted they both looked as Warren wordlessly led her through to the connecting room where Lyndon Johnson lay. Jackie was shocked at how pale he was and how bad the room smelled, a mix of stale sweat and vomit. In a flash, Jackie's long-ago nursing training kicked in. She instinctively reached for his wrist. She felt Lyndon's pulse fluttering underneath his clammy skin.

"Either you get proper medical aid for this man or he's going to be dead within twenty-four hours," Jackie muttered to Lady Bird as she reentered the anteroom. "Listen," she continued in a whisper, "there's a doctor up at the Mayo Clinic who has fantastic success with a special new procedure to remove these stones without surgery." Dr. Gershim

---

is in XYZ Hospital" and in "room number such and such." Though she doesn't give these details, it's safe to assume Stuart's actual note contained all the information she needed.

had cleared up one of Floyd's worse kidney stones, and Jackie suspected the same physician might be able to save Lyndon.

"Lyndon is so stubborn," Lady Bird replied, visibly frustrated that her husband was willing to risk his health rather than disrupt his Senate race. "Make him do something," she implored. "Let's get the Mayo Clinic on the phone."

Jackie called Dr. Gershim, who told her he couldn't get down to Dallas himself. The best option was for her to fly him up to Mayo as soon as possible. Just the thought of transporting a sick friend terrified Jackie.

"What if he dies en route?" she protested.

"Jackie, moving him can't really do that kind of harm," the doctor told her over the phone. "In fact, the movement might make the stones move, and that could only be good."

The doctors in Dallas compiled everything the Rochester, Minnesota, doctors would need to know about the stone's location—it was between his kidney and bladder—and gave Jackie some painkillers she could administer by injection during the flight. Meanwhile, Jackie figured out the flight plan. She decided to leave after midnight. Flying through the colder air, she could get more power out of her engines, just enough that she wouldn't have to stop for fuel en route. She had her maid, Ellen, who'd come to Dallas with her, buy pillowcases, big soft bath towels, a large plastic pan, two quarts of rubbing alcohol, and blankets, and load everything into the plane.

Just before three o'clock in the morning, Jackie returned to the hospital. "Lyndon," Jackie spoke in a low voice, "you're going to the Mayo Clinic, Lyndon. We're taking off now in the middle of the night so no one will see us and no one else will ever know." He squeezed her hand in reply. Dressed in a bathrobe and dragging his feet with every shuffling step, Lyndon bit through his lip on the short trip from the hospital to Jackie's Lodestar. Once on board, she settled him into the bed along one side of the fuselage, strapping him in as tightly as she dared and placing an oxygen mask over his nose and mouth. Ellen took the seat at the table across the aisle while Lady Bird and Warren took seats farther back, where the extra blankets and pillows awaited them. Walking up to the cockpit, Jackie stopped by her patient. "Lyndon, are you okay?" He didn't reply. Settling into her seat next to her young flight engineer, Steve, she was suddenly acutely aware of the lives in her hands.

She took off through a low ceiling, and when the first hint of sunlight appeared through the clouds, she felt calm. She engaged the autopilot. Just as she relaxed into the rhythm of the flight, a blood-curdling scream ripped through the cabin.

"Watch it, boy," Jackie told Steve as she jumped from her seat.

She found Lyndon drenched in sweat. Again her old nursing instinct kicked in as she ripped his clothes off and, with Ellen's help, swabbed him with a sheet soaked in alcohol, hoping the evaporation would cool him off. Then she brusquely turned him onto his stomach to give him another shot of the painkiller, and he was immediately sick all over her. She grabbed the dishpan and shoved it closer to his face to catch any more vomit, checking that Steve was keeping up with everything in the cockpit. The crisis passed, and she and Ellen swaddled Lyndon in clean blankets and waited as he fell asleep. Glancing toward the back of the plane, Jackie was amazed to see Lady Bird and Warren had slept through the whole episode.

After the most stressful six hours she'd ever flown, Rochester came into view. She landed to find an ambulance waiting with a crew who took Lyndon straight to the hospital. There, doctors filled his bladder with water and crushed the stones. After seven days under observation, the candidate was cleared to resume his campaign.

Lyndon won the election with an eighty-seven-vote margin. When a reporter asked "Landslide Lyndon" who had done the most for his campaign, he answered simply, "A woman—not my wife."

CHAPTER 9

Oklahoma City, Fall 1948

"NO COLLEGE."

Jerrie, now a senior at Classen High, should have been thinking about continuing her education. Instead, she was thinking about flying, and her family didn't take the news well. Three voices shouted back, demanding to know what she would do without an education, without school dances or a thriving social milieu. The University of Oklahoma was right next door, and Carolyn could give her an introduction to sorority life. But Jerrie wanted none of that.

"Look," she began, "you all know that I want to get my commercial pilot's license..." Three heads nodded in agreement. "And it takes a lot of money..." Three heads nodded again. "Which means that either I'll have to keep doing a lot of little jobs, or get one big job as soon as I can." There was no nodding this time, just patient waiting to see what would come next. "Well, it just happens that I've been offered a job—one that pays very well. Of course, it means a lot of traveling, with a girls' softball team. The Sooner Queens." Women's softball was a popular spectator sport, but working as an athlete was the last thing Jerrie's family expected of her.

"Unthinkable!" Helena cried. "Now I've heard everything! First horses, now airplanes, now baseball. Geraldyn,"—Jerrie hated the use of her full given name—"you're already seventeen. When are you going to stop thinking about hobbies and games and start thinking about your future?"

"My future!" Now it was Jerrie's turn to explode. "Mother, I've thought about nothing else since I was twelve. All I've ever wanted to do was fly. That is my future. The only reason I want to do this is to earn enough money for my commercial pilot's license." Then she turned to her father, hoping to find an ally. "Dad..."

"Now, Jerrie, the field today is overrun with highly trained men back from the war. A girl doesn't have a chance. Your mother and I don't want to see you break your heart trying to find a place in aviation that isn't there."

"But, Dad, how will I know until I've tried? All I want is a chance!"

The Cobbs reached a compromise. Jerrie was allowed one year for her chance, one year playing with the Sooner Queens before she would have to go to college.

Life with the Sooner Queens was anything but glamorous. Jerrie played as first baseman under the hot sun, dressed in a little satin skirt and blouse. The team traveled from town to town in two station wagons loaded with all their gear. As one of the rookies, she was responsible for lining the base paths, raking and rolling the field, and cleaning the clots from spiked shoes. In spite of some missed signals and failed plays, she earned enough money between her salary and bonuses to cover the additional flight time she needed to qualify for her commercial license, which she earned on her eighteenth birthday.

Unfortunately, Jerrie found that her father was right. When she started looking for work as a pilot, no one was interested in hiring an eighteen-year-old woman. "It's like this, honey," one flight operations manager told her. "I've got pilots running out of my hangar doors. I can actually choose 'em by the services—ex-Navy, ex-Marines, ex-Coast Guard. These," he said, picking up a manila folder, "are all the applications for jobs I don't have. Pilots are a dime a dozen today, and they've had thousands of hours in fighters and bombers—not just a few hundred civilian hours like you puddle jumpers. You'd be about forty-sixth on the list if you still want to apply." It was the same everywhere. Unable to find work flying, she walked like a condemned woman off to the Oklahoma College for Women the following fall, a school she chose because it was close to the Chickasha Municipal Airport.

To stay close to planes, she got a job at the airport as a general flunky and grease monkey. Occasionally she worked with a local crop dusting service, taking a new pilot up for a checkout flight in the company's open-cockpit Stearman biplanes and even flying the odd job herself, though she needed a hand getting the hundred-pound bag of insecticide into the aerial bin. Still, given the choice between three more years of college and a life of crop dusting, she preferred the latter. She knew it, the college knew it, her parents knew it, and her first year of college became her last.

Jerrie returned to the Sooner Queens for the summer season. While traveling around in her little satin skirt she found a plane in Denver, a little maroon-and-yellow Fairchild PT-23 trainer the airport manager assured her was in good shape. The color reminded her of her father's old Taylorcraft, and she absolutely had to have it. The five-hundred-dollar price tag, however, was out of reach for a girl with eleven dollars and change between her purse and bank account. So she put herself into hock to the Sooner Queens: she agreed to play the remainder of the season for the exact sum needed to buy the PT-23.

The deal turned out to be better for Jerrie than the Sooner Queens. She struck out in the last game of the season, keeping her team from going to finals but clearing her to resign. She bought the PT-23 and flew it to Ponca City, Oklahoma, where her parents were now living. She worked at her father's Pontiac-Cadillac dealership part-time while flying occasional missions for an oil company, passing low over pipelines sniffing for leaks. It wasn't much, but the hours built up in her little logbook. She barely made enough to keep the plane fueled, so she worked on maintenance herself. She took apart the instrument panel, much to the delight of the other pilots flying out of her local airport.

"Wanna get in our pool? We're betting on the number of parts you're gonna have left over when you try to put it all together again."

"Hey, Jerrie, is it true that you're going to paint the fuselage with pink polka dots?"

"Need some hairpins from the tool shed, Miss Cobb?"

When she finished, she christened the plane "Par-a-dice Lost," a reference to her love of literature and a nod to the role of chance in her life. She also took immense pride in not having a single part of the instrument panel left over.

CHAPTER 10

Portugal, May 17, 1951

"THE SPLENDID FRENCH FLYING WOMAN DID NOT BEAT MY RECORD, WHICH I obtained with a plane of a different type." Jackie only half believed the official statement she was giving to the media. She was traveling again, this time in Portugal, when she heard the news that Jacqueline Auriol had secured a new speed record of 509.245 miles per hour on a 100-kilometer closed course.

The 100-kilometer closed course was one of the many Fédération Aéronautique Internationale records that pilots could try for. Like with all distance records, the stipulation was that the course be a closed loop—she could fly in a circle, there and back between two points, or any other shape, so long as she passed over preselected checkpoints. Four years earlier Jackie had set the world speed record for the closed 100-kilometer course in a propeller plane, flying at 469.549 miles per hour. So it was true that the "splendid French flying woman" had not beaten her *exact* record, but Jacqueline had nevertheless outstripped Jackie, not because she was a superior flyer but because she'd been flying a jet.

Eleven years Jackie's junior, Jacqueline Auriol was a socialite and daughter-in-law of France's president, Vincent Auriol, making her as well connected to France's upper echelons as Jackie was in the United States. Jacqueline also shared Jackie's indomitable spirit; she was back in a cockpit just two years after a crash that had left her needing twenty-two plastic surgeries to regain any semblance of her previous self. Both women were strikingly beautiful, and they were also a match in flying talent. The rivalry between the two Jackies was a friendly one, but that didn't stop Jackie Cochran from wanting to reclaim the title of fastest woman on Earth.

Keeping a brave face to the media, Jackie poured out her frustrations in a letter to Floyd, whose own busy schedule and worsening arthritis prevented him from flying out to meet her. Instead, he sent a letter that was waiting for her when she arrived in Madrid days later, suggesting a few ways she might take back the speed record from "the French girl." He first suggested she talk to Hoyt Vandenberg, the Air Force chief of staff who also happened to be one of their close friends. She was still on part-time active service with the air force, though women were barred from flying, so he suggested she ask to fly a military jet under the guise of gathering data on how women reacted to jet flight. Another route might be through their good friend Randy Lovelace, who would almost certainly be interested in gathering data on women flying jets; perhaps he could initiate a new research program and hire her as the test pilot. If all else failed, Floyd suggested she could maybe borrow an F-86 jet directly from the manufacturer, Canadair. "So all that remains," Floyd closed the letter, "is to tell you that I love you very much, I miss you, I wish I were there with you, and I hope you are well and happy."

Floyd's suggestions were easier said than done. Jackie started with the most direct solution and talked to Hoyt Vandenberg. Ideally, she wanted to fly a military jet at Edwards Air Force Base in California, where all the hottest test pilots regularly broke records. It was where her friend Chuck Yeager had broken the sound barrier. But this was a nonstarter. As much as Hoyt admired Jackie, his hands were tied.

He couldn't give her permission to fly a jet while denying the same clearance to other women in the reserves. This meant that flying under a medical program with Randy wouldn't work, either.

An American plane out of reach, Floyd stepped in to help her get a Canadian plane. The US Air Force was flying F-86s, an aircraft manufactured under license from the North American Company by Canada's preeminent airplane manufacturer, Canadair, which was itself controlled by General Dynamics. Floyd owned no part of General Dynamics, but he was on the boards of both the Atlas Corporation and Consolidated Vultee Aircraft, and that gave him leverage in the aviation industry. He sold 17 percent of Atlas's holdings in Convair to John Jay Hopkins, the chairman of the board of General Dynamics, who replaced Floyd as chairman at Convair in the deal. Though there was no official mention of her name, the deal also stipulated that Jackie be hired as a flight consultant on General Dynamics' advisory board, which meant Canadair gained a new test pilot. Floyd effectively bought Jackie clearance to fly an F-86 Sabre fitted with a Canadian-made A. V. Roe Orenda engine that could push it to Mach 1. And because the F-86 Sabre was being built for the US Air Force, Jackie was cleared to fly it at Edwards. The arrangement, convoluted though it was, worked for everyone, most of all Jackie.

The last piece of the puzzle was for the Canadian government to okay the transfer of an F-86 to California for a flight test. In the meantime, Jackie started planning her jet flying lessons. She knew this was a whole new kind of flying. Everything happened faster in a jet. The engine reacted differently, and the stresses on her body would be like nothing she'd ever experienced. Even with all her years of propeller flying, all her records and all her awards, she still couldn't just hop into a jet cockpit and fly. It was a high-speed environment that brought as much risk as reward. "Before I fly the Canadair jet for a record, I will need some time in a T-33," she wrote in a personal and confidential memo to Hoyt Vandenberg. She knew test pilots at Edwards who were keen to help, she reminded him, and Randy would be using her data to learn how women fared in high-speed flight conditions. "Remember," she said in closing, "the day may come when you will suddenly need a lot of women for operational flights of jets and won't have any background data."

"I'm not here to talk about my private life." It was close to midnight in early February 1952. Jackie was in a nightclub giving an interview to radio personality Barry Gray, and they were supposed to be talking about Ike Eisenhower, the man the Republican Party hoped would run for president.

The United States was at war again. American soldiers in South Korea were fighting the Soviet-backed North Koreans in an effort to stop the spread of communism. Unsuccessful peace talks did little but damage Democratic President Truman's reputation, and the Republican Party was planning to use the president's flagging approval rating to retake the White House in the 1952 presidential election. What the GOP needed was a candidate, and it set its sights on Ike Eisenhower. Ike, however, from his home in Paris where he was commanding NATO forces in Europe, had not declared his candidacy. He maintained he was too busy or else referenced an army regulation that said a serviceman may accept a nomination for public office but couldn't solicit it. In truth, he was nervous. He was a career army man, and one of four men in America's history to earn the rank of five-star general. Losing the election could mean putting his military career on the line. On the other hand, Ike couldn't deny the allure of the country's most powerful office.

The Citizens for Eisenhower movement thus took up the task of showing Ike just how much his country needed him, and this grassroots mission to revitalize the Republican Party had sought Jackie's help. The party's finance committee chairman, John Hay "Jock" Whitney, knew Jackie had some interest in politics and had offered her the chance to test the waters by helping radio entertainer Tex McCrary and his wife, Jinx Falkenburg, put on a rally for Eisenhower at New York's Madison Square Garden. Her celebrity could prove useful in drawing attention—not to mention she and Floyd were friends with the Eisenhowers and had money that the campaign badly needed.

"I know. We just want a little bit, honey, just a little..." Barry Gray's condescending tone irked Jackie, especially since she was there to talk about Ike, not herself.

"Well, *honey*," Jackie replied, patronizing him right back, "if I'm going to do that I'm going to get paid for it. Come on."

"May I ask what business you're in?" Barry switched back to an air of politeness.

"Well, I have a very large business. Plug. All right," Jackie gave in, frustrated but keen to get the conversation moving. "I own Jacqueline Cochran Cosmetics, I own Charbert, Nina Ricci, a luggage company, a moving picture exchange for foreign distribution, and I've built it up on my own and it's a pretty good sized company employing some seven or eight hundred people, and I've never shouted about it, so you wanted a plug, so you have it. And I still started working in a beauty shop." Jackie's empire was indeed growing. The Jacqueline Cochran Personal Analysis Chart was a staple in department stores, helping women find their ideal cosmetic routine that almost always included her signature lotion, Flowing Velvet. Her name appeared in *Vogue* as often as it did in any flying magazine.

"That's wonderful," Barry said.

"That is America," Jackie replied, earning a round of applause from the audience.

"And you don't even look like a pilot."

"When they say, 'you don't look like a pilot,' what are people supposed to look like?" Jackie's frustration intensified. "I think that Miss Cochran, Jacqueline Cochran, looks like an intelligent gal who can fly a plane and do anything else that she made up her mind to do."

As the interview came to a close, Jinx told Barry and his listeners that Jackie would be flying footage from the rally over to Ike in Europe.

"I don't know how I'm mixed up in this thing," Jackie said with mock humility, "but I'm very proud to be along."

The next night, Friday, February 8, the Citizens for Eisenhower moved into Madison Square Garden as a prizefight was wrapping up. They had hoped boxing fans might linger and grow their number, but they needn't have worried. Though no political heavyweights or professional organizers were involved, the crowd still swelled to more than 15,000 people, so far over the Garden's capacity that people spilled out into the street. It was a mix of new voters, established voters, Republican and Democrat voters, all of whom wanted to see Eisenhower become president. Before the assembled masses, Jackie took the microphone. "I am for Dwight D. Eisenhower because this nation never needed an experienced pilot at the controls more than it does now. As a candidate he would not divide this nation, he would unite us. He would be a president of all the people." This wasn't pandering. Jackie truly believed that her friend could be the next Abraham Lincoln.

The following Tuesday, Jackie met with Ike at the Supreme Headquarters of the Allied Powers in Europe building in Paris. Sitting in the projection room, she told him she was carrying a message from the common people of America that was of the utmost importance to the country. Then she played him the reel. By all accounts, the rally had been an unprofessional mess. Critics had gone so far as to suggest Jackie dump the footage into the Atlantic to spare Ike the shame of seeing such a poorly organized event put on in his name. But detractors hadn't taken the emotional element into account. Jackie watched Ike watching the film reel and saw tears roll silently down his face. She knew how deeply he was touched by the raw outpouring of support from his fellow Americans.

Days later, Ike Eisenhower announced he would run as a Republican if nominated, and the Democratic Party asked Jackie if she would be interested in running for Congress. She had to decline. Politically, she and Floyd agreed they were Republicans, and after publicly backing Eisenhower she couldn't very well run for the opposing party. Not to mention she was preparing for the 100-kilometer closed-course speed record and had no intention of putting that on the back burner. Instead, she threw herself into the Eisenhower campaign. At Ike's behest, Jackie proudly stood on stage alongside his wife, Mamie, and a handful of others as he accepted the presidential nomination. Throughout the year she flew around the country appearing at events on his behalf. She and Floyd were at the Ranch on election night, following the news broadcasts as their friend was elected president.

A little over two weeks after the election, Jackie got the news she'd been waiting for: the minister of defense in Canada and the chief of the air staff in the United States both approved her contract with Canadair, which meant she was officially cleared to fly the F-86 Sabre jet at Edwards Air Force Base. She had three months beginning on March 10, 1953, to make a series of test flights for Canadair to check the stability, control, airspeed calibration, and static thrust of the plane before it would be sent back to the manufacturer. That was also her window to secure the 100-kilometer closed-course speed record. But there was something else. With a jet, she could take on a speed record no woman had achieved: flying faster than sound. It was time to go back out west to learn to fly a high-performance jet.

"If you want to fly this program you're going to be on time!" Just past six o'clock in the morning, Chuck Yeager's furious voice echoed down the hallway at Edwards Air Force Base. "You've got fifteen people out here working at four in the morning to preflight your airplane and get your gear ready while you, a single pilot, can't get here on time. Look at the man time you've already wasted for the air force, not to mention the guys who are busting their tails for you."

Jackie knew Chuck was right to be angry. He wasn't some race official she needed to keep waiting while she touched up her lipstick in a power play. In the seven years since their first meeting, Chuck had become one of her closest friends and something of an adopted son to her and Floyd. He and his wife, Glennis, were frequent guests at the Ranch. Theirs was a relationship based on mutual respect and adoration, neither of which she'd shown that morning. Nor had she shown that courtesy to Floyd, who had paid to have her insured in the Sabre, and paid the Fédération Aéronautique Internationale to send representatives to verify her record attempts. He was also paying Chuck $1,200[15] a month to take a leave of absence from the air force for her benefit with a promised $500[16] bonus for every record she broke including the sound barrier as well as covering any incidental costs.

Jackie had to admit she'd been cavalier and that Chuck had been right to put her in her place. She vowed to take their lessons more seriously, and by the end of 1952 had made fifty-six successful jet flights in a Lockheed T-33 without another episode of tardiness. With every flight, she grew more comfortable with the controls and the speed of jet flying, and both student and teacher learned to trust each other in this high-speed, high-risk environment. Chuck had one cardinal rule: "If I tell you to do something, you do it immediately and don't ask why." He drilled it into her head until he was confident that

---

15 About $11,300 in 2019.
16 About $4,700 in 2019.

she would push the plane until it melted if that's what he asked from her, and she knew to trust his judgment.

The media, on the other hand, was far less celebratory about Jackie's progress. Once the story of a woman flying at Edwards got out, rumors swirled about Floyd's finagling to get his wife in a jet, leading many journalists to dismiss her project as little more than vanity. Some men within the air force were unhappy to learn that a woman had clearance to fly a jet, a sentiment that in turn caused the Canadian government to consider withdrawing its consent for Jackie's test flying; its relationship with the US Air Force was more important than her contract. Canadair stepped in to suggest playing up the angle that Jackie's flight was a joint Canadian-American effort to secure a speed record away from France, but the media persisted with a negative slant. Stories insinuated that the air force was embarrassed by Jackie selfishly taking up flying time at a military base or else frustrated that a Canadair test program was being done on American soil when it could just as well be flown in Canadian airspace. But the people who mattered and supported Jackie put these rumors to rest. Air Secretary Thomas Finletter affirmed the air force had no objections to the Canadair flights at Edwards or with Jackie serving as the pilot, and Chuck assured her she was getting sufficiently adept at jet flying that she was likely to break both the sound barrier and the 100-kilometer speed record. Jackie knew the rumors were unfounded attempts to damage her reputation. She also knew Chuck wouldn't lie to her, he wouldn't tell her she was ready to fly if she wasn't, though his support brought an incredible amount of pressure. She had to succeed, or it would fall back on him as her instructor.

Jackie's stress evaporated when the Sabre finally arrived at Edwards in March of 1953, starting a ninety-day countdown before it had to be back in Canada. She had three months to fly the jet as much as she could on any kind of record-breaking flight she wanted. The pressure was on for her to make the most of the opportunity, especially as new reports had surfaced saying a British woman was eyeing the sound barrier as well. Jackie wasn't about to lose that race. She moved into the Yeagers' house to be closer to Edwards, and Chuck moved into officers' quarters on base. Glennis Yeager, who had never been as close to Jackie as her husband was, vacated to the Cochran-Odlum Ranch.

On the morning of May 12, Jackie donned a pressure suit. Rubber tubing woven throughout the garment pressed on her legs and abdomen, ready to expand at altitude to stop the blood rushing from her head and causing a loss of consciousness. Her blond hair was tucked neatly inside a white knit brimmed hat with a red pom-pom on the top. The Sabre fitted with the Orenda engine sat waiting for her on the tarmac. She checked out the plane, then exchanged her knit cap for a helmet with "Jacqueline" curving over the straight and bold "Cochran" on the front. At long last, she climbed into the cockpit of the Sabre. She rolled down the runway and took off through clear skies with Chuck flying close behind her.

Jackie immediately felt that the Orenda engine was more powerful than anything she'd ever flown. She also felt that the cockpit was uncomfortably hot.

"This cockpit is burning me alive!" she called to Chuck over the radio.

Closing the distance between them, Chuck could see that someone had inadvertently turned her defroster on. It was heating the cabin in an attempt to dispel nonexistent ice.

The smell of gasoline soon flooded Jackie's cabin. Mixing with the sweaty heat, she started to feel nauseated. Chuck calmly instructed her on how to shut down the heater and guided her to a landing so technicians could make sure there was nothing else wrong with the Sabre. The checkout revealed nothing. The plane was in perfect order, though it now smelled horrific. Unable to stomach the sweaty odor, Jackie sprayed the cockpit liberally with her perfume and climbed right back inside. Chuck shook his head and laughed; he would have preferred the gasoline to perfume so overpowering that the cockpit smelled like a French whorehouse.

Jackie only had one other incident during her familiarization flights with the Sabre. On a high-speed run, she and Chuck flew low over a nearby farm. The sound of their jets spooked the chickens, making them all run to one side of the paddock, where more than 130 were smothered to death. The farmers filed a complaint with Edwards AFB and asked that the flight path be changed so no more chickens had to die. Jackie just wondered why dead chickens seemed to keep popping up in her career.

On May 18, one week after her forty-seventh birthday, Jackie arrived at Edwards Air Force Base on another clear and crisp morning. She opted for a simple flight suit rather than the pressure suits she had been wearing; she figured if the boys didn't wear them, she didn't have to, either. Again she switched her white knit cap for a helmet as she climbed into the Sabre, then she closed the cockpit. Alone in the now-familiar jet, she took off with Chuck on her wing. Together the pair headed south, turning back just before the Mexican border, rising higher into the sky. Flying over the Ranch, Jackie hoped Floyd might be outside to see the contrail of ice crystals she left in her wake, but she was far too high to be seen from the ground. By the time she neared Edwards, she reached her peak altitude of 45,000 feet. Flying that high, there was nowhere to go but down.

Out of the corner of her eye, Jackie noticed that the blue sky had turned nearly black. Tearing her eyes from her instruments, Jackie looked out the canopy window to marvel at the sight. It was barely noon, but she could see the stars. For a brief moment, she felt how small she was in her little jet plane, and felt as though she were teetering precariously on the horizon close to the gates of heaven. For a moment she was entranced, but soon realized she wasn't there to stargaze. Forcing her focus back into the cockpit, Jackie put the Sabre into an S-dive.

In an instant, she was losing altitude so fast the needle on her altimeter was a blur, but it was the machmeter she was focused on. She called the numbers to Chuck over the radio.

"Mach .96. . . .97. . . .98. . ."

She was in the turbulent transonic zone. The air passing over her wings was traveling faster than sound and the air underneath them was not. Air molecules couldn't get out of the way of her plane fast enough so she slammed into them. Her right wing dipped as it dug into this compressed air, then the plane lurched, and the other wing dipped as she fought to keep it level. Then she felt the nose try to tuck under as though the plane wanted to loop over itself, but she held it steady with all the strength in her hands.

"Mach .99. . ."

Even with a helmet and radio in her ear, she could hear the plane rattling.

"Tell me what you're feeling," Chuck called as he dived alongside her. Jackie vaguely registered that she could see the shock waves rolling over the canopy; she told Chuck it looked almost like a delicate film of water trailing off the window. Then, without warning, the turbulence stopped. The shock waves disappeared. The rattling was replaced by an unearthly silence. She was through the sound barrier; the turbulent air and shock waves were behind her and the noise couldn't catch her. For a fleeting moment, Jackie felt a spiritual connection with something greater than herself. She didn't feel scared, just confident and keenly aware of the plane's every movement.

A split second later, the stillness was replaced with the sound of rattling metal and rushing air as Jackie pulled the Sabre out of its dive at 18,000 feet and slowed to below the speed of sound. The heavy air closer to the ground pressed on her plane as though trying to rip it apart as the drag slowed her further. Then she did it again. She flew high enough to see the stars then dove through the sound barrier. She would have gone a third time, but she was running low on fuel.

Jackie landed to hearty congratulations from the other pilots and ground staff at Edwards. Though she'd just become the first woman to break the sound barrier, she was disappointed to learn that no one on the ground had registered her sonic booms. Unwilling to let the day end without a record of her achievement, she refueled the Sabre and got right back up in the sky to break the sound barrier again, this time taking care to make sure she was at the right point for everyone at Edwards to hear her boom.

Five days later, Jackie took on the flight record that had started her foray into jet flying: the 100-kilometer closed course.

Setting up for a record run had taken a lot of planning. Floyd helped design the course. Rather than picking a point fifty kilometers away and flying a straight path with a slow turn at one end, he found that if she passed over twelve pylons arranged in a circle she could stay in a tight turn the whole time and fly the course faster. So Jackie had the pylons installed in an area that, as per regulations, was free of power lines and property lines. She planned to fly in a thirty-degree bank about 300 feet above the desert floor. To make sure she didn't cut any corners, two observers were stationed at each pylon as well as at the start line. Those official observers were backed up by two cameras stationed 1,200 feet from the course and four feet off the ground to capture the moment she passed over each checkpoint; the photographs could be used to correct the time based on the plane's position with known landmarks if there was some dispute. In the air were two official observation planes with sealed barographs on board just like the one in Jackie's plane: two backups to her official instrument. There were also two Fédération Aéronautique Internationale judges on hand armed with automatic electric timers.

Leading the group was Charles S. Logsdon, the FAI's chairman of the Contest Committee. Floyd was there, manning one of the sighting devices so he could find his wife in the sky. Randy was behind a sighting camera, too, as interested in Jackie's record as her medical data. Chuck was also stationed at a sighting camera, with nothing left to do but watch his student. They all waited in the hot desert sun. Most hid under hats or the shade of a tarpaulin while Chuck opted simply to take off his shirt. No one could leave their post. Jackie had enough fuel on board for two flights around the course. She was going to come at it fast, and they had to be ready.

The quiet desert day was broken by the sound of the Orenda engine in the distance. Then Jackie appeared, a blur in the sky tearing toward the course. After so many practice flights, the controls felt natural in her hands as she maneuvered the plane from checkpoint to checkpoint. She shattered Jacqueline Auriol's record; her official time was 675.471 miles per hour, 166.226 miles per hour faster than the French flyer. For the second time that month, Jackie landed to cheers and accolades from the test pilots at Edwards.

Jackie's time flying in the California desert ended with a small celebratory dinner in her honor. Place cards featuring a picture of the Sabre told everyone where to sit. The guest list included test pilots and their wives, including Chuck and Glennis Yeager, as well as Randy and Mary Lovelace and Lyndon Johnson. Jackie dined at the head table with Floyd before a floral centerpiece also in the shape of the Sabre.

The weeks of high-speed flying left Jackie emotionally thrilled but physically in pain. Every time she'd shot toward the desert floor then pulled the plane up sharply, she'd felt the g-forces on her body but hadn't known how much damage she was doing to her insides. The complication from her teenage appendectomy, the scar tissue in her abdomen, had come back to haunt her. Her high-speed flying had pressed her intestines in the perfectly wrong way that left her digestive tract adhered to that scar tissue. At first, Jackie felt some painful cramps, but when they didn't go away, she knew it was more serious. She needed Randy to operate and break down the adhesions, something he'd done for her once before. Telling Floyd she was visiting one of Randy's patients so he wouldn't worry, she flew herself to the Lovelace Clinic, where he operated on her once more.

Stuck in the hospital unable to eat or drink just days after securing an international speed record was not how Jackie wanted to celebrate. After four days, she hit her breaking point. She pulled out her own nasogastric tube and intravenous line, then called her secretary to bring her a bottle of whiskey and get the car. She was going home.

Back at the Ranch, Jackie tried to keep up with the letters that came pouring in. Air force brass commended her on her flights and her determination. Jacqueline Auriol sent her congratulations and noted that the twelve turning points had indeed been a wise choice. "When preparing for a new record later on," she wrote, "I will adopt your procedure of the twelve turning points. You are right—the gain in time should be considerable." She also learned that her supersonic runs had earned her her fifth Harmon Trophy win, an honor she'd be sharing with Chuck, who won the male trophy.

But the standout letter was a personal note from President Eisenhower. "Your letter, congratulating me because I became the first woman to go through the sonic barrier, touched me deeply," Jackie wrote in reply to the president. "The letter has become one of my most prized possessions. With this splurge of flying behind me, I am seriously thinking about entering politics. You may see me showing up in Congress before you are out of office."

CHAPTER 11

Summer of 1953

"DURING THE WASP PROGRAM, MANY PRECONCEPTIONS CONCERNING women were proven wrong. The general belief at the start of the program was that women were handicapped as pilots, due to menses." Jackie knew it was an altogether ridiculous notion, but she also knew she couldn't not mention menstruation in her medical report for Randy Lovelace and Walter Boothby. The doctors, both of whom saw no reason why women shouldn't fly jets if the need arose, were keen to include a female perspective in a new manual on high-altitude respiratory physiology for the US Air Force's School of Aviation Medicine. "This belief found no support in experience," Jackie finished her appropriately short comments on menses. "The conclusion was that in properly selected women, menstruation is not a handicap to flying or dependable performance of duty."

With supersonic jet flying under her belt, Jackie was the only woman in the country keeping pace with male aviators, and as far as she could tell the only real difference between the sexes was physical size and strength. Height for height, women were less muscular than men, but newer planes demanded less muscle and more skill. Hydraulic controls were becoming much more common, allowing a pilot to control her flight surfaces with small inputs from the cockpit; she didn't need to muscle a plane through bad weather or out of a dive. In her own supersonic flights, Jackie hadn't felt a dependence on anything physical, she just felt confident in her abilities.

Aware that her career had reached a new height and with one eye on a potential seat in Congress, Jackie decided it was time the world knew just how extraordinary she was. Taking ownership of her story in her memoirs, she knew, would cement her reputation as America's standout female pilot and show the vulnerability voters liked in their representatives. She would, of course, maintain her orphan origin story; she'd been telling it for years and couldn't risk being caught in the lie now. So she began to write.

I am a refugee from Sawdust Road...

Little had changed in the professional landscape for women by the summer of 1953. If women chose to work, they were expected to be secretaries, waitresses, or some other appropriately feminine pursuit. Flying was decidedly a man's job, as Jerrie was reminded every time she applied for an open position.

"A pilot? You want to work for us as a pilot? Are you sure you don't mean stewardess?"

Jerrie caught a glimpse of herself reflected in a glass partition. Her medium-length blond hair was styled into a loose pageboy, her plain tan dress accentuated her lean five-foot-seven frame, and though she had a slight tan, it didn't hide her freckles. The overall effect was that she looked far younger than her twenty-two years. Instinctively, she sat up a little straighter, hoping the man interviewing her might not notice. She'd been rejected from countless small-time flying jobs and was desperate for someone to take her seriously.

"Well, why not," the man continued with a note of resignation in his voice. "Maybe a woman pilot is what this outfit needs. It'll be novel anyhow, good publicity—and she can double as a stewardess."

Jerrie couldn't believe it. She had just been hired as a pilot for the brand-new Trans-International Airlines. She would be flying paying customers, something unheard of for a woman. She immediately threw herself into training to prove she could keep up with the airline's male pilots. She learned the ins and outs of the propeller-driven DC-3s she would be flying, everything from the hydraulic and electrical systems to the right way to offer guests milk with their tea or coffee. But Trans-International Airlines had some trouble pulling together enough funding to buy airplanes. Time wore on, and the pilots had to get creative to keep their jobs. Male pilots painted hangars while Jerrie worked as a typist for a dollar an hour. Finally, one DC-3 arrived. Jerrie flew it on a checkout flight, but then the airline's financial backer pulled out, forcing the company into bankruptcy.

Deflated, Jerrie resumed sifting through the classifieds looking for any job that would bring her close to an airplane. She finally found a position at the Miami International Airport in the customer service department of Aerodex Incorporated. The good part was that she dealt more with airplanes than with customers, putting through work orders and checking up on ongoing repairs. The bad part was she never got to see the inside of a cockpit.

One October morning, Jerrie arrived at Aerodex at seven o'clock, a full hour before the office opened for the day. Her cherished quiet time among the airplanes in the adjacent hangar was interrupted by an unshaven, rumpled, and irate man staring through the office window. Behind him, she saw an airplane parked on a ramp.

"Whodyhaftaknow to get something done around here?" he snapped at Jerrie the moment he saw her.

"Perhaps I can help you, sir," she replied, leading him into the office. "Now, what's the problem?"

The problem turned out to be his plane, a de Havilland Beaver. He was flying it from Canada to South America, where the Colombian Air Force was eagerly awaiting its arrival; Colombia had bought the old trainer plane to build up its air power. It needed some maintenance, and every minute it sat on the ground was costing him money, which explained why he was angry, but not why he was directing it at Jerrie. Mustering her good manners, she offered to put the work order through for him first thing when the office opened. He barked a rough "thanks" then left with his sense of entitlement apparently intact. Jerrie shook off the exchange as the workday started and began tackling the mountain of paperwork on her desk.

A few hours later, Jerrie noticed that the girls in the office were tittering. She looked up to see the rude man was back, though he was clean-shaven now and wearing a pressed suit. Jerrie had to admit it made quite the difference; he was actually quite handsome with his dark, wavy hair falling loosely in front of his face, drawing attention to deep eyes that were alight with energy. She watched him walk through the room, unfazed by the stir he was causing as though he was quite used to it, and straight into her boss's office. Keen to avoid another exchange with him, she busied herself with work orders, but the sudden appearance of her boss looming over her commanded her attention.

"Miss Cobb," he said, gesturing to the rude man standing next to him, "Mr. Ford very much appreciated your putting that work order through so early in the morning. He'd like to buy you a cup of coffee." Jerrie looked from her boss to the rude man, who was now smiling at her, the picture of civility. She began to protest, saying there was work to be done and that she simply couldn't leave her desk, but her boss insisted. Jerrie rose from her seat, only vaguely aware of the firm but gentle arm directing her along to the coffee shop.

Mr. Ford bought her an iced tea and placed it before her on the small table before sitting down himself. Jerrie could hardly believe he was the same man from earlier. He was charming and kind as he thanked her again for rushing the work order through and apologized for his attitude. He told her to please call him Jack and asked if he might call her Jerrie rather than Geraldyn. As they talked, she learned that he was more than just a pilot.

Jack had spent the Second World War overseas flying B-17s and B-24s as an Army Air Force heavy bombardment pilot. While in England, he'd met and married another American pilot, an ATA girl flying under Jackie Cochran named Mary Zerbel Hooper. Jack was Mary's second wartime husband; her first, Roy Hooper, had gone missing in action months after their marriage and was presumed dead. Once the war ended, Jack and Mary—who was now going by Mary Zerbel Hooper Ford—co-founded a company called Fleetway that ferried planes around the world from its headquarters in Burbank, California.

Now, Jack told Jerrie, he was having a hard time finding willing pilots. Most of the planes Fleetway transported were single-engine trainers, American planes sold to other countries to build up their air power. Experienced ferry pilots were less than keen to fly them on routes that covered oceans and mountains. Jerrie understood why: a single engine meant there was no redundancy; if it failed, the pilot had nowhere to go but down. But rather than registering the danger, she saw an opportunity.

"I know a girl pilot who would fly any single-engine airplane over any ocean or mountain," she blurted out before she could lose her nerve.

"And who might that be?" Jack asked.

"Me," Jerrie answered without missing a beat. She could sense he was amused rather than impressed, but she wasn't backing down. It had been far too long since she'd flown either for work or pleasure, and she was tired of people not taking a girl pilot seriously.

"You're a woman, and flying is a man's job. I'm not saying that women shouldn't fly—for fun or sport,"—he made this small allowance—"but as a career, it's no good. They haven't got the stamina or the temperament.

It's a rare woman who can compete with the men in this field." Flying meant living out of a bag, never knowing what city or country you'd end up in at night. Flying meant erratic schedules and long hours alone in a cockpit. Flying meant no creature comforts. It was a hard, dangerous, and demanding job he told her was best left to men.

Jerrie felt her scalp prickle, she was so riled up. Through gritted teeth, she laid out her qualifications, licenses, ratings, and her 3,000 hours in the air—all of it, she added, gained without a man supporting her. She was self-sufficient and determined, but Jack wasn't convinced she could handle full-time flying.

"For how long?" he asked. "Be honest. Don't you think it's just barely possible that you're still a youngster? Flying full-time isn't easy."

"Mr. Ford." She steadied herself. "You talk as though men have a monopoly on flying, not only commercially but emotionally as well! I fly because I can't not fly, same as you."

"Jerrie, you still haven't answered my question. For how long."

"'Til I buy the farm!"

Jerrie laid everything on the line, but all Jack did was shake his head. It pushed Jerrie to her breaking point. She thanked him for the iced tea and rose to leave.

"Jerrie!" She paused and turned back to face him at the sound of her name. "You put up a good fight," he said. Jerrie smiled in spite of herself.

"If you run out of manpower, I know a girl pilot . . ." With that, she left Jack to his plane and his South American destination.

The next week, Jerrie was at work when the phone on her desk rang.

"Jerrie Cobb? This is Jack Ford. Touché." She didn't understand. "Touché," he repeated. "You win." He had a fleet of T-6s in New Jersey that needed to be in Peru as quickly as possible, but there wasn't a male pilot around who would fly the single-engine trainer over water. He'd fired them all and needed someone with real moxie to take on the job, even if that someone was a girl. And so he was calling to offer her the job—that is, if everything she'd said over iced tea hadn't just been talk. She knew she was his last resort, but she didn't care.

"You've got yourself a pilot," she told him.

Days later, Jerrie met Jack at the Miami International Airport at five o'clock in the morning. There were two T-6s on the tarmac, one for each of them, both painted in the colors of the Peruvian Air Force with bomb racks and machine guns already installed. Their destination was Las Palmas. Jack gave her plane a once-over, then handed Jerrie a set of maps with the instruction to keep her radio on and stay in formation. As an afterthought, he asked how she was feeling; privately, Jerrie was sure he cared more about his plane's safety than hers. Then he climbed into his cockpit and signaled to Jerrie to do the same, but once inside, she couldn't figure out how to get it started; she'd never been inside a T-6. He angrily walked back to her cockpit and pointed out the energizer pedal on the floor she had to press to start the propeller, then stalked back to his own plane. Luckily Jerrie was a fast learner, and before long managed to get the propeller going, and she was off. But as soon as they were airborne, she lost Jack in the clouds. He'd assumed she was familiar with formation flying, which she wasn't. Still, she held it together and made it the two and a half hours to Camaguey, Cuba. They stopped to refuel, and Jack took advantage of the opportunity to chew her out for not following his scant instructions to the letter. He wasn't sure she'd make it to Kingston, Jamaica, in one piece.

Back in the air, Jerrie again lost Jack in the clouds and somehow managed to land a full ten minutes before him. Far from impressed, he was visibly irritated with her as they left the airport and checked in to a hotel for the night.

Their next leg the following morning was to Barranquilla, Colombia, for another refueling stop. Then it was on to Cali, where Jerrie landed first again. She was still on the runway when Jack came in for a landing, but he hit a chuckhole and ground-looped—the plane lost balance and spun around so violently it sustained significant damage. It was clear the plane would need to stay in Cali for a few days for repairs, and the pilot would be stuck with it. As Jack walked away from the accident unhurt, Jerrie was amazed to see he looked sheepish about it, but he soon snapped back to anger, this time frustrated over losing money while the plane was grounded.

"Take my plane and at least get that one delivered," she offered.

For the first time since they'd left Miami, Jack actually smiled at her. "No, it's your baby. You've brought it this far in good shape. I'm sure you can make it the rest of the route." He patted her on the back as she climbed back into her cockpit, a small gesture Jerrie took to mean he was not only confident in her but also they were becoming friends. "You deliver it." Leaving him behind, Jerrie took off toward a foreign land completely alone.

Navigating around the peaks of the Andes, Jerrie felt at peace. The landscape stretching out below her was serene and beautiful. She made it to her next refueling stop in Guayaquil, Ecuador, easily, but almost as soon as her wheels hit the runway, she saw what looked like the whole Ecuadorian army running toward her with rifles and pistols at the ready. What little Spanish she'd learned in high school vanished from her mind as, hands shaking, she reached for her flight case that held her immigration and clearance documents. She found the ones she needed for her stop in Ecuador, forms cosigned by the US State Department and the Ecuadorian embassy in Washington, and handed them to the man who seemed to be in charge. It was then she remembered that her plane had a Peruvian military paint job and bomb racks that she knew spoke louder than any documentation.

She was escorted from the plane and loaded into a car. As she watched the city go by, Jerrie's hope that she was being taken to a government building vanished when she was driven into what was unmistakably an army camp; she was halfway around the world, but the barracks looked just like the ones she remembered from watching her father's military exercises. The car finally stopped at a rudimentary wooden building. The door opened, and Jerrie was led into a small room with a wooden table and two chairs, one of which she took. A man joined her and sat in the other chair, then asked her again why she was in Ecuador. She again attempted to explain she was a ferry pilot and tried to show him her travel papers, but her efforts were fruitless. Unbeknownst to her, Peru and Ecuador were in the midst of one of their frequent skirmishes, and the appearance of a Peruvian plane had become a military event. She was moved into a drab but clean room with a dresser and rudimentary bathroom, a hurriedly assembled cell for the jail's first-ever female prisoner. The door was shut behind her with the unmistakable sound of a lock clicking into place.

Days went by. Jerrie was kept confined to her small cell, the monotony broken at regular intervals by further questioning from military and government officials clearly still trying to figure out why she was in Ecuador. Three times a day, an unidentifiable gruel arrived from the mess hall that Jerrie couldn't bring herself to eat. A sympathetic guard eventually brought her a far more palatable meal of rice and gravy and taught her rudimentary Spanish to pass the time.

Without warning, after twelve days in her cell, a guard came to release her. She was told that the Ecuadorian government had spoken to American officials throughout her imprisonment and determined that she was an American citizen ferrying an American plane; though Peru had paid for it, the T-6 hadn't yet been delivered, so it wasn't yet Peruvian property. There had been no legal grounds to hold her in the first place, and keeping her in custody risked becoming an international incident. She could only imagine that Jack, learning she hadn't made it to Peru, had traced her last known location and realized she had been stopped in Ecuador. Whether he cared about freeing her or getting the plane to its destination, she couldn't say. She just wanted to get on her way.

Jerrie hurriedly signed a Spanish-language document that translators assured her was merely a statement testifying she hadn't been mistreated. Then she returned to the airport, checked out her plane, and left before anyone could change their mind. Safely in the air, she relished in the feeling of the wind whipping around her face and through her hair, and drank in the view of the forest stretching beneath her. Flying always felt freeing, but it never felt as good as it did that day.

She landed in Las Palmas a hero, the female pilot with a brand-new plane. To the locals, she was like something out of a modern romance story.

The months after that eventful first flight passed in a wonderful blur. With every ferrying mission to South America, the foreign cities with exotic names became more familiar. Jerrie got to know the small islands and densely forested countries from the air as well as she'd known the Oklahoman fields and the wide spaces in Texas. The language barrier melted away, too. Her Spanish steadily improved, and before long she was able to communicate with the hoteliers in small towns who helped her find rooms that were sometimes palatial suites and other times cubicles with four-foot-tall walls offering a modicum of privacy. She delighted in learning the local customs, often rallying her energy for late-night fiestas lest she insult a host. She learned to carry a tarpaulin on every trip to keep out the rain if she was forced to spend a night in the cockpit. But what stood out above everything else were the friendly smiles that greeted her and the hospitality of the South American people, their generosity and eagerness to make her comfortable regardless of how much or little they themselves had. Everywhere she went, she was greeted by crowds amazed to see a female pilot in a captain's uniform dismounting from the plane, often taking pictures of her before she took off again.

Within six months, Jack put Jerrie, the woman he'd doubted could fly at all, in charge of Fleetway's South American ferry operations. Far from a desk job, this managerial position had Jerrie organizing the other pilots as well as flying the routes herself, often with Jack on her wing delivering a second plane.

Before long, an unforeseen challenge arose. It happened without Jerrie even realizing it. As she spent weeks exploring foreign lands with Jack on those trips, she started thinking of him as more than a friend. But she couldn't imagine her feelings would be reciprocated. With his movie-star good looks he could have passed for Cary Grant's brother, and he was notorious for dating movie stars when he was home in Burbank. The roguish, handsome pilot had his pick of women Jerrie knew were far more glamorous than she was. So she put her feelings aside.

Jerrie and Jack were flying formation somewhere over the Caribbean Sea one day delivering a pair of T-6s to Peru. The flight was perfectly routine until, out of nowhere, a loud pop caught Jerrie's attention. Almost immediately she saw oil spewing from her engine and coating the canopy. She'd lost a propeller seal. "Can't see outside," she called to Jack over the radio, not wanting to waste time on needless words. "I'm flying by instruments now."

"I'm on your left wing," his reply came crackling through, "coming closer now... Looks like we'll have to ditch. We're too far to return to Cuba, and Jamaica's another eighty miles. I doubt we can make it."

Despite her slowly mounting panic, the relentless stream of data she was relaying to Jack about her engine pressure, and his instructions to keep her on their current flight path, she hadn't missed his use of the word "*we*." "What do you mean, *we'll* have to ditch?" she demanded. There was no reason for Jack to risk himself. Trying to land on water blind without breaking the plane apart was going to be hard. She needed him in the sky guiding her, not following her on a suicide mission.

Jack ignored Jerrie's question and continued with his instructions. "Pull your raft out from under your parachute," he instructed. "Check your Mae West straps," he called and she checked that her life jacket was secure. "Get everything ready to ditch. No loose items in the cockpit that can turn into projectiles in a crash. You'll need help landing in those waves blind. I'm going with you." Jerrie couldn't believe it. This wasn't the Jack she knew, the Jack who cared more about a plane than its pilot, the Jack who wouldn't dare risk his own perfectly good airplane just because hers was going down.

Oil started leaking into the cockpit, covering Jerrie's instrument panel. "I'll need your help with headings, speeds, altitudes as all the instruments are covered in oil," she called. "Don't go in with me. I know about rafts, and I'm a strong swimmer. I'll make it through."

"Forget what the book says, slide your canopy all the way back and jam it with something. Or it'll slam shut on impact, and you'll drown because you won't be able to get out. I'm with you all the way. Airplanes are expendable, good pilots are not." He paused for a moment. "I didn't wait thirty-six years to lose the love of my life now."

The disbelief that hit Jerrie seemed to stop the drama unfolding in her cockpit momentarily. In spite of everything, a grin spread across her face. Suddenly, the prospect of ditching blind into the ocean seemed easy; the oil coating her canopy seemed like a minor nuisance. She had the love of her life at her side. Anything was possible.

"You still there?" Jack's voice broke through her reverie. "Call out oil pressure. We're going to try for a long glide to the Jamaican coast. I can just see it now on the horizon."

Still slightly dazed, Jerrie listened as Jack called to the ground that their two-plane formation would be making an emergency landing in Montego Bay. She flew by the sound of his voice. He matched her pace to tell her her airspeed, altitude, and heading, guiding her unpowered plane to a runway she couldn't see. They both knew she only had one chance to make it. "Nose up," he called, and she pulled the control stick back in response. Jerrie braced herself for the sound of a crash and the feeling of her tail kicking up in the air as her plane flipped end over end, but it never came. Instead, she heard the unmistakable squeal of one, then both tires hitting the tarmac and finding their grip on the rough surface. "Right rudder," came Jack's voice, "easy on the brakes... a little bit to your left..." Her T-6 rolled to a stop.

Still sitting in the cockpit, Jerrie felt an almighty lurch that told her Jack had jumped up on the wing. A thumping sound seconds later told her he'd slipped in the oil and slammed into the canopy. She heard him fumble with the slick surface before he managed to wrench it open. His arms reached in, found her, and lifted her easily from her seat and down to the tarmac. He wiped a tear from her face leaving a smear of oil in its place, then lifted her in an embrace, swinging her feet off the oily tarmac and spinning her in circles. She barely registered the emergency vehicles congregating around them and the damaged plane as he set her down and guided her wordlessly aboard the tow truck hauling her plane to the hangar.

Once inside, Jack went straight for the phone while Jerrie sat on the floor, still in shock. As though from a distance, she heard Jack on the phone talking to parties unknown demanding all prop seals be replaced on all the T-6s, and that someone get down to Montego immediately to fix Jerrie's plane—he wasn't about to lose money with it sitting in the hangar longer than it had to. "And one more thing," he said before hanging up, "I'm unreachable the next few days."

Jack replaced the antique phone's mouthpiece and walked over to pull her to her feet. "After we get this oil off us we're going to go stretch out on a beach, okay?" Jerrie smiled. Words felt unnecessary.

The pair tried to clean themselves up as best they could, sopping up the oil and grease with newspaper, but that only added ink to the mess. Jerrie tried in vain to wash the gunk out of her hair in the airport's makeshift shower but gave up. When she emerged, her blond hair turning slightly green from the mix of oil and ink, Jack had managed to find a local with a car and inside knowledge of the most romantic and secluded beaches on the island. He climbed in front while Jerrie took a seat in the back, still fussing with her hair. After what felt like hours on the road, they reached the promised beach. Without a moment's hesitation, Jack grabbed Jerrie's hand, and they ran into the surf, clothes and all. Jack dove in, taking Jerrie with him.

"See, I told you we were going into the ocean together," he said as they came up for air.

"My love," Jerrie replied without even thinking, "that's when I knew we were kindred spirits, that we would always be together."

Right there in the surf, still dressed in their greasy, oily clothes, Jack leaned down, pulled Jerrie in close, and kissed her. Jerrie felt the full weight of the day dissolve as though she were floating away into Jack's arms.

All too soon their island escape came to an end. Two days later, Jerrie's T-6 was repaired and ready to fly, and they couldn't afford to keep it on the ground any longer. "But we'll return again soon," Jack told her as they set out to finish the flight they'd started.

CHAPTER 12

Ponca City, Oklahoma, Christmas 1954

JERRIE WAS EXHAUSTED. SHE'D PICKED UP A LAST-MINUTE FERRYING JOB to Miami after returning from Paris and still somehow managed to get home in time for her parents' country club Christmas Eve dinner. Tired as she was, she couldn't deny her parents the chance to show off their jet-setting daughter to their friends. As she dressed, the phone rang. It was Jack calling from the Wichita airport, less than two hours away.

"How wonderful!" Jerrie was thrilled he could make it to this big family event. They'd been together a little over a year, and as Jack had already charmed her parents, she was sure they would love to show him off at the Christmas party, too. "I'll be right out to get you."

"Honey, I can't stay. I've got to take that plane out tonight. Come with me to get it, and I'll drop you off on my way back."

"Oh, darling, I can't do that." Jerrie felt her smile falter. "You've no idea what this evening means to Mother and Dad. They've had their hearts set on my going with them. Isn't there any way you could delay that flight and come with us?"

"Impossible," Jack told her. "Even leaving tonight, I'll be lucky to make delivery on time in this weather. Look, Jerrie, I'll have you back in a couple of hours. Please, honey."

"Jack, I'd love to but I just can't. They see so little of me as it is. I couldn't disappoint them now."

"What about us, Jerrie?" he pleaded. "This is all the Christmas we can have together. It's a long time between trips."

"Jack, please don't put me on the spot this way. I just can't do it."

"Okay, baby. Have it your way. Merry Christmas." Without a trace of merriness in his voice, he hung up the phone.

Jerrie put on a brave face as she played the part of the happy daughter all night, regaling family friends with stories of her European and South American adventures. But inside, she felt miserable.

Hours later, the Cobbs arrived home to find the hulking form of Jack Ford bundled up on the doorstep. He'd said he could have her home within a couple of hours, so instead, he'd delivered himself: he'd picked up the plane in Wichita, flown back to Ponca City, then taken a taxi to her house.

"Jack," Jerrie whispered. Standing alone together outside, the tears she'd been fighting back all night finally spilled out. "I'm sorry about tonight."

"I know, sweetheart. I'm sorry, too. I was wrong. Forgive me."

"Oh, of course, you big dope. You didn't have to fly down here to ask me that."

"I know. I didn't. I brought you your Christmas present. I wanted you to have it before—well, New Year's." He handed her a small box. She opened it to find an aquamarine ring. He'd found the raw stone in South America and had it set in California. That the ring had gone halfway around the world to reach her finger simply because he thought she'd like it brought the tears back to her eyes.

"It's lovely," she managed, then the tears came as he put his arms around her.

"C'mon, madam pilot. That's no way for a grown-up girl to act."

"Oh, darling, please don't let's ever fight again." Jerrie smiled.

"Never again. That's a promise."

Jerrie's life as a professional ferry pilot was anything but conventional. In addition to frequent flights to South America, she also started ferrying in Europe, delivering big bombers across the Atlantic Ocean. The odd night that her travel path crossed with other Fleetway pilots, all of whom were men, they stayed in the same hotel, pooling their laundry with the understanding that the last to leave in the morning would bring any leftover clothes back to the company's headquarters in Burbank. Her relationship with Jack was similarly unorthodox.

One night in London, the phone in her hotel room rang. The operator told her it was a long-distance connection from San Francisco, and she knew immediately whose voice she would hear on the other end.

"Hi, honey," Jack's deep, warm voice crackled over the transatlantic line. "When did you get in?"

"This afternoon. How did you know I was here?" she asked, surprised.

"Talked to the office. Just got in from Japan. Look, I have to pick up an airplane in Burbank and come on over."

"Darling," she replied, "I'm leaving in the morning. Commercial, back to New York."

"I know—" She could hear the smile in his voice. "I'm meeting you there tomorrow night. Got something for you."

Jack would often finagle his way to wherever Jerrie had a layover, always bringing her a foreign coin or some trinket from a country she hadn't yet visited to add to her collection. When their flight paths crossed, they took mini-vacations wherever they met. They did return to the beach in Jamaica, just as Jack had promised. They met in Prestwick for a quick game of golf, still dressed in their flight suits to make the most of their short time together. When they overlapped in London or Paris, they took the opportunity to get dressed up and go out on the town, pretending that they were regular tourists on vacation, even if the fantasy only lasted the night. Normality was rare, but when their schedules coincided near Oklahoma, Jack consented to take some time off and visit the Cobbs. Jerrie was determined that her parents should get to know and like Jack, and of course they adored him.

As the months wore on, Jerrie and Jack's times together remained brief but perfect. They saw each other so infrequently they didn't have time to fight or fall into the pitfalls of jealousy or pettiness. But Jerrie knew that she would want more before long. Ferrying was an incredible adventure, but she ultimately wanted a home, a husband, maybe even a family, all things she couldn't have if she was constantly on the go.

In the summer of 1955, Jerrie got a taste of what a more conventional life might look like. Fleetway had a ramp full of airplanes at the Burbank airport that needed to be checked out, so she volunteered for the job. Day after day, she flew those planes hard enough to reveal all the bugs and nagging little issues. It was exhausting work, far more draining than a long but level flight across the Atlantic, and less of an adventure, but she was flying unique missions every day, and working out of one of Fleetway's main airports meant she saw Jack more than once in a blue moon. Jerrie was happy, but Jack couldn't understand.

"How are you bearing up with this sedentary life?" he asked her one day.

"Sedentary?" she shot back. "Are you kidding? Or haven't you noticed that I spend all day, every day, flying airplanes inside out and backwards."

"Rough," he replied, shaking his head. "Well, don't worry, hon, we'll have you out of this rut before long."

"It's no rut, Jack. I like it."

"Like it? How can you? The same thing, the same place, day in, day out. That's drudgery. Boy, it would drive me nuts. I couldn't stand it for more than a week."

Jerrie was all too aware that Jack's simple statement spoke volumes. If he couldn't stand the "drudgery" of test flying out of one airport, she doubted he could withstand the routine of marriage. International courtship was one thing; it was romantic, exciting, and unexpected in the best ways. International marriage was something completely different, and Jerrie knew the glamour of it would eventually give way to frustration and anxiety. She never questioned Jack's fidelity. It wasn't the pull of another woman she was worried about, it was the pull of another airplane. Flying was his mistress, and she didn't know if she could prove as tempting.

Jerrie started to realize there were two Jacks. Jack the man would change his pattern and sacrifice his freedom so they could build a life together. But Jack the pilot was like Peter Pan, who'd never really want to grow up or settle down. It was a conflict she knew he didn't recognize as he spouted off clichés in defense of their unconventional romance. "Absence makes the heart grow fonder," he'd tell her. "Familiarity breeds contempt."

With Jerrie in Burbank, the incompatibilities in their relationship began to show. They discussed marriage; Jack wanted to be married by a justice of the peace while Jerrie wanted a church wedding. They agreed on taking a honeymoon, of course, but while Jerrie wanted a long, luxurious, lazy holiday on a beach somewhere, Jack was content to go wherever their ferrying schedule took them. It was as though everything Jack had warned her of not two years ago in the Aerodex coffee shop—the long hours, the erratic schedule, the constant movement—was indeed starting to prove too much for her.

Maybe he had been right when he said women just didn't have the right temperament for a life of flying. Perhaps she'd been wrong when she'd told him marriage didn't mean retirement for all female pilots; she had to admit her ambitions were changing now that she was in love. Or maybe Jerrie did have the right temperament for a life of flying, just not one that jelled with Jack's idea of a flying life. The tone of their phone calls changed. Jack grew impatient, and Jerrie grew silent as what used to be loving check-ins became arguments about nothing with each playing the role of the aggressor in turn.

"Where were you when I called?"

"At the airport."

"So late?"

"Yes—why the cross-examination?"

"Maybe I'm all tensed up," Jerrie told Jack one night. "I haven't had a vacation in two years. I'll get some rest; that will do us both some good."

"It's not you, honey. It's me," Jack replied, shouldering the blame. "I'm just trying to do too many things at once. You take a couple of weeks off, and by that time I'll be in the clear. A new man!"

Jerrie took a leave of absence from Fleetway, and with distance, she could see they were tearing each other apart. Jack was the love of her life, but marrying him would be akin to caging a bird and clipping its wings. Slowly, she admitted to herself that she had to set him free.

On July 29, 1955, White House press secretary James C. Hagerty sat before a microphone at a small table under the glare of bright TV lights. On his right were representatives from the National Science Foundation, on his left were representatives of the National Academy of Sciences. They were all better equipped to answer the inevitable barrage of questions that James knew were well beyond his comfort zone. The small group of men faced a room packed with eager journalists who fell silent as he leaned forward toward the microphone.

"On behalf of the president," he began, "I am now announcing that the president has approved plans by this country for going ahead with the launching of a small unmanned Earth-circling satellite as part of the United States participation in the International Geophysical Year." Few people in the room had ever heard anything like it. The IGY wasn't new; reports had been circulating for over a year about this international, eighteen-month-long scientific investigation into solar and geophysical activity. Akin to the International Polar Years that had seen great scientific collaboration in the early 1880s and the early 1930s, this one would run from July of 1957 to December of 1958, continuing the tradition of cooperation as international scientists gathered data during a period of peak solar activity. It was the mention of satellites that gave the gathered journalists pause. Rockets and satellites were fodder for science fiction, but the White House had just committed the United States to launching a real-life space mission as part of its IGY program.

The news reached the 6th International Astronautical Congress in Copenhagen, Denmark, four days later. When Soviet academician Leonid Sedov heard the news, he knew what the Americans were thinking—studying the Sun and the Earth in space meant no atmospheric disturbances—and knew his own nation's scientists had to keep up. That same day he held a press conference and announced to the world that "the realization of the Soviet satellite project can be expected in the near future."

In the days that followed, no other country announced its intention to launch a satellite as part of the IGY. It was as though James Hagerty, in the simple act of reading a presidential statement, had unwittingly fired a starter's pistol in a race with the Soviet Union to be first in space.

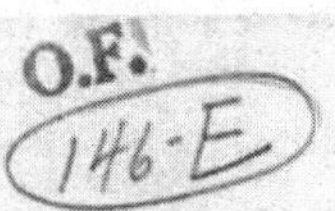

IMMEDIATE RELEASE July 29, 1955

James C. Hagerty, Press Secretary to the President

- -

THE WHITE HOUSE

Statement by James C. Hagerty

On behalf of the President, I am now announcing that the President has approved plans by this country for going ahead with the launching of small unmanned earth-circling satellites as part of the United States participation in the International Geophysical Year which takes place between July 1957 and December 1958. This program will for the first time in history enable scientists throughout the world to make sustained observations in the regions beyond the earth's atmosphere.

The President expressed personal gratification that the American program will provide scientists of all nations this important and unique opportunity for the advancement of science.

# # # #

Eisenhower Presidential Library

CHAPTER 13

Palm Springs, California, January 1956

"WHY I WANT TO RUN FOR CONGRESS. THE ANSWER IS SIMPLICITY ITSELF. I started life in desperately poor circumstances. But the political freedom and rights we have in America let me succeed. I owe my country a lot. I wish to serve it in return." Jackie read her prepared statement before the Palm Springs Republican Assembly, the pages typed in a font large enough she could read without her glasses. "If I go to Congress, I do not expect to have a nicely packaged plan for the solution of every problem. I would expect to be an active member among the four hundred thirty-five members of the House of Representatives." Weeks later, Jacqueline Cochran-Odlum was officially announced as one of the six Republican candidates running in California's 29th district.

Now fifty, Jackie wasn't looking for an avenue away from high-speed flying, though it had been three years since she'd secured a new speed record. She'd developed an interest in politics working on the Eisenhower campaign, albeit more for the power and influence that came with an elected position than her need to right wrongs she saw affecting citizens. With the president a personal friend, she was well positioned to take on this new challenge as a Republican in a region that skewed Democratic. It just so happened that a congressional seat in her district was up for grabs, but it wouldn't be an easy win. Her district was one of the largest in the country, spanning 11,000 square miles from the eastern edge of Los Angeles to the border of Arizona and from the Mexican border north for about 200 miles. Her constituent base was largely Mexican farmers concerned with education opportunities and agriculture. None of this made Jackie the likely candidate, so she leaned on her celebrity to court voters.

By now her story was well known—at least the version she wanted known. Her memoir *The Stars at Noon* had hit shelves two years earlier, telling the story of an orphaned girl forced into the cotton mill by her foster family, scraping together meager meals and fighting for union rights before striking out on her own to make a living in the beauty industry. "I know what it means to work with my hands—what a good job and security or lack of security mean to the individual," she'd orate in coffee shops and small rallies before promising to represent the common man in Washington. She walked up and down the streets as spring temperatures soared to 120 degrees, promoting a free, strong, and secure America until her feet were blistered and her hands swollen. She liberally referenced her military service as evidence of her commitment to veterans' rights. "I do not believe this country can have second-class citizenship for any of its people," she declared before vowing to serve her country with the highest loyalty. She peppered speeches with names of her well-positioned friends in Washington whom she vowed would support her campaign promises.

Jackie might have been saying all the right things, but her actions spoke louder than words. She flew around the district in her Lodestar decorated with a larger-than-life portrait of herself, then set up a table right on the tarmac where she'd glad-hand voters while collecting their signatures in support of her candidacy. Jackie's identity was so wrapped up in flying it seemed natural to her to make her plane a centerpiece to her campaign, but to voters, it underscored how much she couldn't relate to the common man. She came off as boisterous, and her campaign seemed amateurish, almost as though she thought she could buy votes by dazzling constituents. Critics openly questioned her motivations, suggesting Jackie was more interested in benefitting her wealthy friends than the low-income voters who made up the majority of her district. And for once Floyd couldn't help; the public saw him as a mean old man who used his money to buy his wife anything she wanted.

While her name was dragged through the mud in the papers, Jackie fought to turn criticisms into positives. She maintained that her flying campaign was a demonstration of her willingness to visit any corner of her district on a dime and proudly posed for pictures with Floyd in matching ranch outfits looking every bit the devoted couple. Ultimately, Jackie hoped her endorsement from President Eisenhower would put her over the edge for all voters and decided to put her name on both the Republican and Democratic ballots. If she won both, there would be no second election; she'd win the seat outright. She could only hope her appeal would transcend political parties.

On Sunday, May 27, Jackie threw her energies into one final bid for votes, a "Salute to Jackie" rally held at the Riverside County Fairgrounds. The event kicked off with a barbecue at four o'clock followed by two hours of entertainment beginning at five, headlined by her friends, comedian Bob Hope, actress Rosalind Russell, and even the famous pooch Rin Tin Tin. Speaking from the podium for the final time in her campaign, she promised voters that she would "represent all of the people and all of the district!" Whether for Jackie or the star-studded stage show, more than 7,000 people braved a dust storm to attend.

A little over a week later, on June 6, the results were in. Jackie won the Republican nomination by just 4,000 votes. Now she had to take on Dalip Singh Saund, the Democratic candidate, in a second campaign that promised to be far harder.

Jerrie's leave from Fleetway wasn't a two-week reset—it became permanent. She felt lost, empty of energy, ideas, and inspiration. Rather than move to a new city to find another flying job and begin her life anew, she returned to her parents' house in Ponca City. She wanted to be cared for, to have decisions made for her. But Jack wasn't giving up without a fight. He called, pleading with her to work things out, but Jerrie remained resolute in her decision. By the summer of 1956, the physical toll of ending her relationship was evident. Her weight fell to a dangerous ninety-eight pounds on her five foot seven frame as she struggled against the bleak feeling that came with having no idea what to do next.

Harvey and Helena were beside themselves with worry. Desperate to get their daughter out of her depression, they finally found something they thought might help: an all-woman international air race from Hamilton, Ontario, to Havana, Cuba, sponsored by the Ninety-Nines. Of course, knowing the young woman as they did, they knew they couldn't outright suggest she do something. The idea had to be planted discreetly in her mind. So Harvey brought it up very casually one day.

"I guess you're not up to it."

"Not up to what?" Jerrie asked.

"Not up to entering the race next month—the International."

"Oh," Jerrie yawned her reply. "I guess I could. But I don't have a plane, and I don't have a sponsor. Maybe next year." But her parents weren't giving up that easily. A few days later, Helena reopened the discussion.

"Too bad, too bad," she muttered within earshot of her daughter.

"What's too bad?" Jerrie asked.

"Your father tells me that race starts in Hamilton, Ontario. I'd kind of like to see my Connecticut friends. If you were flying in it I'd go up to Canada with you and then go on to Connecticut."

"You would?" Jerrie was surprised.

"I just said I would, didn't I?" Helena heaved a sigh she hoped would play on Jerrie's conscience.

The final phase of their assault on their daughter's depression came when Harvey announced that the Ponca City Oil Company was willing to sponsor with the loan of the company's twin-engine Piper Apache. Jerrie couldn't deny her parents or herself. The race did sound fun.

"We'll have to fatten you up a bit first," Helena said. "After all, I don't want a sick pilot for my first long trip by air."

"We have to get some color in your cheeks, too," Harvey added. "They won't let you across the border in your condition!"

Her parents' unconditional love and support worked wonders. A year after hiding in the safety of her parents' home, after the calls from Jack finally stopped, Jerrie was ready to return to the air and fly the international air race.

As the congressional race in California's 29th district picked up steam in the summer of 1956, the country quickly took notice. It was hard not to be enthralled by the battle between American-born famed aviatrix Jacqueline Cochran-Odlum and Dalip Singh Saund, whose Sikh religion felt as foreign as his native India. The candidates were so strikingly different, the contrast was irresistible. Jackie was always immaculately made up from her hair to her pressed dresses, while the stockier, bushy-haired Dalip wore the same suit to multiple appearances. Jackie piloted her Lodestar between events or was driven in her air-conditioned Oldsmobile with several changes of clothes in tow. Dalip drove more than 26,000 miles in his six-year-old Buick accompanied by his wife, son, two daughters, and other family members. Jackie hired four professional campaign organizers who managed her every move. Dalip, the hardworking family man, relied on relatives to do the same.

Jackie's team warned her that Dalip was likely to take aim at her wealth, her lavish campaign style, and even Floyd's business interests in an attempt to chip away at her credibility. Her best bet, they advised, was for her to scale things back, to connect with voters as one hardworking, patriotic American to another. But Jackie didn't like that approach. She maintained that her celebrity was her most valuable asset, so she turned to her famous friends for help.

That help came in the form of endorsements. Several congressmen avowed that Jackie would be an asset in Washington. Second World War air ace turned special assistant to the Air Force chief of staff Jimmy Doolittle told her she represented "the type of patriotic, courageous, honest, intelligent person" America needed in Congress. President Eisenhower's brother Edgar Newton Eisenhower, a noted lawyer, confirmed that she had the support of the whole Eisenhower family, including Ike. Retired USAF General George C. Kenney lauded her "courage, integrity and good old-fashioned common sense" as valuable traits. Jackie took these words of encouragement and turned them into promotional material. When President Eisenhower paid a visit to the Ranch to take a picture and write a formal letter of endorsement, Jackie promptly created leaflets. Who she knew became a driving force in her campaign, which soon took a nasty turn.

While Jackie was meeting and greeting voters, Floyd busied himself digging into Dalip's past. He pored over Dalip's memoir *My Mother India* and curated a list of excerpts that could be read as un-Americanism. He looked for Communist ties and searched for laws that would make Dalip ineligible for office since he hadn't been born in the United States and had been in the country on an expired visa for years before declaring his candidacy. Floyd even tried to find evidence that Dalip's abandoning his birth name, Piyush, was proof he was hiding something. Incredibly, neither Dalip's team nor dogged journalists managed to find evidence of Jackie's own birth name, and by extension never tracked down living relatives on the East Coast.

Everyone readily accepted the story in her memoirs that she'd picked it out of a phone book.

Dalip's team fought back with retaliatory personal attacks that amounted to a stunning smear campaign. He accused Jackie of running for office to further Floyd's business interests, of failing to register as a voter until after the war, and of spending more than quadruple what he had spent campaigning. When Jackie referred to Dalip as "that Hindu," he shot back, dismissing her as "that woman" or the "wealthy Mrs. Odlum." He maintained that she was so far removed from her poor upbringing that she couldn't possibly understand the hardships faced by laborers. He accused Jackie of racism in an attempt to dissuade black and Mexican voters from supporting her. He cited her lack of formal education as evidence that she was unfit for politics, to which she replied that at least she spoke clean and unaccented English. When Dalip accused her of having no relevant experience, Jackie pointed to her wartime service and Distinguished Service Medal.

Jackie had her defenders. Military brass with whom she'd worked during the war and former WASPs wrote editorials expressing their dismay over these false characterizations, but the negative press coverage started taking a toll as the campaign wore on. After reading such horrific things about herself every day for weeks on end, Jackie couldn't help but take it personally. She took an overly defensive stance in public appearances while privately she alternated between feeling sorry for herself and feeling unappreciated. Eventually, she turned her anger toward her staff, lashing out about how poorly the campaign was going, until her campaign manager warned her she ought to be nice if she didn't want a fistful of resignation slips in her hand. The machine of her political dream was crumbling.

When the votes came in in the early morning hours of November 7, Jackie's worst fears were confirmed. She captured 48.5 percent of the district's votes to Dalip's 51.5 percent. Absentee ballots would come later, but Jackie wasn't optimistic they would affect her standing. She admitted defeat. "I doubt that the absentee vote will change the presently known results," she wrote to Dalip in a telegram. "If you have been elected as our representative in Congress I warmly congratulate you." She issued a similarly simple press release offering Dalip her congratulations and her hope that during his time in office, the "interests of our 29th District will be well served." Campaign activities wound to a close as letters of condolence poured in. "Your loss is to me a personal one for I know so well what fine work you would have done in the Congress," Ike Eisenhower wrote. "My heartfelt regrets go to you and Floyd."

Gradually, Jackie felt the full weight of the loss. It was like the WASPs' deactivation all over again, a very public failure rooted in her own ego. If she had listened to her campaign managers and downplayed her celebrity, things might have been different, but she hadn't. She was too wrapped up in being Jacqueline Cochran to see how her reputation was proving to be more of a hindrance than a help. She finally admitted to herself that running for Congress had been a mistake, particularly as a Republican in a Democratic region. Jackie was a pilot first and businesswoman second, not a politician, which meant whatever her next big campaign or undertaking, it would be something in the realm of flight.

The international air race was one of the worst of Jerrie's career. The first leg from Hamilton to Buffalo was fine, but as she approached the airport to land and refuel she was told to enter a holding pattern. Another contestant had made an emergency, gear-up landing. Jerrie's handicap was high, and the hour she spent circling while airport attendants cleared the runway from the crash killed her time. There was no way she could recover.

Losing the race so early on was one thing, but losing out on a reunion with other pilots was another. Now that she was back in the air, she couldn't miss a chance to reconnect with her sisters in flight, so she flew to her next stop in Marathon, Florida, then on to Havana. High in the air, she unconsciously began making plans. She had loved Miami before she'd joined Fleetway. Why not move back and look for a job there? Or maybe somewhere else in the Caribbean, since she loved the area? It was time to rejoin the flying world.

Unfortunately, Jerrie found that little had changed in the aviation landscape during her ferrying years; airline managers and airplane companies alike remained reluctant to hire female pilots. She wrote hundreds of letters to dozens of companies and airports looking for any job until she found one in Kansas City, Missouri, with the Executive Aircraft Company, running ground school, charter scheduling, and occasionally flying. But being back at work wasn't enough. She wanted more from her career. She wanted something spectacular. Something like an international record.

CHAPTER 14

Oklahoma, Spring 1957

"JERRIE?"

Jerrie recognized the voice on the other end of the line as belonging to Ivy Coffey. A journalist for the *Ponca City News*, Ivy had been one of the few people patient enough to press through Jerrie's natural shyness and extract enough details about her life as a ferry pilot for a story. Jerrie had been thrilled that the resulting article hadn't painted her as a cross between a barnstormer and a bearded lady. As a thank-you gesture, Jerrie had taken Ivy to lunch, and the two had struck up a strong friendship.

"Do you want to set a world's record?"

Jerrie had confided this private dream to Ivy, so figured she wasn't just making conversation. "Do you want a Pulitzer Prize?" she replied with what she thought was a similarly obvious question.

"Do you think you could set two world records?" Ivy asked, upping the ante.

"Ivy Coffey, tell me what you're talking about! This minute!"

"Okay," Ivy began, "the Oklahoma Semi-Centennial Exposition Committee agrees with me that a great way to publicize the state and to draw national attention to the semi-centennial would be for you to set a world's record in an Oklahoma-built airplane."

"Wow." Jerrie was stunned. Her dream had just fallen into her lap. "Ivy, when you say Oklahoma-built airplane, do you mean an Aero Commander?" Jerrie had developed a fondness for this propeller-driven plane back in her ferrying days, and it was one of the only notable Oklahoma-built planes in the air.

"What else?" Ivy replied. "Jerrie, they all want to see you and talk about the possibilities."

Jerrie took the next day off from Executive Aircraft to meet with the Aero Design and Engineering Company in Oklahoma City. They talked about the plane and how she intended to go about her record-setting flights. They agreed that Jerrie would aim for two international propeller plane records in the 1,750- to 3,000-kilogram weight class, one for nonstop distance and the other for altitude. The semi-centennial committee would pay all expenses, Aero Design would provide the plane, and the pilot would earn one dollar for the year. The arrangement suited Jerrie just fine—she wasn't doing it for the money—and so she set about planning.

Jerrie tackled the distance record first. Oklahoma City was the obvious endpoint for the flight, and the existing distance record held by a male Soviet pilot was 1,236.64 miles. To beat the record, she picked Guatemala City as her starting point. Her flown distance would be 1,504 miles, enough for the record, and she'd score one for the United States over the Soviet Union in the process. The problem was that Oklahoma City wasn't an official entry point to the United States. To make the landing both official and legal, she worked with the United States Customs, Immigration, and Public Health Services Bureau to make Oklahoma City a temporary immigration point for her flight only. Scores of people from the right department had to be moved through a sea of red tape just to be in Oklahoma for one day.

On May 25, Jerrie and her Aero Commander, christened *Boomtown I*, were in Guatemala City. Representatives from the Fédération Aéronautique Internationale and the Federal Aviation Administration were there, too. The officials checked that the specially installed barographs were in good working order; they would be recording all her flight data since Jerrie couldn't take a passenger on a record flight. Not only was it against the rules, but the added weight would also slow her down. Jerrie did her final checks. She knew her routes, she'd studied the weather patterns. She climbed into the cockpit and, ready, sped down the runway. As she took off, she waved goodbye to the newsmen following her for the story in a second plane, *Boomtown II*, then tore off over the mountains surrounding the city.

She kept the plane steady as she covered the 750-mile stretch over the Gulf of Mexico. Over Texas, she fought her way through and around cumulonimbus clouds carrying tornadoes, hail, and slashing rains, but she didn't deviate from her planned course more than she had to. As she approached the Texas border, she began the familiar dance unique to female pilots: racing along at 190 miles per hour against the Soviet record without the benefit of an autopilot system, she slithered out of her flight suit and wiggled into the gingham dress, stockings, and high heels she'd brought with her, all without a flying or fashion disaster.

Eight hours and five minutes after taking off, she landed appropriately dressed with the Soviet record broken. She waited a half hour for the journalists to arrive in *Boomtown II*, then posed for pictures. One newsman noted she looked as fresh as though she had just dressed for a date. The barographs were sent to Washington to be unsealed and the flight verified by the United States Bureau of Standards, but no one had doubts she'd done it. Jerrie had one record down and one to go.

Preparing for the altitude flight was a different beast. For this record, Jerrie would have to fly vertically rather than horizontally, pushing the Aero Commander beyond its tested limits in the process. By design, the Commander could reach 27,000 feet; any higher and the air became too thin for the propellers to keep the plane aloft, but she would have to force it higher if she wanted to secure the record. There was also the change in temperature to consider. On the tarmac in Oklahoma, it would be about ninety degrees, but at her peak altitude, it would be around ten below. But the most challenging part of the flight for Jerrie wasn't the technical part, it was the fact that she'd never gone that high. She had no idea how she'd feel—she might be excited, or she might mistake the onset of hypoxia for excitement and put herself in a potentially fatal situation. All in all, though Jerrie had more experience in the air than most people her age had in cars, this was one of the most dangerous flights of her career. And it was going to be a public one. The semi-centennial committee wanted the altitude record to open the state's celebratory exhibition, so if anything happened, it would be in front of an audience.

The day of the flight arrived. On June 13, 1957, oxygen mask in place over her nose and mouth, Jerrie took off from Aero Design's runway just before three-thirty in the afternoon in her twin-engine Commander and started her upward trajectory. She flew steadily higher, pushing every ounce of power she could get out of her engines to hit a record altitude in her class of 30,330 feet. She could barely breathe, but not for lack of oxygen—the mask was still in place. The tranquil beauty of the deepest, bluest sky she'd ever seen, unblemished by a horizon, took her breath away. Looking out the cockpit window, she saw no boundaries, no limitations. The Sun felt closer, and the stars seemed brighter, so bright she felt as though she could reach out and touch them. As the laboring plane hung in the thin air, Jerrie wanted nothing more than to continue her upward climb all the way into space, an urge that gripped her as nothing had before.

The Commander couldn't stay at that altitude; the propellers needed air. Jerrie guided the plane back down through the thicker atmosphere, and less than two hours after taking off she landed among a horde of reporters waiting by the runway. Again the barographs were removed and sent to Washington for verification as journalists asked her about the flight. "I think I'll just take it easy and stick with straight flying for a while," she told the gathered journalists. She didn't mention anything about the pull of the dark sky, but it had left its mark.

"Check under the hood?"

John Herschel Glenn Jr. gave a wry laugh. "Don't bother, I'm sort of in a hurry this morning."

Hurry was an understatement. It was July 16, 1957, and John was high over New Mexico, traveling 300 miles an hour as he deftly lined up his US Navy Chance Vought F8U Crusader probe into the waiting tanker's funnel-like drogue. He held his position as the tanker transferred fuel into his jet's tanks. It was the first of three midair refueling stops on his world record attempt at a high-altitude, supersonic, cross-country speed record, though flying as fast as he was, it was more like a race car driver making a pit stop without actually stopping.

A career military pilot with combat experience in both the Second World War and the Korean War, John was now a test pilot working with the navy's Bureau of Aeronautics in Washington. He'd been working with the Crusader for months, now testing its Pratt & Whitney J-57 jet engine. While it had an excellent record, there was some question as to whether it had logged enough hours at combat power and high altitude to be cleared for installation in military jets. John's plan was to test the engine on the supersonic cross-country flight, drawing attention to the new plane at the same time. After all, it was taxpayer money making those planes fly, so publicity couldn't hurt. When he'd realized the Crusader would be flying faster than a bullet out of the muzzle of a gun, he'd named the flight Project Bullet.

"Come and see us again. I'll mail your Green Stamps in the morning," the tanker pilot joked as John, the refueling finished, disengaged from the drogue. He put a safe distance between himself and the tanker before returning to the Crusader's cruising altitude of 51,000 feet and its cruising speed of Mach 1.48, well over the speed of sound. Nestled inside the cockpit, everything seemed eerily still. There was no sound save the static over his radio, and he was so high above the ground the countryside below looked almost stationary. Everything felt so smooth and calm, almost like driving down an empty country road at a leisurely thirty miles an hour.

As he crossed the country, John's flight path took him within ten miles of his hometown of New Concord, Ohio. His parents were outside in the yard with friends, trying to find a glimpse of the plane overhead. As the sonic booms that reached the ground in John's wake shook all the windows in his parents' neighborhood, a boy ran out into the street yelling, "Mrs. Glenn, Johnnie dropped a bomb!"

Finally, New York's Floyd Bennett Field came into view. John passed the towers that marked the finish line where timers stood to log the end of the flight. John had covered the country in 3 hours, 23 minutes, 8.4 seconds with an average speed of 723 miles per hour, sixty-three miles faster than the speed of sound at that altitude. He idled the plane's throttle, pulled up into a sharp turn, and glided back around to make a smooth landing on the runway. He had just enough fuel to taxi to where a welcome party waited. He quickly spotted his wife, Annie, in the crowd wearing an off-the-shoulder dress topped with a choker necklace and a white hat, the woman he'd known and loved since they'd met as toddlers. Standing next to her was ten-year-old Lyn in a striped dress and white gloves and twelve-year-old Dave in a white shirt and pleated pants. As the band played "Anchors Aweigh" followed by the "Marine Hymn," John pulled Lyn's cat brooch and

Dave's Boy Scout knife out of his pocket and handed these supersonic souvenirs to his children. The family posed for photographers, and the following day their picture appeared in the *New York Times*. The accompanying article pointed out that at the age of thirty-six, "Major Glenn is reaching the practical age limit for piloting complicated pieces of machinery through the air."

The Glenns stayed in New York for a flurry of press conferences and public appearances, but it wasn't all work. They spent time shopping, wandering through the big-city department stores where the award-winning Jacqueline Cochran Cosmetic displays encouraged women to consult with a beautician to learn her skin type and how to treat it. During a jaunt into Macy's, John noticed a woman staring at him and David. He'd quickly grown used to the public attention so thought nothing of it, but the woman eventually approached them.

She didn't want an autograph. Rather, she wanted to know if he'd ever heard of the game show *Name That Tune*.

John Glenn made his television debut on the game show weeks later, partnered with child actor Eddie Hodges. Every time they correctly identified a song from a few notes, they won $5,000 and moved on to the next round. They made it all the way to the final round, the Golden Medley. If they named every tune in the Medley, they would split the grand prize of $25,000.

Viewers nationwide tuned in to watch the final round on October 4. Before the Medley began, host George DeWitt asked John what he would do with the money if they won.

"Well, it would help pay for Dave and Lyn's education, and if there was any left over, it would go towards Lyn's wedding," John told the studio audience, which included his now-blushing daughter. George turned his attention to Eddie, who admitted to having some trouble with a girl at school. Seeking advice, George asked John if he'd had any trouble with girls when he was Eddie's age.

"I did," John admitted. "There was a girl in school who wouldn't talk to me because all the other boys liked her, too."

"Did you ever get to talk to her?" George asked.

"I did. I got to marry her," John replied, smiling at Annie as the audience around her clapped. America couldn't ask for a better hero in that moment. The handsome young marine with his beautiful wife and two children were the very picture of American family values.

After a moment of silence in the studio, the Golden Medley began. John and Eddie guessed the first four songs correctly, then had nine seconds on the clock as the fifth and final song started. John listened for a moment, pressed the buzzer, then leaned down to whisper the answer in Eddie's ear.

"'Amaryllis'?" the boy guessed.

"'Amaryllis'...is right!" George exclaimed, proclaiming the duo grand prize winners as they hugged in celebration.

The final strains of the Golden Medley had barely faded from the airwaves when, at 8:07 p.m. Eastern Standard Time, a very different sound was picked up on an RCA receiving station at Riverhead, New York, then relayed to the NBC radio studio in Manhattan. The sound was recorded then broadcast for anyone with a radio or television to hear. It was a slightly distorted, intermittent elongated beeping tone the special reports announced was coming from an artificial satellite called *Sputnik*.

Lyndon Johnson stepped outside after dinner, accompanied by the guests he and Lady Bird had hosted that evening at their ranch. It was his custom to walk at night since suffering a heart attack two years earlier. But tonight, as he looked up at the familiar Texas sky, it felt different, almost alien. Spaceships and rockets to the Moon were no longer the stuff of stories told by the fire on a rainy evening. That future was just over the horizon, and so were the powerful Soviet missiles that had launched the small satellite. He'd never imagined the Soviets would succeed in launching the first satellite, and he knew America had no equivalent technology. Lyndon realized this nascent era in man's conquest of the natural world marked a grim development in the Cold War, but it also brought a unique opportunity.

Since becoming Senate majority leader in 1954, Lyndon Johnson's political mastery had made his a career to watch. He skillfully united the liberal northern and conservative southern Democrats on crucial issues, but the Democratic Party as a whole was struggling to find something that it could leverage to regain the White House. That night at his ranch, Lyndon thought that if he could take charge of space policy and mold the nation's reaction to *Sputnik*, he might be able to launch himself to the presidency.

AFTER SPUTNIK, OCTOBER 1957: A SPACE INTERLUDE

SPUTNIK DOMINATED THE NEWS CYCLE IN THE DAYS THAT FOLLOWED. IT made headlines in the *New York Times* that landed on Jackie's breakfast tray and in Jerrie's local *Daily Oklahoman*. The nearly unanimous reaction among Americans was fear. No one was sure whether *Sputnik* was, as the *Los Angeles Times* called it, a "dove of peace or the sword of Damocles hanging by a thread over the free world." The panic that gripped the nation was more extreme than anyone in Eisenhower's administration had anticipated. The president's science and political advisors knew *Sputnik* wasn't a threat; the 184-pound sphere couldn't do anything more than beep from orbit. The R-7 missile that launched it, on the other hand, was concerning. If it could put a small satellite into orbit, it could easily launch something heavier, like a nuclear warhead, from Russia to the United States. The president was thus armed with everything his staff had been able to dig up on missile development when, at ten-thirty in the morning on October 9, he prepared to address the nation for the first time in the space age.

Sitting behind the large wooden desk in his executive office, he seemed calm, almost apathetic at the prospect of losing face to the Soviet Union on an international scale. In front of him were journalists eagerly awaiting his words as well as television cameras that would bring the press conference to Americans who needed their president to allay their fears.

"We congratulate Soviet scientists upon putting a satellite into orbit," the president read from his prepared statement. He went on to remind Americans that both the United States and the Soviet Union knew the other was developing a satellite capability as part of the International Geophysical Year. America's IGY satellite program, he said, was the Navy's Vanguard, a civilian program deliberately separated from all other military activity to underscore the scientific nature of the program and prevent it from interfering with high-priority military programs. "Our satellite program has never been conducted as a race with other nations," he said. "I consider our country's satellite program well designed and properly scheduled to achieve the scientific purposes for which it was initiated."

Lyndon didn't think much of Ike Eisenhower's promise of a superior US satellite. Neither did his aide George Reedy, who stayed up late into the night reading everything he could about space until he too became fixated on the idea of leveraging the issue into a Johnson presidency. George suggested LBJ take the stance of a nonpartisan senator and advise Congress to conduct an inquiry into *Sputnik*'s importance as well as the leadership decisions that had kept American satellites grounded. The US Army's missile program was sufficiently advanced that it might have beaten *Sputnik* into orbit, but orders from Washington stopped it. Decisions like these were the key issues for an inquiry, in George's mind. Lyndon agreed, not only because he knew space would be critical for the foreseeable future but also because taking control of new space policy was an appealing way to raise his public profile.

The senator started laying the foundation for his inquiry the following Friday at the twentieth annual Texas Rose Festival in Tyler. "The real meaning of the satellite is that we can no longer consider the Soviet Union to be a nation years behind us in scientific research and industrial ability," Lyndon orated before the gathered crowd. "The mere fact that the Soviets can put a satellite into the sky—even one that goes beep—does not alter the world balance of power. But it does mean that they are in a position to alter the balance of power." To stave off that power shift, he called for the establishment of a Senate Armed Services Committee that he promised would get to the bottom of America's flagging performance. This warning that America needed to act became his public refrain while he privately planned how to tackle Congress.

Before LBJ had a chance to address Congress, the Soviets struck again on November 3. *Sputnik 2* was more than five times heavier than the first and carried a dog named Laika on board, which meant the Soviets had a working, if rudimentary, life support system. It was clear they were planning to put a man in space. Appalled by America's continued inaction, Lyndon issued a press release the following day calling for "bold, new thinking in defense and foreign policy" to regain American technological preeminence.

"We meet today in the atmosphere of another Pearl Harbor," LBJ addressed the new Senate Preparedness Subcommittee on satellite and missile programs in the caucus room of the Senate office building on November 25, 1957. "There were no Republicans or Democrats in this country the day after Pearl Harbor. There were no isolationists or internationalists. And, above all, there were no defeatists of any stripe. There were just Americans anxious to roll up their sleeves and wade into the enemy." He didn't care if he was being overly dramatic. He knew the press would lap up every over-the-top statement and his speech would make headlines.

After this ostentatious start, Lyndon led the Preparedness Subcommittee in cross-examining a rotating cast of witnesses from the military, aviation industry, and science arenas. Going into the Christmas season, Americans learned that the country had no planned timetable to catch up with *Sputnik*. The hearings quickly became a platform for every voice that opposed the president's approach to space, among them Wernher von Braun.

The tall, broad-shouldered, soft-spoken former Nazi engineer had found notoriety during the Second World War as the lead designer behind the V-2 rocket. His simple liquid-fueled guided rocket had been the first system able to deliver bombs to relatively precise locations from a considerable distance. The technology was so appealing that, when he orchestrated his team's surrender to the United States in the closing months of the war, he was welcomed with slightly wary open arms. However frightening his weapons, America preferred to have him on its side than working for the Soviets.

Now, more than a decade later, Wernher was an American citizen leading the Army Ballistic Missile Agency in developing the Redstone intermediate-range ballistic missiles that, with the slight modification of a fourth stage, could launch a satellite into space. This team could have beaten *Sputnik*, but the president had forbidden it; he didn't want America's first satellite to come from former Nazis. The result, Wernher testified before LBJ's subcommittee, was a failing American program that could only recover with the creation of a dedicated and well-funded space agency.

The Preparedness Subcommittee hearings were still going on when the navy attempted to launch America's first satellite, *Vanguard*, from Cape Canaveral on December 6. People throughout the country gathered around their televisions for the live broadcast, their hopes pinned on its success. The whole world watched as the rocket lifted off the ground, paused as it struggled to gain height, then settled back on the launch pad where it collapsed under its own weight. The rocket erupted into a fireball that engulfed the launch pad and sent thick, black smoke curling high into the clear Florida day. The three-pound satellite, meanwhile, had sensed a moment of weightlessness and punted itself free. Thinking it was in orbit, *Vanguard* proudly emitted its beep from a nearby puddle for the world to hear.

The international press had a field day giving *Vanguard* creative nicknames like *Flopnik* and *Stayputnik*. In Washington, the very public failure forced Ike Eisenhower's hand. He gave Wernher von Braun's army team sixty days to get their satellite up while the Preparedness Subcommittee took advantage of the media attention to try its most outspoken witnesses. LBJ allowed witnesses to speak freely about their issues with the Eisenhower administration, and though the senator himself cast no stones, the president came out battered, and the majority leader emerged as a politician concerned with the nation's welfare above all else. The Preparedness Subcommittee's inquiry ended with the unanimous agreement that decisive action was vital to accelerate an American space program. To avoid a destructive war, America needed a space program, powerful rockets, fast-tracked "crash programs" for its intercontinental and intermediate-range ballistic missiles, and above all, some structure over the warring service branches trying to gain dominance in space.

At ten-thirty on the last night of January 1958, Wernher von Braun paced relentlessly in the communications room at the Pentagon. Being in Washington was hell when his rocket was 800 miles away at the Air Force's Cape Canaveral launch site in Florida, but he couldn't disobey an order. If the launch was successful, he would go straight to the Academy of Sciences building for a celebratory press conference. If it wasn't, he'd put on the dark sunglasses hidden in the breast pocket of his dark suit jacket and escape to an even darker movie theater.

At 10:48, telemetry from the rocket reached the Pentagon, and Wernher could see that the launch had been good, but he couldn't celebrate until he got confirmation that the satellite was in orbit. An hour passed with more pacing, coffee drinking, and cigarette smoking. One hundred six minutes after launch, Wernher knew the satellite should be over the west coast, its signal audible to local tracking stations. Unable to wait, the men in Washington called the men who had built the satellite at Caltech's Jet Propulsion Laboratory in California. Both ends of the phone call were tense until JPL got the signal confirming the satellite was indeed in orbit. At 12:51 in the morning, exactly seventeen weeks after *Sputnik*'s launch, America officially entered the space age.

Less than a week later, Lyndon pushed a Senate resolution through to establish the Special Committee on Space and Astronautics that would oversee space policy as well as direct its use of resources, and against a lone dissenter, he was elected chairman. Funding was forthcoming and LBJ would be in control of it.

It wasn't long before word got around Washington that the National Advisory Committee for Aeronautics was getting into the space game. America's preeminent aviation research organization, the NACA was a strictly civilian agency that worked closely with the military and had played a part in virtually every major development over the last four decades. Now it was spearheading the move into space. The NACA's Research Center at Langley Air Force Base in Virginia was looking for a few test pilots to fly simulated missions and help figure out what a future manned spacecraft would look like. When John Glenn heard about it, he immediately volunteered.

Langley introduced John to concepts he'd never heard of and sensations he'd never experienced. He learned how to adjust a spacecraft's orbit in flight and what it took to make a manual reentry. Sitting in a cabin at the end of a fifty-foot arm, he tried to control the axes of pitch, yaw, and roll with a simple hand controller while the centrifuge spun him fast enough to feel crushing g-forces. On a visit to the McDonnell Aircraft plant in St. Louis, he saw the first mockup of the spacecraft. It was nothing like the sleek aircraft pilots flew through the sound barrier. This was a squat-looking truncated cone with a rounded bottom designed to disperse the heat it would generate falling through the atmosphere during reentry. Though it was mostly research and rumors, John knew he was seeing the very beginning of America's manned spaceflight program. He also knew that the second they called for people to fly in space, he would be first in line.

Lyndon was frustrated as President Eisenhower remained at loggerheads with his science advisors. The president wanted America's new space program to fall under the Department of Defense whereas his advisors were pushing for a civilian space agency that could foster creativity and international cooperation. The president finally relented and sent a note to Congress recommending that the new agency form with the NACA as its nucleus and the goal of managing everything but the nation's military needs in space. LBJ sprung into action.

On April 14, Senate resolution S. 3609 was presented to the House. "Be it enacted by the Senate and House of Representatives of the United States of America in Congress assembled," the resolution read, "that this Act may be cited as the National Aeronautics and Space Act of 1958." It went on to stipulate that the agency's focus would be on expanding man's knowledge of atmospheric phenomena and the space environment as well as improving the usefulness, performance, safety, and efficiency of aircraft. Organizationally, it would be headed by a director, appointed by the president, who would, in turn, be advised by the sixteen-person National Aeronautics and Space Board.

A month later, Lyndon began the hearings of the Senate Special Committee on Space and Astronautics that would turn that bill into law. "I believe," he said in his opening statement, "it is entirely fair to say that seldom if ever, has a Congress and an administration faced a more challenging task. We are dealing with a dimension—not a force. We are dealing with the unknown—not the known. The challenge of the space age, at the beginning now, is to open a new frontier to permit its use for peace."

LBJ, with his trademark panache, laid the foundation for a civilian space agency.

Randy Lovelace knew America had little time to waste in figuring out who America's new space agency should eventually send into space. As a leading authority in space medicine and an active member in the Fédération Aéronautique Internationale, he had traveled to Russia a number of times over the years and knew just how advanced the Soviets were in their quest to conquer space. It was enough to make him seriously concerned for America's future. He was thus thrilled at the opportunity to shape America's future astronaut program as chairman of the Working Group on Human Factors and Training for the Special Committee on Aeronautics.

The eight-man committee met toward the end of April 1958 at the High Speed Flight Station at Edwards Air Force Base in California. Committee member Scott Crossfield, a former NACA test pilot who was now working with North American Aviation on the X-15 rocket plane, had firsthand knowledge of the challenges facing pilots in space. Also on the committee was Dr. Don Flickinger, an air force flight surgeon who was working on the service's own manned spaceflight program called Man in Space Soonest, a program using blunt capsules like the ones John Glenn had seen at Langley. Tapping into the experiences of its members, the working group determined that pilots would be the best people to make the first flight into space.

"I have today signed the National Aeronautics and Space Act." The president read his prepared statement at the White House on July 29, 1958, before gathered politicians and journalists, including Lyndon Johnson. LBJ was proud that his bill had been passed so swiftly and that he'd stayed in the headlines, but he knew the real work in space hadn't even started. His own presidential ambition, meanwhile, oscillated by the hour; he'd wake up in the morning desperate for the top office only to be repulsed by the idea by lunchtime. Regardless of his next political move, he remained confident that his close affiliation with the space agency would be vital going forward.

The NACA's successor organization, the National Aeronautics and Space Administration, opened its doors two months later on October 1. The eight thousand employees at research centers around the country who went to work that Tuesday morning suddenly found themselves on the forefront of the space age. NACA president Hugh Dryden stayed on as deputy administrator to Ike Eisenhower's pick of T. Keith Glennan as administrator. The new agency moved into the Dolley Madison House in Washington's Lafayette Square.

Almost as soon as it opened for business, the Space Task Group with T. Keith Glennan as its chairman began laying out the qualifications for the astronauts NASA would soon hire. Building off the lessons learned from earlier aviation medicine and experimental high-altitude programs, the agency decided to release a public call for "research astronaut-candidates" as part of "Project Astronaut."

Applicants had to be American citizens between the ages of thirty-five and forty by the date their application was filed, be in excellent physical condition, and stand no taller than five feet eleven inches. Candidates were also required to hold a bachelor's degree in physical, mathematical, biological, medical, psychological sciences or engineering as well as have three years of relevant technical or operational experience. Though astronauts would go through extensive training and the capsule would be largely automated, the announcement read, "he will contribute by monitoring the cabin environment and by making necessary adjustments. He will have continuous displays of his position and attitude and other instrument readings, and will have the capability of operating the reaction controls, and of initiating the descent from orbit... will make research observations that cannot be made by instruments; these include physiological, astronomical and meteorological observations."

These qualifications were ultimately deemed too broad. The president needed a measure of security in the program and advised that the existing pool of military test pilots was a large enough group from which to select astronauts. Administrator T. Keith Glennan agreed. The qualifications remained the same with the addition that an applicant must have graduated from a test pilot school, have a minimum of 1,500 hours flight time, and be a qualified jet pilot. This made sense for NASA in light of the unknowns of spaceflight. Everything was new, the whole venture an experiment. The closest analogue, really, was military test flying. These pilots possessed the fast thinking and even faster reaction times needed to keep themselves alive in high-risk environments while also serving as the engineers' eyes and ears in the air. For the foreseeable future, while NASA figured out how to live, work, and fly in space, astronauts would need to play the same role. It was the only way the agency could get off the ground.

Even with these qualifications in place, there were more than a hundred servicemen NASA could test. Since it didn't need that many astronauts to begin with—a half dozen, maybe—it could be exhaustive in its testing to pick the absolute best men for the job.

Orders marked "Top Secret" made their way to John Glenn's desk at the Bureau of Aeronautics in Washington asking him to report to the Pentagon for a briefing. He arrived to find a room full of his fellow test pilots and two NASA officials who confirmed the rumors he'd been hearing: NASA was looking for pilots to send into space. It would be a new kind of flying that had never been done. It was a volunteer position, and anyone could reconsider their choice to participate at any time, but everyone in that room met the basic qualifications. John volunteered without hesitation.

There were some 110 elite aviators nationwide who received the same orders as John. Each had more than 1,500 hours of flying time, extensive jet piloting experience, and had graduated from a test pilot school. But they weren't all as excited about the opportunity as John. Some career military pilots didn't like the idea of flying with a civilian agency, while others thought the gamble of flying in space risked sidetracking their careers and removed themselves from contention. The pool narrowed to just over eighty-five candidates. Several further screenings followed. The candidate pool was whittled down based on the men's experience, their motivations, personality, communication skills, and less tangible qualities like dependability and adaptability. The pilots worked through a long series of interviews and exacting technical quizzes. Psychologists probed for odd personalities: anyone too shy or too strident and forceful in their personal ambitions was pushed aside.

Over the course of these preliminary tests, the pool narrowed, and John was deselected. He did not have the required academic degree. The Japanese had attacked Pearl Harbor midway through his sophomore year, and he'd promptly dropped out of Muskingum College to enlist with the navy. As a career pilot, he'd taken more than enough classes and gained enough experience to make up for the missing degree, but it was still a vital requirement he didn't have. He tried to convince his alma mater to take his career into consideration and award him the degree, but it refused.

Unbeknownst to John, his commanding officer, Jake Dill, stepped in. Jake took John's combat and academic records as well as his technical flight reports to NASA and sat down with the selection board to make the case that John was more than qualified. NASA, satisfied with John's background, made an exception. In spite of his missing qualification, the agency sent him a letter saying that if he was still interested in becoming an astronaut to please present himself at the Lovelace Clinic for physical testing.

Dozens of highly qualified pilots were removed from consideration, and the pool was down to the final thirty-two candidates when John got to the Lovelace Clinic for the final phase of testing: exhausting physical checks. John spent a week in New Mexico submitting to the most detailed medical tests he'd ever experienced before submitting to an equally exhaustive examination into his psyche at the Wright-Patterson Air Force Base. Throughout it all, he stayed positive, knowing that any show of displeasure could lead to his disqualification on account of attitude.

The men were ranked, the tiniest variation from the norm affecting their standing. NASA took the top eighteen candidates and winnowed the list to seven, all men with the experience to jump right into NASA's cutting-edge test flying environment. It was, after all, an extension of the work they'd been doing for years.

John had been back at work for a week when the phone on his desk rang. The voice on the other end belonged to Charles Donlan from NASA.

"Major Glenn, you've been through all the tests. Are you still interested in the program?"

"Yes, I am. Very much," he answered, then held his breath for what he hoped was coming next.

"Well, congratulations. You've made it."

John barely registered the rest of the conversation. All he knew was that he was going to be an astronaut. Hanging up the phone, he noted that it was April 6, his fifteenth wedding anniversary with Annie. He couldn't help but smile, sure that it was a good omen.

CHAPTER 15

Oklahoma City, End of 1958

JERRIE HAD THOUGHT THAT BREAKING TWO WORLD RECORDS WOULD MAKE it easier to find a flying job, but the opposite was proving true. Her name had become synonymous with Aero Design, the manufacturer behind her record-setting Commanders, and no company was willing to hire a pilot who was so closely linked to its competitor. She finally went to the source of her troubles and met with Tom Harris, Aero Design's new sales manager working out of its Oklahoma offices.

"Mr. Harris," Jerrie announced, "I want to set some more records with the Aero Commander. I'd like to go to work for you right now. Today."

"Miss Cobb, I'd like to have you with us. You've shown what you can do with the airplane, and there's no question about your ability as a pilot. But,"—she knew that *but* was always the harbinger of bad news—"you're a luxury we can't afford."

A luxury? She didn't understand what he meant. How could she be a luxury?

"You'd be a luxury because first of all we don't need another pilot. And second, because this is a man's industry. Front offices are womanless. And I'm afraid it's going to continue that way."

Jerrie had to respect his honesty, but she also wasn't prepared to let that be the end of the conversation. She returned to his office with dozens of ideas, from speed and distance flights, to a flight around the world from pole to pole, a never-before-done circumnavigation route that would be a record for her while also showing just what an Aero plane could do. Tom killed every one of Jerrie's suggestions with ifs—*if* she could get clearances from the nations she'd be landing in, *if* the manufacturers allowed her to take on a dangerous flight, *if* she got some assurance from the weather bureau that the temperatures she'd meet were survivable. Every proposal hit an impasse until she came up with something she was sure would win Tom over.

"Mr. Harris, I have an idea." She was back in his office for yet another meeting toward the end of 1958. "You're probably planning to have an Aero Commander in the show at Las Vegas."

Las Vegas was the site of the Fédération Aéronautique Internationale's First World Congress of Flight, and it was entirely Jackie's doing in her new capacity as the FAI's first female president. The five-day-long event of meetings and flight demonstrations showcasing the latest developments in aviation promised to bring together the top people from every corner of the aviation industry, from military to commercial and even the fledgling space world. Delegates were coming from dozens of countries including Spain, Yugoslavia, Holland, Belgium, Portugal, Israel, Austria, and the Soviet Union to attend a handful of FAI subcommittee meetings. Coordinating with Charles Logsdon, the chairman of the FAI's contest committee who had verified her speed runs in 1953, Jackie arranged for the European officials to fly via Military Air Transport Service from Frankfurt to Las Vegas to make sure everyone arrived on time. For Jackie, the event was her chance to lead the FAI to the cutting edge of space. For Jerrie, it meant she'd have the ideal audience for a record-breaking flight.

"How about letting me try for a world's speed record there?" She wanted to go for the 2,000-kilometer closed-course record in a twin-engine propeller plane, one currently held by a male Soviet pilot. "You'll have a captive audience of the top people in aviation." Jerrie paused, letting Tom think through the prospect of her flight yielding immediate sales.

Tom couldn't deny this great idea. He hired an elated Jerrie on a four-month contract that would last just long enough to try for the record in Las Vegas in April.

Jerrie wasted no time getting to work, and she quickly found the 2,000-kilometer record one of the hardest flights she'd ever undertaken.

First, she had to figure out her route. The most direct and obvious path distance-wise was flying a there-and-back course between Las Vegas and some point in Canada or Mexico, but Jerrie thought there might be a better way. She studied a half century of weather bureau data and found that strong winds coming off the Pacific Ocean could give her an added speed boost in the air, so she settled on a triangular course. From her starting point at McCarran Field in Las Vegas, she'd fly to Reno, then to San Francisco, where she'd pass over a beacon at Pescadero that would serve as her official checkpoint. From there, she'd turn south toward another beacon at Lindbergh tower in San Diego, her second checkpoint. Then she'd make a straight shot back to McCarran. On paper, it was a perfect route, but in reality, there was one big problem: it meant flying over the atomic proving grounds out in the desert. This wasn't just restricted airspace, it was prohibited. No plane—not commercial, private, or even military—could fly over the area. Jerrie's heart sank. If she had to bypass the site, it would kill her time before she even started. Hoping for a miracle, she asked the Atomic Energy Commission for clearance, but the answer was no. She got every person she could think of with enough importance to ask a favor from the Atomic Energy Commission, but still, the answer was no. Another if—if she could get clearance—was standing in her way.

Then one day, without warning, the answer changed to "okay." No one offered an explanation for the change of heart and Jerrie didn't dare ask questions. She just hopped in the plane and flew right over the proving grounds on a trial run.

The flight was a disaster. Not only did the winds blow opposite from all the weather data she had, but foul weather also forced her down to refuel in San Francisco. She wouldn't set any nonstop records if she had to land.

Undaunted, she set about preparing her plane as per National Aeronautic Association guidelines. The rules said she needed to be visually identifiable, so she painted "COBB" in four-foot-high dayglow letters on the underside of the fuselage. She also bored a hole in the floor next to the pilot's seat and installed a tube through which she could fire a Very pistol at every checkpoint; the flare would signal to the judges that it was indeed her flying overhead. Then came the business of lightening the plane. To qualify for her weight class, she and the Aero Design team needed to strip out at least 887 pounds. They removed everything they deemed superfluous, including the heater, every seat except the pilot's, and all the instruments except a transceiver so she could talk to the checkpoints and the automatic direction finder so she would know what direction she was flying. The only weight they added was ten two-gallon emergency cans of gasoline. Refueling while flying wouldn't be easy, but it was a better option than landing.

The final piece of the puzzle was organizing the Fédération Aéronautique Internationale representatives to verify the run; teams would be stationed at each checkpoint with Charles Logsdon manning the start and finish lines in Las Vegas.

Finally ready, Jerrie, her stripped-down Commander, and the team from Aero Design set off for Las Vegas before the World Congress of Flight officially started. Jackie was already in town when they arrived, staying in a bungalow in the brand-new Stardust Hotel at the north end of the Strip where she could entertain guests in the evening.

Everything was set. In a few days, the weekend would kick off with a press preview.

Sequestered behind a curtain in the ballroom at NASA's headquarters on the afternoon of April 9, John Glenn could hear what sounded like a veritable horde of journalists and photographers filling the room. He milled around with six other men, military pilots all slightly self-conscious in civilian suits. The agency's tall, stoop-shouldered head of public affairs, Walter Bonney, said he'd never seen such an excited crowd.

Just before two o'clock, Walter pulled the curtain back. John was momentarily blinded by the glaring television lights as he walked along a table covered by a blue cloth. He took the seat behind his nameplate as the others did the same, putting them in alphabetical order by surname. On the floor were two models, one of an Atlas rocket and the other of the conical Mercury spacecraft. Behind the table hung a large red, white, and blue NASA insignia.

"Gentlemen," Walter addressed the audience from the podium, "these are your astronaut volunteers. Take your pictures as you will." The sound of furious shutter-clicking joined the low babble as press kits were handed out and journalists with afternoon deadlines ran for the telephones against the back wall to call their editors. After a few minutes, administrator T. Keith Glennan took the podium.

"Today we are introducing to you and to the world these seven men who have been selected to begin training for orbital space flight." A few people shuffled in their seats. "Which of these men will be first to orbit the Earth, I cannot tell you. He won't know himself until the day of the flight." Then came the moment everyone was waiting for. "It is my pleasure to introduce to you, from your right,

Malcolm S. Carpenter,"—Scott stood as his name was called—"Leroy G. Cooper,"—Gordo also stood—"John H. Glenn Jr.; Virgil I. Grissom; Walter M. Schirra Jr.; Alan B. Shepard Jr.; and Donald K. Slayton—the nation's Mercury Astronauts." The room gave a round of applause as the men regained their seats, shooting sidelong looks at one another as though confirming they were all slightly uncomfortable at being on display. Deke Slayton pulled out a cigarette and started smoking almost for something to do.

Randy Lovelace took the podium next. He was there because NASA had asked him to run the medical testing phase of the astronaut selection program, trusting him to be both thorough and discreet, and now he was presented to the press as the man who picked the most physically prepared astronauts. "I just hope they never give me a physical examination," he joked of the thorough nature of his medical examinations. "It has been a rough, long period that they have been through. I can tell you that you pick highly intelligent, highly motivated, and intelligent men, and every one is that type of a person," he spoke of the astronauts as a group. "I am not worried about their stability, their powers of observation, or their powers to accomplish the task which they are given." US Air Force physician Don Flickinger echoed Randy's remarks, adding that the high caliber of men they tested spoke volumes of the military for turning out such remarkable candidates.

Then the journalists got their chance.

"I would like to ask Lieutenant Carpenter if his wife has anything to say about this, or his four children?"

"They are all as enthusiastic about the program as I am." Scott's answer was straight and to the point.

"How about the others?" a voice called out from the crowd. "Same question." The men went down the line, leaning forward into the microphones to give answers that were as concise as Scott's. Until it was John's turn. "I don't think any of us could really go on with something like this if we didn't have pretty good backing at home, really," he began, using "us" as though speaking for the group. "My wife's attitude towards this has been the same as it has been all along through all my flying. If it is what I want to do, she is behind it, and the kids are too, a hundred percent."

When each man was asked to discuss why space was appealing, John waxed poetic as he spoke for the group. "Every one of us would feel guilty I think if we didn't make the fullest use of our talents in volunteering for something that is as important as this is to our country and the world in general right now."

"Could I ask for a show of hands from you seven as to how many are confident that they will come back from outer space?" Another question from the audience. All seven raised their hands—John and Wally Schirra raised both—as the room laughed.

"Were any civilian test pilots considered, or is this confined to military career pilots, and why?" asked another journalist.

"The answer to that is that the selection process was limited to military test pilots," Walter Bonney answered as the voice of NASA. "It was a purely arbitrary decision because we knew that the records on these people were available. We could run them through the machines and very quickly make first-cut selections from an elite group."

As the press conference wore on, the media fell in love with John. He was verbose where his colleagues were almost monosyllabic. He gave emotional answers about his motivations, his family's support, and his Presbyterian faith. He charmed the room describing the medical tests they'd all been subjected to. "If you figure out how many openings there are on a human body, and how far you can go into any one of them,"—he paused for effect—"you answer which one would be the toughest for you!" The whole room laughed as he smiled.

When it was over, John knew he'd marked himself apart from the group. While his colleagues were the epitome of laid back and cool, just as a test pilot should be, he'd offered up too much information about himself and his family. But he figured there was nothing wrong with a little publicity, especially since flying in space wasn't a popularity contest.

Four days later, the astronauts were still headline news all across America as Jerrie rolled down the tarmac at Las Vegas's McCarran Field in her Aero Commander at ten o'clock in the morning. The World Congress of Flight was beginning its second day with her world record attempt. A small crowd had come to the airport to watch her take off while attendees in the convention center could watch on closed-circuit television. She gained speed, and the second she was airborne the timer started. She could only pray this flight would be better than her practice run.

Barely off the ground, she noticed the needle on her compass seemed to be reversed. Unable to trust it until she got to Reno and had a known checkpoint against which to check it, she had to consider it useless and rely solely on her backup navigation method: a map balanced on her knee.

As she ascended to her cruising altitude of 10,000 feet, Jerrie noticed the faint smell of gasoline. It took her a second to realize that it wasn't coming from a leak in the fuel tank but from the canisters she'd brought with her—the lower atmospheric pressure had popped their lids right off, and fumes were filling the cabin. She snapped open her vents to get a flow of fresh, freezing air. It had been ninety degrees on the runway that morning, and she hadn't thought about the missing heater when she'd left her sweater in the hotel. Faced with the choice between cold hands and deadly gas fumes, she chose cold hands.

Nearing Reno, Jerrie got her first piece of good news. The compass was working perfectly in reverse, so she could rely on the feather end of the needle to guide her rather than the point. Pleased with this stroke of luck, she reached for the Very pistol, slid the nozzle into the tube next to her seat, and fired the flare. The gun fired, but the flare got stuck in the tube. A quarter of the way through a world record attempt and she was flying an aircraft full of gas fumes with a lit flare sticking out the bottom of it. *Well,* she thought to herself, *if God wants me to die in a fiery explosion today, then that's His plan*. She put the mounting danger from her mind and turned toward San Francisco. Then her radio died, cutting her off from the ground. Without it, she couldn't verbally confirm the official timers at her checkpoints had seen her. She had to hope someone in San Francisco had figured out what was going on and warned San Diego she'd be coming in silent. She turned south and soon met low clouds so dense they formed an almost solid ceiling. She had no choice but to descend low over San Diego so officials could see the "COBB" painted on her underside, though it would cost her precious time. Frustrated, she dipped below the clouds as long as she dared then darted back up to her cruising altitude as she turned toward Las Vegas.

On the final leg of the most stressful flight of her career, a new problem arose: she was running out of gas. Luckily, she was prepared with her extra canisters, but without an autopilot system, she had to get creative. Jerrie got the plane flying straight and level, then clambered out of her seat to grab a gas canister and pour it into the fuel tank. The moment it was empty, she scrambled her way back to her seat, adjusted the plane's heading, gulped some fresh air, then maneuvered her way back to do it all over again. She repeated this dance ten times to give her engines every possible extra drop of gas. With just fifty miles to go, she pushed the plane as hard as she could. Finally, McCarran Field came into view, and to her horror, she saw the judges walking away from the finish line. She tried to squeeze a last little bit of speed out of the Commander, willing the sound of the engine to call the judges back to their station, and mercifully saw them stop and rush back to the finish line in time to log the moment her wheels passed over it.

Jerrie landed exhausted but tense. Her flight was done, but the judges still had to verify her time. Minutes passed as the Las Vegas team called each checkpoint, and the longer they waited, the more the Aero team felt her chances slipping away. Then, after twenty minutes that felt like two hours, Charles Logsdon gave Jerrie the verdict.

"You made it," he told her. She'd beaten the Russian record by just twenty-six seconds, making it the closest record he'd ever measured, but it was enough for the win. Tom Harris and the Aero Design team were ecstatic. Jerrie was, too, and hoped that she might parlay this success into a job with Aero Design, but for the moment she focused on celebrating. The small team took Jerrie out for a private celebratory night with steak and baked Alaska on the Las Vegas Strip away from the World Congress of Flight.

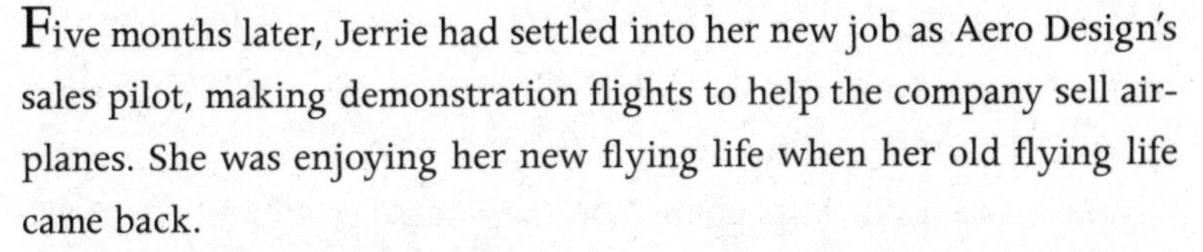

Five months later, Jerrie had settled into her new job as Aero Design's sales pilot, making demonstration flights to help the company sell airplanes. She was enjoying her new flying life when her old flying life came back.

In early September, she heard that Jack Ford, her old love from her Fleetway days, was dead; he and his navigator–radio operator were both killed in an explosion shortly after taking off from Wake Island on a ferrying mission to Tokyo. The obituaries Jerrie read felt cold and impersonal. Jack was described as a capable "goodwill aviation ambassador" and a "completely dedicated pilot." It had been years since she and Jack had spoken, but Jerrie felt the loss deeply. She also felt she could have added so much more to the stories of his death, preserving the legacy of the man she'd loved.

But Jack was her past. Walking along a nearly deserted stretch of beach in Miami at seven o'clock in the morning with Tom Harris, Jerrie was focused on the future. She and Tom were in town for an Air Force Association Conference, and the pair was taking advantage of the quiet morning to plan for the afternoon's events. As they walked along the surf, two men emerged from the water and, recognizing Tom, waved and walked over.

"Dr. Lovelace, General Flickinger," Tom made the necessary introductions, "I'd like you to meet Miss Jerrie Cobb." Turning to Jerrie, he added, "These gentlemen just flew in last night from Moscow. They were at a meeting of space scientists."

Jerrie looked on as the men caught up about their respective projects. When the conversation turned to Russian aeronautics, Jerrie jumped in with a minor comment about the problems one Russian plane was known to cause its pilots. The doctors turned to her, their interests piqued.

"Are you a pilot?"

"Oh, yes, I've been flying for sixteen years," Jerrie began but soon grew shy at the looks of disbelief on the men's faces.

"Jerrie has more than seven thousand hours in her pilot log," Tom jumped in, making sure she didn't sell herself short. "She's set three speed and altitude records for us besides."

"Have you really, Miss Cobb?" Don was amazed. Right there on the beach, he told her that they'd been interested in the effects of flying on women for a long time and had even designed a pressure suit for Jacqueline Auriol to wear on a recent speed record attempt to steal the crown back from Jackie Cochran.

"You'd better make one of those pressure suits for Jerrie." Tom laughed. "She's liable to try for a record in space next."

A heavy pause hung in the air before Randy broke the silence. "Matter of fact, we had indications at the Moscow meeting that the Russians plan to put women on space flights." A chill ran down Jerrie's spine despite the warm morning air. "Look," Randy continued, "let's get together later and talk some more."

That evening, Jerrie met with Randy and Don in one of the Fontainebleau Hotel's elaborately decorated meeting rooms. They asked her pointed questions about her flying peers, their ages, physical fitness levels, and whether they were typically licensed the same as men. Then they got to the real reason for the meeting.

"Are you familiar with the Mercury astronaut tests given to select the first seven space pilots at the Lovelace Foundation in Albuquerque?"

Of course she knew about the astronauts; it was impossible not to see them in newspapers and magazines. *LIFE* magazine even had an exclusive contract with the men so their stories appeared under their own bylines regularly. And if Jerrie was being honest with herself, she'd been wondering why there were no women pilots in their number, but she didn't ask that question. Instead, she simply responded with, "I've read about them, yes."

Slowly, the conversation revealed the doctors' motivation. After running the medical testing program that had helped select the Mercury astronauts, Randy knew better than anyone what kind of person was physically fit for the rigors of spaceflight. He also knew that women were smaller, lighter, and consumed fewer resources, all of which made them potentially better medically suited to spaceflight. His own curiosity was driving him to find out whether women were as physically and psychologically resilient as the Mercury candidates had proved to be. There was nothing stipulating that he test women pilots; really, he could test any women. But pilots promised a similar baseline as the Mercury astronauts, and having a group of medically cleared female pilots, he thought, could be an asset to NASA should it broaden its astronaut corps. But for the moment this was purely a pet interest for him, albeit one he knew could yield significant results.

Though Jerrie had no jet flying experience, she was close in age to the men he'd tested and had ample flying experience. He thought she might be just the woman he needed.

"What can I do?" she asked. "How can I help?"

"Would you be willing to be a test subject for the first research on women astronauts?"

It was as though Jerrie's whole flying life flashed before her eyes. She vividly recalled the thrill of that first plane ride with her father all those years ago, the glorious freedom that came from flying alone for the first time, the adventures it had brought her. More than anything, she remembered her altitude record flight and the incomparable, mysterious beauty of the blackness of space dotted by bright stars. And now it sounded like they were asking if she wanted to take the first step to make that journey. It was all she could do to keep tears from spilling down her face.

"I would," she said.

Jerrie returned to Oklahoma City and the routine of her job with Aero Design after the conference with the unshakable feeling that something momentous was waiting for her just over the horizon. Every day she checked the mail hoping for a letter saying her flying and medical background was sound and she was cleared to begin the testing, and every day she was disappointed to find nothing from Randy.

In the meantime, she stayed in touch with Don. Together, they pored over the records of women pilots whose background information was available and who could eventually fill out the ranks of the female medical testing program. What neither knew was what that program would look like. Don had hoped the Air Force's Air Research and Development Command might pick up the project, but the service had moved away from the idea of testing women for fear of negative publicity. NASA, of course, was just getting started training its seven men. For the moment, it looked like Randy Lovelace would be the lone physician gathering data on how women might perform as astronauts as part of his own private program.

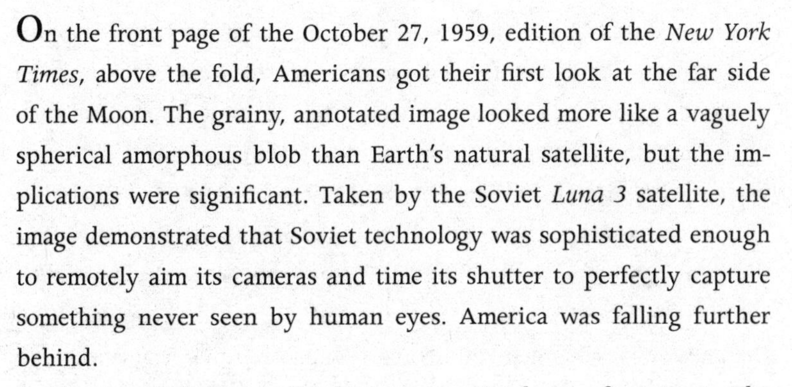

On the front page of the October 27, 1959, edition of the *New York Times*, above the fold, Americans got their first look at the far side of the Moon. The grainy, annotated image looked more like a vaguely spherical amorphous blob than Earth's natural satellite, but the implications were significant. Taken by the Soviet *Luna 3* satellite, the image demonstrated that Soviet technology was sophisticated enough to remotely aim its cameras and time its shutter to perfectly capture something never seen by human eyes. America was falling further behind.

The overall feeling in Washington remained one of wariness. The Soviets were still outpacing American technology, and there was some speculation that Premier Nikita Khrushchev would propose a cooperative mission in space that would turn out to be some kind of booby trap. Whatever the case, it was clear that space would be a major issue in the 1960 election, which meant Lyndon Johnson was in a prime position. He was still considered "Mr. Space," still the go-to senator for a comment on the Soviets' every move.

At Jackie's urging, and to keep the record straight on America's legacy in aviation on the cusp of the space age, he had Congress declare December 17, 1959, "Wright Brothers Day"; no one, he least of all, wanted to see the Russians referring to the Wrightske brothers. When the date came, Lyndon gave a speech at the Wright Memorial Dinner in Washington as Jackie watched from the audience. "Are we in a competitive space race with the Russians? An unqualified yes!" He restated the position laid out in the National Aeronautics and Space Act and reiterated America's need to act if it was going to be an international leader in space. The one thing Lyndon remained mum about as the year drew to a close was his seeking the Democratic presidential nomination.

"Lyndon would make the ablest president of any of us running, but he can't be elected." The sentiment of Senator John F. Kennedy from Massachusetts was the same as the Democratic Party's. LBJ's detractors maintained that he was too closely connected with the south to appeal to northern liberal voters, and Lyndon worried that as soon as he declared himself a candidate, he would be the front-runner and then by default the target of a "stop Johnson" campaign. And if he was honest with himself, he knew it would be successful. The charisma, charm, and political rhetoric that he was so adept at in small meetings didn't always come across in front of large crowds, particularly when those crowds were part of televised events. He couldn't compete with the likes of Jack Kennedy and Hubert Humphrey in a beauty contest. No, Lyndon thought his best bet was to stay quiet. Let the other candidates declare themselves then tear each other apart in the preliminary debates. Then the party would have no choice but to come to him at the last minute before the convention.

Just in time for Christmas, Jerrie got the letter she'd been waiting for. The Lovelace Clinic had looked into her professional and medical backgrounds and determined that she was fit for testing. The last thing she needed was Tom Harris's approval. She'd need to take a leave of absence from Aero for the tests, and if she passed, she thought she might be away even longer for more tests. It was up to Tom whether she could take time off from her job or if she would have to leave the company. Either way, she would be taking the astronaut medical tests.

Somewhat apprehensively, she walked into Tom's office.

"What's cooking this morning?" he asked. "Still thinking of flying to the Moon?"

"Yes," she answered, then told him about the letter.

"How much does trying to qualify for the space program mean to you? Have you thought how you'd feel if you failed, or if nothing came of this?" Jerrie knew he wasn't trying to dissuade her, he was being realistic. She had thought about it; she'd thought of little else while waiting for the Lovelace Clinic's verdict on her candidacy. The idea of flying in space had gripped her so completely that even the faintest possibility of these tests leading to a mission meant she couldn't not try. Even if she made only the smallest contribution, it would still be worthwhile. This wasn't some girlish lark, she told him, and Tom knew it. "All right, Jerrie, all right."

With Tom's blessing to arrange her work schedule around her Lovelace Clinic testing, Jerrie began a strict, self-administered fitness regime. She got a stationary bicycle and pedaled every night. Her diet consisted of steak, chops, and other lean meats with a side of fresh fruit and vegetables. As her endurance improved, she started running every morning before work, sweating so much that her hair would be wet and stringy when she arrived at the office. Bewildered coworkers questioned whether she was swimming in the middle of winter, and her hairdresser shuddered over her ratty ponytail, fixing it up only to see it completely destroyed by their next appointment. Jerrie didn't care.

The February 2 edition of *Look* magazine hit newsstands with Betty Skelton on the cover dressed in a silver spacesuit and standing next to a Mercury capsule against a blue sky. The tantalizing headline asked, "Should a girl be first in space?" The article featured images of Betty on a tilt table, Betty sitting in the cockpit of a jet talking to astronaut Wally Schirra, Betty laughing over a coin toss with six of the seven Mercury astronauts. The photos suggested she was training alongside her male counterparts, but the accompanying text was far from a scientific look at the relative merits of putting a man or woman into space. Though Betty was a pilot, the article told readers, she had only been invited to look at the spaceflight from a woman's perspective at *Look*'s invitation. For the moment, "there is no announced program to put women into space."

America didn't know that Jerrie was hoping to change that.

CHAPTER 16

Albuquerque, New Mexico, Sunday, February 14, 1960

JERRIE CHECKED IN TO THE BIRD OF PARADISE MOTEL ON SUNDAY NIGHT and found it left much to be desired. The room was the very definition of spartan. The staff only changed the sheets on Wednesdays, so she would be sleeping on the previous guest's—or guests'—dirty linens for the next three nights. There were certainly nicer options for lodging in Albuquerque, but what the functional roadside motel lacked in charm and cleanliness it made up for in location: it sat right across the street from the Lovelace Clinic. Jerrie looked out the window at the distant mountains on the fringes of the city, tinged purple from the haze. Her mind wandered to what those mountains would look like from space, wondering if she'd be the first woman to see them from that orbital vantage point.

The phone rang, pulling her from her reverie.

"Miss Cobb, this is Dr. Secrest. How do you feel?"

"Never better," she assured him.

In a quick conversation, Dr. Robert Secrest confirmed their meeting in the morning, then told her to abstain from eating, drinking, or smoking until she was given permission to do so the following day. He urged her to get a good night's rest and hung up. Abstaining from cigarettes was easy for a nonsmoker, but her stomach was growling. She'd hoped to catch up with Randy Lovelace that night over a hot meal since he was the only person she knew in Albuquerque, but he was out of town. Feeling suddenly very alone, Jerrie reminded herself there were no restaurants in space and went to bed.

The next morning, Jerrie skipped breakfast as per her instructions and presented herself to Vivian at the front desk of the Lassetter Laboratory Building promptly at eight o'clock. She felt a heavy weight on her shoulders. She daydreamed that her performance over the next five days would determine her eligibility for further testing. Lost in thoughts of what it would mean to open space for her sex, she imagined how it would feel if her success triggered the start of a real women's astronaut program. Knowing every pilot tested at the clinic was given a number to protect their anonymity, Jerrie privately gave herself the only number she thought was appropriate: unit one, female.

"Good luck," Vivian said as she handed Jerrie a schedule, "and I mean it."

No one could say for sure what would happen to a human body in space, what kind of havoc weightlessness would wreak on the body's vestibular and digestive systems, or whether muscle loss would render astronauts immobile. It might be impossible to swallow without gravity, or eyeballs might distort so badly astronauts wouldn't be able to see. To give the first space voyagers the best chance of surviving whatever damage the mission brought, NASA figured it should send up the most physically fit individuals. It had tasked Randy with developing the medical tests to help find those individuals. Now Jerrie was willingly submitting to the same tests.

Doctors at the Lovelace Clinic began with a series of lab tests to get a baseline on Jerrie's vitals. This was the easy part, drawing blood for a complete blood count, blood smear, blood sugar test, nonprotein nitrogen test, serology, sedimentation rate, cholesterol test, Rhesus factor test, and a urine sample for a complete urinalysis. Then the demanding work began.

For five days, Jerrie was poked and prodded by a rotating cast of doctors.

She blew into a tube to see how high she could raise a metal disk while doctors listened to the changing pressures in her heart chambers, checking for any otherwise undetectable murmurs. Strapped to a slablike board with a footrest, she was wired for an electrocardiogram and had a blood pressure cuff fixed around her arm before the table tilted from supine to nearly upright and back again; this was a sneaky way of checking for circulation and cardiovascular issues that could make an astronaut black out in orbit. Ear, nose, and throat specialists tested everything from her hearing to how clear her speech was over a radio. Ophthalmologists shone a bright light in her eyes, plunged her into darkness, then measured how long it took her eyes to adjust enough to see a horseshoe shape on the wall. Dr. Kilgore seated her in a chair with her head tilted all the way back and injected supercooled water into her ear to freeze her inner ear bone and induce nystagmus; the light on the ceiling began to spin and her hand fell off the armrest as vertigo set in. A nurse timed how long it took for her to regain her equilibrium. One test had her fly to Los Alamos for a human radioactivity counter to measure total body radiation. Stool samples were collected. Everything that could be x-rayed was x-rayed. Across the board, the physical tests were the same as those the Mercury astronauts had done except that in the place of a semen sample for motility testing, Jerrie was given a full gynecological exam.

On Friday, her final day of tests, Jerrie was led into a brightly lit room in the physiology department. In the center of the room sat a stationary bicycle. It was covered in wires that led to a bank of instruments and had a big green plastic bag attached to the front end.

"Miss Cobb, this test is to see how your body reacts to hard physical work," the attending physician, Dr. Luft, said. "When we tell you to start, Miss Cobb, keep time with the metronome. Just keep pedaling to its beat until I tell you to stop."

It seemed simple enough. Jerrie hopped up on the seat, and almost as soon as she sat down, the technicians descended. They wrapped another blood pressure cuff around her arm, put sensors on her head and torso, closed a clamp over her nose, and fitted an oxygen tube in her mouth that connected to the green bag. Wired and ready, she nodded as best she could, and on Dr. Luft's signal started to pedal.

Jerrie dimly registered the sensation of resistance being added to her back wheel, but she remained focused on the metronome. The resistance kept increasing, she kept pedaling, and soon, her heart rate rose to 180 beats per minute. Vaguely aware of the doctors talking off to the side, Jerrie tried to blink the sweat dripping down her face out of her eyes. Her eyes burned, her legs burned, her lungs burned, but she kept going as Dr. Luft nodded his encouragement in her peripheral vision. Jerrie said a silent prayer as she bit down on the mouthpiece and willed her legs to keep pedaling. Then, finally, the metronome's relentless ticking ceased.

"All right, Jerrie, you can stop," Dr. Luft said.

The wall in front of her was blurry as Jerrie slumped over the handlebars. She had absolutely nothing left. One of the nurses dabbed the sweat off her forehead as her right foot slid to the floor.

On Saturday morning, Jerrie followed the final instruction on her schedule and reported to Dr. Secrest's office at nine o'clock. This was the appointment she had been most worried about, the meeting where she would find out how she'd done. With a mix of excitement and anxiety, Jerrie walked into his office to find the doctor smiling.

"Let me sum this up quickly, Miss Cobb. You're a remarkable physical specimen. I wish there were more women like you..." He paused to look over the papers on his desk before continuing. There were only two issues with her tests. Jerrie had some slight hearing loss in her left ear, which wasn't unheard of for a pilot, and had poor circulation that led to a condition known as cold feet. But on the whole, he was satisfied. "We've gained valuable firsthand information on a woman's performance. Thank you."

"No, thank *you*. Thank *you* very much."

Jerrie's final instructions were to tell no one about her results. Randy wanted to present her data in a scientific paper at a medical conference. Until then, only the doctors at the clinic, Jerrie, and a handful of reporters from *LIFE* magazine who had an exclusive look at Jerrie's testing could know what had happened that week in Albuquerque.

Back in Oklahoma and her day job with Aero Design, Jerrie relished the feeling of holding on to a secret. She hoped that these medical tests would turn out to be the first steps on a path that would lead her to orbit.

LBJ's supporters were desperate for the senator to declare his candidacy, but he refused to make a commitment. Watching the ongoing campaigning from afar, Lyndon felt nothing but disdain for Jack Kennedy, whom he regarded as little more than an inexperienced boy. Nonetheless, the majority leader refused to commit to running.

He was still undeclared when thousands of people streamed into the Los Angeles Convention Center on the afternoon of July 13, among them the 1,520 delegates whose job it was to select the Democratic Party's presidential nominee. LBJ wasn't there. He was at his home in Texas watching the proceedings on TV. He was watching when Minnesota governor Orville Freeman nominated John F. Kennedy. Then he watched House Speaker Sam Rayburn take the stage to nominate Lyndon Baines Johnson. "I have been a member of the Congress of the United States for nearly half a century. I have worked beside more than three thousand members of Congress from every nook and cranny of America. Every giant of the past half century I have known personally. I think I know a great leader when I see him. This is a man for all Americans."

The delegates, apparently, felt differently. JFK won on the first ballot, pulling nearly double the votes that LBJ brought in. Lyndon shut the TV off and spent the rest of the evening lounging around in his pajamas. Senate majority leader was a perfectly powerful position for a man with big political ambitions, he decided. He'd done a lot in this role, and he could continue to do more. But Jack Kennedy had other ideas.

"The son of a bitch will do us a lot less harm as vice president than he will as majority leader," was Jack's refrain on LBJ that night. Jack was an Irish Catholic, liberal, and the darling of organized labor and civil rights organizations, which made him a lock in the north. But the south and southwest needed to find something equally likable about him if he wanted to turn voters away from Richard Nixon, the Republican candidate, and he thought that something was Lyndon Johnson. Jack, the son of the wealthy Joseph P. Kennedy, could appeal to the upper classes while Lyndon's poor roots made him an appealing candidate to the lower classes. While the Kennedy team wasn't thrilled that Jack was eyeing Lyndon as his running mate, they all agreed it would be better to have a Kennedy in the White House with LBJ as vice president than a Nixon victory. So the Kennedy team reached out to the Johnson team, and everyone agreed Lyndon should accept the nomination if Jack made the offer.

"Lyndon!" LBJ was shaken awake by Lady Bird the following morning, phone receiver in her hand. "It's Senator Kennedy, and he wants to talk to you."

An hour later, Robert Kennedy and Lyndon sat in the Johnsons' bedroom while the whole JFK team waited in LBJ's suite. Bobby didn't beat around the bush. In a ten-minute conversation, he told Lyndon that Jack wanted him to run as vice president. LBJ didn't immediately accept, but he did advise that the most important issues should be organized labor, big-city bosses, and the needs of black voters.

Around ten o'clock Jack Kennedy called and read to Lyndon over the phone the press release he had prepared, the one that listed Johnson as his running mate.

"Do you really want me?" Lyndon asked.

Jack assured him that he did, and that was it. Whatever the Kennedy team thought, they now had to deal with the gruff Texan if they wanted to win. For LBJ, it meant he could plan on campaigning as himself; he could be as liberal and progressive as he wanted in the north and northwest while still appealing to the southern constituents Jack needed him to carry. He would also be able to prove his detractors wrong; he could prove that he wasn't just a provincial phenomenon, that he could appeal at a national level.

In the days after the official announcement, congratulatory telegrams and letters poured in, but few meant as much to Lyndon as those from Floyd Odlum expressing his confidence in the senator's vice presidential potential. Lyndon had long looked up to and admired Floyd, and facing the prospect of a new position in Washington he hoped he would continue to merit the confidence of his longtime friend.

On August 18, as America geared up for another presidential election, half a world away in Stockholm, Sweden, Randy Lovelace presented his paper about the Mercury astronauts' exams and the one woman who had taken the same medical tests to the delegates of the Space and Naval Medical Congress. He emphasized in his talk, and reiterated to the press, that there was no female astronaut training program as of yet and this was nothing more than a private medical program. But the words "woman" and "astronaut" in the same sentence were all anyone heard. Newswires stretching across continents and oceans started buzzing with the name Jerrie Cobb.

CHAPTER 17

New York City, Friday, August 19, 1960

A LITTLE BEFORE FOUR O'CLOCK IN THE MORNING, A PIERCING RING CUT through the relative quiet of Jerrie's friend's New York City apartment. From the guest room, Jerrie heard her host get up to answer.

"Who? What? Say, do you know what time it is?" Jerrie's friend's grumpy voice said into the phone, then, after a pause, "Yes, she is a friend. Yes, she was here. No, she's not here now." Jerrie had an idea of who might be on the line. She knew Randy Lovelace had announced the results of her medical testing that day in Sweden and suspected a journalist was trying to track her down, but she was under strict orders not to speak to the press until the Time-Life press conference on Tuesday and so had to stay hidden. After another pause, Jerrie heard her friend say, "No, I don't know where she is. For all I know she could be driving back to Oklahoma." Jerrie heard the receiver being placed back in its cradle then. She was pleased with her friend's fast thinking; the two girls went back to bed.

Within minutes, 1,400 miles away in Ponca City, Oklahoma, the phone rang in Harvey and Helena Cobb's house. It was the same Associated Press reporter still in search of Jerrie.

Jackie Cochran chats with Alexander P. de Seversky in front of her 1938 Bendix plane, a Seversky P-35. *(Courtesy of the Eisenhower Presidential Library)*

A young Jackie Cochran and Floyd Odlum early on in their courtship, likely sometime in the mid-1930s *(Courtesy of the Eisenhower Presidential Library)*

Jackie and Floyd not long after their marriage *(Courtesy of the Eisenhower Presidential Library)*

Jackie Cochran (right) and Amelia Earhart sitting on the diving board at the Cochran-Odlum Ranch in Indio *(Courtesy of the Eisenhower Presidential Library)*

Randy Lovelace holds a version of the mask he developed with Drs. Harry Armstrong and Walter M. Boothby, the same early model Jackie used during her 1938 Bendix race. *(Courtesy of the Museum of History & Industry, Seattle)*

Jackie Cochran shakes hands with President Roosevelt during the Collier Trophy ceremony on December 17, 1940. Behind them, in the center, is Dr. Randy Lovelace. *(Courtesy of the Eisenhower Presidential Library)*

Jackie Cochran in the foyer of her Manhattan apartment, the entryway decorated with a compass inlay in the floor and flight scenes on the walls, and lined with her various trophies and awards. *(Courtesy of the Eisenhower Presidential Library)*

Jackie Cochran smiles at the wheel of a Jeep full of WASPs during the Second World War. *(Courtesy of the Eisenhower Presidential Library)*

Jackie sits next to Cary Grant at a dinner party in the early 1940s; at the time, Floyd owned RKO Pictures. *(RKO Pictures/Courtesy of the Eisenhower Presidential Library)*

Jerrie Cobb in flight at the controls of her favored Aero Commander *(Courtesy of the National Air and Space Museum Archives)*

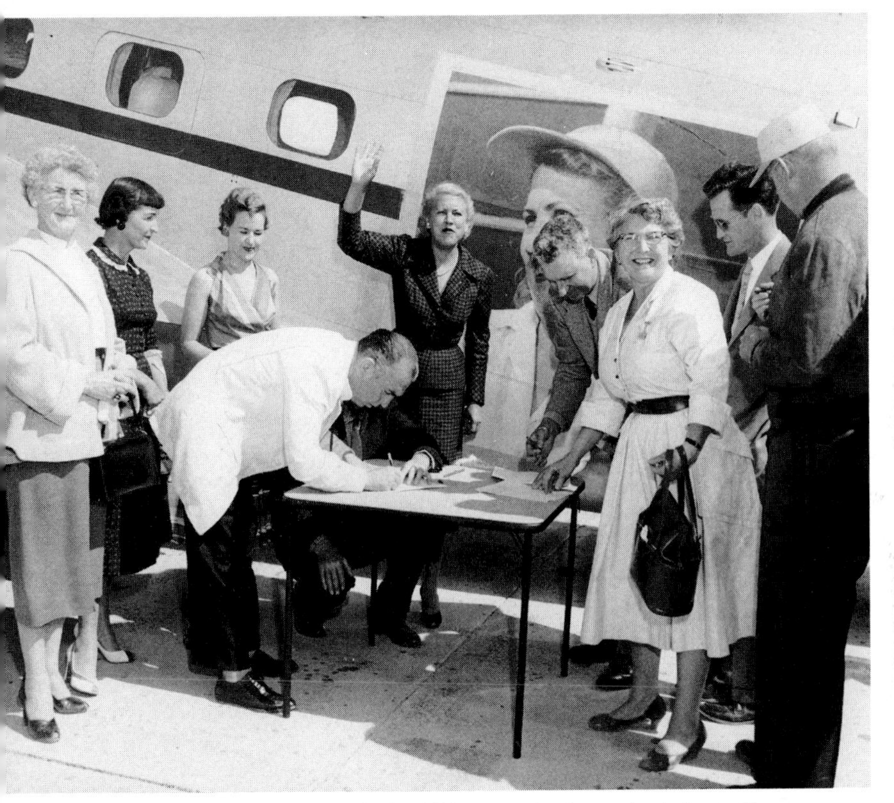

Jackie Cochran canvasses voters for her 1956 congressional campaign in front of her personal Lodestar emblazoned with her larger-than-life portrait. *(Courtesy of the Eisenhower Presidential Library)*

A nurse at the Lovelace Clinic wipes Jerrie Cobb's forehead after her exhausting bicycle stress test, the last medical test she describes taking as part of her medical evaluation in February 1960. *(Photo by Ralph Crane/The LIFE Picture Collection, courtesy of Getty Images)*

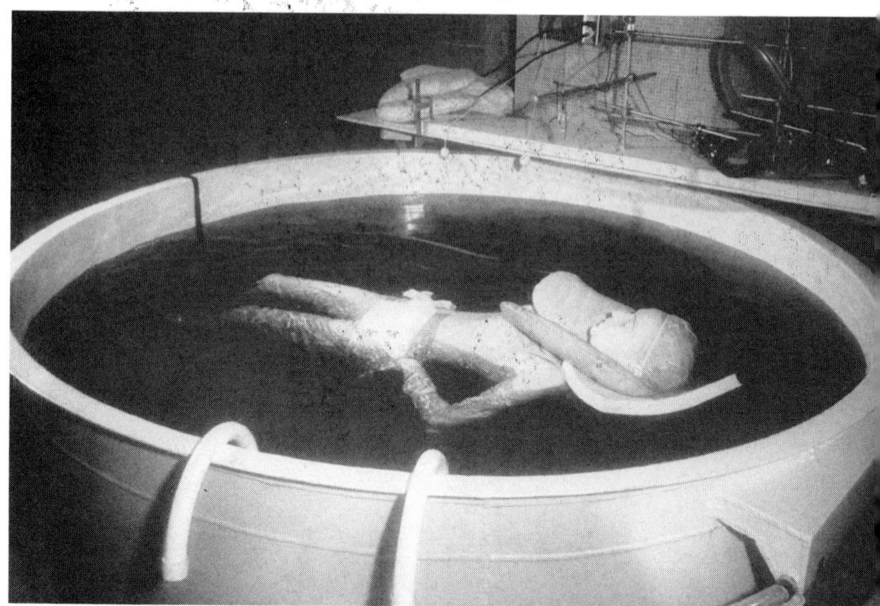

An infrared picture of Jerrie Cobb floating in Dr. Jay Shurley's isolation tank. This was during her second run, the one staged for the cameras. *(Photo by A. Y. Owen/The LIFE Picture Collection, courtesy of Getty Images)*

A multiple exposure of Jerrie Cobb during the tilt table test, the one designed to check for circulatory and cardiovascular issues *(Photo by Ralph Crane/ The LIFE Picture Collection, courtesy of Getty Images)*

Jerrie Cobb at the Peaceful Uses of Space conference in Oklahoma, posing with a Mercury capsule *(Courtesy of NASA)*

Astronaut John Glenn relaxes on the deck of the USS *Noa* after his *Friendship 7* orbital flight on February 20, 1962. He's debriefing into a tape recorder, getting his experience on record before any memories fade. *(Courtesy of NASA)*

Jackie Cochran pauses to touch up her lipstick before climbing into a T-38 jet. Behind her, hanging his head on the jet's ladder, is Chuck Yeager. *(Courtesy of the Eisenhower Presidential Library)*

Jerrie Cobb and Janey Hart sit at the witness table during the first day of the subcommittee hearings on qualifications for astronauts, July 17, 1962. *(Bettmann, courtesy of Getty Images)*

Jackie Cochran with her speedy rival, Jacqueline Auriole. The two Jackies engaged in a friendly competition to outfly each other for more than a decade. *(Agence Intercontinentale/Courtesy of the Eisenhower Presidential Library)*

The Lovelace women attending Eileen Collins's first shuttle launch in 1995. From the left: Gene Nora Jessen (née Stumbough), Wally Funk, Jerrie Cobb, Jerri Truhill (formerly Sloan), Sarah Ratley (née Gorelick), K Cagle, and B Steadman. Though some of the women met through air races or other aviation groups before and after their Lovelace testing, this was among the few times they met as a group. They weren't all in attendance; some of the women couldn't make it to Florida for the launch, and by then both Marion Dietrich and Jean Hixson had died of cancer. *(Courtesy of NASA)*

The next morning, the Associated Press's three-paragraph story about Jerrie appeared in the *New York Times* under the headline "A Woman Passes Tests Given to 7 Astronauts." The short piece, accompanied by an old publicity shot of Jerrie in a dress and pearls smiling as she turned back from the pilot's seat to face the camera, gave little more than her name, age, and home state, but it was enough information for the press to start hunting her down in earnest. The phones in Aero Design's Oklahoma offices and on the desks of the company's representatives in New York started ringing in search of the elusive Miss Cobb. Coworkers lied, saying she was on a trip to South America, out of town vacationing with friends, or simply not in the office. In actuality, she had moved to a hotel for the weekend, waiting out the media storm feeling as though she had jumped to the top of the FBI's most wanted list.

Jerrie emerged on Monday to prepare for her first press conference, starting with an appointment at the Savoy-Hilton's salon. Sitting in the stylist's chair, she watched in the mirror as he considered her blond locks. After a moment, he proposed he style her with a French twist.

"No...no twist. I usually wear it with just a little softness in the front and a ponytail in back."

"A ponee-tail!" the stylist replied, apparently offended. "Oh, madam..." He implored her to consider something a little more sophisticated, rattling off names of celebrities whose hair he'd styled, hoping to change her mind. "A simple chignon? Very chic?"

"Ponytail," she insisted.

"Ah," he sighed. "Ponee-tail."

Jerrie watched in the mirror as the stylist flitted around behind her, his eyes jumping to the mirror occasionally to check how it looked from the front. On one of these mirror checks, her hair half finished, he looked at her face more closely. "You know, I think I know you from someplace." Jerrie smiled at him in the mirror but said nothing, privately entertained that this man now had the dubious honor of having styled a celebrity with a pedestrian *ponee-tail*.

The following morning, with clean hair and frazzled nerves, Jerrie arrived at the towering Time-Life Building in Manhattan. She walked through the lobby and took the elevator up to the eighth-floor auditorium, but as soon as she saw the doorway, she froze. Every muscle in her body tensed up, and her feet felt glued to the floor. She had sudden visions of ravenous reporters, angry that she'd been hiding all weekend, expecting her to rattle off perfect answers like a teletype. But the rational voice in her head urged her forward. *They'll only expect you to answer their questions*. Gathering her courage, she forced herself inside.

The room was filled with news teams setting up bright lights and adjusting wires to give their cameras and microphones a clear shot of the stage. Journalists milled around waiting, some chatting happily with one another, all oblivious that the woman of the hour had just entered their midst. The one welcome element Jerrie noticed in the room was the smell of coffee and doughnuts.

Before she had a chance to relax, Jerrie was ushered onto the stage where she sat with a camera trained on her face in the first of a series of one-on-one interviews. Then came the free-for-all. She sat on stage alone as journalists fired questions at her from every corner, hazarding guesses as best she could without any real experience in space.

"Miss Cobb, aren't you afraid of spaceflight as an unknown, unexplored quantity?"

"No, not really," she answered.

"Are you afraid of anything?"

"Yes, of course," she said without offering any details.

"Specifically what?"

"Grasshoppers." The room laughed, but it was true. Grasshoppers were the one insect she could never really get used to.

"What advantages would a woman astronaut have?"

"Well, by her physical makeup alone, she would require less oxygen and food. The weight factor—and high cost for each pound of space payload—is terribly critical in our space effort. Women have a built-in advantage being smaller and lighter." Jerrie tried to explain the science from what Randy Lovelace had told her, but the press had other concerns.

"Can you cook?" one woman asked. This seemed like an odd question when the issue at hand was space travel, but the room leaned forward to hear the answer.

"Yes, I cook, when I have time."

"What do you cook?"

Was it really that important? Or interesting? Jerrie thought for a minute. "Chickasaw Indian dishes are my favorites... steamed beef, dried corn, squaw bread, yonkapins."

"How do you spell that last—and what is it?" As Jerrie explained how to cook these water lily roots, the room fell silent save for the sounds of pencils scratching on paper. She hoped they were done talking about food.

"Why do you want to beat a man into space?"

Jerrie stared at the man who'd asked the question, stunned. No one had been in space yet, and here was a reporter challenging her intentions. "I don't want to beat a man into space. I want to go into space for the same reasons men want to. Women can do a useful job in space."

"Will there be more women tested?"

"There will, and now hopefully the government will set up a formal test program for women astronauts. Research on me is being done as a private, pilot project." In truth she didn't know what was coming next. Randy Lovelace wanted to add more female data points to his research, and Jerrie herself was starting to feel that if more women got involved, it could bring weight to her growing desire to fly in space. So for the moment, her own fervent wish about a woman-in-space program was the best answer she could give.

The press conference was just the beginning. When the *LIFE* article came out in the August 29 edition, the title declared that "A Lady Proves She's Fit for Space Flight." Readers got to see Jerrie in her element, at the helm of an Aero Commander and wired for testing at the Lovelace Clinic. Her more formal publicity photos, meanwhile, graced newspapers accompanied by articles calling the "five foot seven inches, 122 pounds, with measurements of 36-26-34" pilot the front-runner in a test program to put women into space, what she called "the greatest adventure a pilot could experience—any person could." Of course, most articles included her fear of grasshoppers, but not one mentioned that her medical testing was a private program nor offered a discussion of the broader requirements needed to fly in space. What press coverage did bring was sponsorship offers from companies wanting her to endorse this brand of mattress or smoke that brand of cigarette in a commercial. She declined every offer. She was serious about her intention to pursue any path that looked like it might lead to spaceflight, and was determined that any woman-in-space program would be a serious one, not one based in novelty or publicity hype.

When Jerrie got back to Oklahoma the following week, she found her desk at Aero Design buried under a pile of fan mail.

The response to Jerrie's story wasn't universally positive. Media reports of a female astronaut program made it seem like Randy Lovelace was working on NASA's behalf, and the agency was finally forced to set the record straight. In a hastily arranged press conference, an agency spokesman made it clear that NASA has never "had a plan to put a woman into space, it doesn't have one today, and it doesn't expect to have any in the foreseeable future." NASA had good reason for selecting test pilots as astronauts, and for the moment it wasn't interested in relaxing its strict requirements to include women. Any story to the contrary was simply not true.

There were some within NASA whose bias against women went beyond a disinterest in female astronauts. Clark T. Randt, director of NASA's Office of Life Science Programs, wrote Randy a letter congratulating him on his research but confessed he couldn't see the value in testing women. "Perhaps I am just one of the old school who favors keeping them barefoot and pregnant!" he wrote.

Randy knew NASA's stance on women as astronauts, but he was nonetheless surprised that more colleagues didn't share his scientific curiosity. But Randy didn't work for NASA, he was merely a consultant, so he was free to pursue further testing on women to satisfy his own interests. With Jerrie's help, he began compiling a list of possible candidates. They started with baseline requirements of age and experience similar to the male pilots he'd tested for NASA. He wanted women with at least a thousand hours in the air—a tall order, Jerrie pointed out, since women couldn't fly for the military and were rarely hired by airlines. As for age, he wanted women in their early thirties, though he was willing to make exceptions for younger women if they demonstrated a singular focus on their careers. He knew he couldn't stipulate the women have college degrees or test pilot experience, qualifications that had been vital for the men: it was uncommon for women to earn science degrees, and since they could not fly for the military, they could not be test pilots. Nevertheless, there were hundreds of candidates, many of whom Jerrie had earmarked with Don Flickinger months earlier and knew through societies or the air race circuit. Though there was no woman-in-space program, rumors of some investigation into women as astronauts was starting to buzz in the ears of female pilots all over the country.

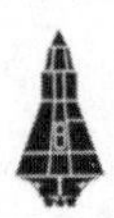

In Long Beach, California, Jan Dietrich followed the stories about Jerrie's medical testing with keen interest. The slim thirty-four-year-old with dark brown hair and high cheekbones had been enamored with flying since she was a little girl. Both she and her twin sister, Marion, had loved listening to their father's stories of flying during the First World War when they were girls and always made him bring back autographs of pilots and stewardesses when he traveled on business. Jan always thought being a stewardess would be the way to fly for a living, but had made her way into the cockpit. Now, she had 8,000 hours in the air, an airline transport rating, and even a degree from the University of California at Berkeley. Not just a casual flyer, she was one of the lucky women to succeed as a career pilot; she worked as a flight instructor and occasional pilot for a construction company. Reading about Jerrie, Jan thought the whole idea of women astronauts was pretty interesting, so she took matters into her own hands. On September 8, she called the Lovelace Clinic she saw mentioned in all the articles. The employee she reached took her credentials over the phone and told her the clinic would follow up with a letter asking for more information. From there, providing her medical background withstood scrutiny, it would consider her as a candidate for the medical testing program.

Only after she'd started the ball rolling with the Lovelace Clinic did Jan stop to think about the implications of her being involved in whatever this testing became. If it did lead to something, some kind of further testing, she would probably have to quit her job to participate. If she did and nothing came of it, she might never get another flying job. As she realized this medical testing could be the end of her flying career, she began to have serious reservations.

Jerri Sloan tossed her mail onto the kitchen table at the end of a long day, though it had been no more frantic than usual. Unlike her suburban neighbors, Jerri was a career pilot. She had more than 1,200 flying hours under her belt, a commercial pilot's license, multi-engine rating, had flown a number of air races, and even had experience flying B-25 bombers. Determined to fly for a living, she had founded her own aviation school, the Air Services of Dallas, but as a woman, she'd needed a man's signature on the bank paperwork for the loan; her friend Joe Truhill had thus become her business partner. She built up the business and now had a small roster of pilots flying for her. Her own ongoing flight projects, meanwhile, included a contract with Texas Instruments testing top-secret terrain-following radar that demanded she fly low over water at night. But Jerri wasn't just a pilot, she was a wife and mother of three, so on top of her own work, she got her children to and from school, fed, and ready for bed.

Now with a quiet moment, she started sorting through the usual assortment of bills and letters. When she came across one that bore the return address of the Lovelace Clinic, she stopped cold. She knew who Dr. Lovelace was from reports about the Mercury astronauts' testing. She also had a vague idea why he might be writing to her; her friend Jerrie Cobb had mentioned that she and the doctor were working on something together. Jerri ripped open the letter and found herself holding an invitation to take the astronauts' medical tests. "These examination procedures take approximately one week and are done on a voluntary basis," the letter said, adding that the tests "do not commit you in any further part in the Women-in-Space program" unless she wanted to continue. Jerri could hardly believe it. She'd been too young to fly with the WASPs in the Second World War, and now a chance to participate in what sounded like an early spaceflight program had landed in her lap. Before she'd even finished reading Randy's letter, Jerri had made up her mind. She wasn't about to turn down a chance to take part in an aviation program like this, especially one that might give her the opportunity to serve her country.

She was still sitting at the table when her nine-year-old son, David, came into the kitchen and saw the letter. He immediately ran outside and started yelling into the night that his mother was going to the Moon.

Jan Dietrich's identical twin sister, Marion, got home after a dinner date midway through September and decided to tackle her pile of mail before bed. Sifting through letter after letter, she hit one that just about made her fall off her chair. She stared at the page, rereading the opening line to make sure she wasn't seeing things, but it was true. The letter really was inviting her to volunteer to take the astronaut medical tests at the Lovelace Clinic. She'd heard about the testing from her sister, but still, she could hardly believe it. To be invited to participate in a program like this on account of her skills as a pilot was incredible.

Like her twin, Marion was also a pilot. Marion vividly remembered the day sometime in grade school they went to the airport to watch planes land and take off. It was a common pastime, but that day they'd seen a sign: "Learn to fly—$50." To the girls, it was an invitation to the world they so desperately wanted to be a part of. Before turning sixteen, they gathered their paperwork for their student-pilot exams and even convinced their high school to let them into the typically all-male preflight class. In a matter of months, both earned their private licenses just two days apart. She and Jan had since pursued advanced licenses together and co-entered a number of air races, but knowing that aviation jobs were few and far between for women, Marion had taken a different route than her sister. She'd turned to writing, first a small aviation column that she leveraged into a full-time position as a reporter for the *Oakland Tribune*. Occasionally she got an assignment that required her to fly, but more often than not, she only spent her weekend hours in the air. All the same, she felt more at home in her plane than her car.

Marion had no hesitations about the tests but knew Jan had mixed feelings. The swell of emotion she felt reading the letter made her realize she couldn't let her sister miss out. The girls had always encouraged and supported each other in their shared love of flying, and she wasn't going to let this be an exception. "Jan," Marion penned an imploring letter. "We are poised on the edge of the most exciting and important adventure man has ever known. Most must watch. A few are privileged to record. Only a handful may participate and feel above all others attuned with their time. To take part in this adventure, no matter how small, I consider the most important thing we have ever done. To be ASKED to participate, the greatest honor. To accept, an absolute duty. So, go, Jan, go. And take your part, even as a statistic, in man's great adventure."

Marion's emotional appeal worked. When the girls both got letters from the Lovelace Clinic saying they'd been accepted as candidates for the medical program, both twins accepted. As they awaited the next letter with further instructions, the pair had a doctor who'd worked with Olympic athletes set them a diet of protein-heavy breakfasts, light lunches, and even lighter dinners. They got in the habit of swimming fifty laps at a time to increase their endurance and pedaling hard on rented stationary bicycles every night. They were determined to be ready for whatever was coming next.

On a Wednesday nearly a month after the world learned Jerrie had taken the astronauts' medical tests at the Lovelace Clinic, she left work at noon. No one at Aero Design imagined that she was doing anything more interesting than her routine physical exams to maintain her advanced pilot ratings. Only Tom Harris knew that she was off to begin her psychiatric evaluation.

Everyone involved with the Mercury astronauts' selection knew there was more to being an astronaut than physical fitness. Astronauts needed mental fortitude, too. The same way no one knew what would happen to a human body in space, there were some questions about what would happen to an astronaut mentally and emotionally when seeing the Earth from orbit for the first time. For the men, this testing had been done at the Wright-Patterson Air Force Base in Ohio, but the air force had no operational need for a woman, especially a civilian woman, taking up time in a military laboratory. So Jerrie had taken it upon herself to find comparable testing.

Her searching uncovered Dr. Jay Talmadge Shurley at the Oklahoma City Veterans Administration Hospital. Jay Shurley, founder and director of the hospital's behavioral science laboratory and professor of psychiatry and behavioral science at the University of Oklahoma's College of Medicine, had been interested in the question of astronaut psychology since reading about the Mercury astronauts' testing. Specifically, he was interested in how sensory isolation on Earth could approximate spaceflight. He hadn't thought too highly of the men's isolation test; sitting in a dark room at a writing desk for a couple of hours was, in his professional opinion, a poor approximation of the probable emotional stresses of flying in space. Out of curiosity, he'd developed a water flotation tank that he thought would create a more complete isolation experience and more accurately test a candidate's mental fortitude for something as isolating as spaceflight. When Jerrie asked whether he would be willing to run some psychological exams on women pilots, giving him a chance to gather data in his tank, his own medical curiosity was piqued. After consulting with Randy Lovelace, Jay Shurley consented to run some psychological tests for the women who went through the medical checks at the Lovelace Clinic, starting with Jerrie.

Over three afternoons, Jerrie submitted herself to a battery of tests, including the Wechsler adult intelligent scale IQ test, Rorschach tests, draw-a-person tests, and sentence completion tests. The Minnesota Multiphasic Personality Inventory was a set of 561 personal questions designed to uncover character or psychological problems. Electroencephalogram and neurological examinations went hand in hand with informal interviews to uncover her childhood, her adolescence, and her occupational attitudes. All in all, it totaled some thirty hours of talk, a challenge for naturally shy Jerrie. On Friday, Jay told her she had demonstrated enough mental and emotional fortitude to try his isolation tank.

Jerrie woke up at six-thirty on Saturday morning. Warm September sunlight streamed into the den at her parents' house where she knelt in prayer, asking for strength. This test, she was sure, would make or break her dreams of qualifying for spaceflight. "Profound sensory deprivation," Dr. Shurley called it, describing the sensation as akin to being struck blind and deaf without a sense of taste, smell, or touch. Jerrie had heard that such complete sensory isolation could cause hallucinations and nonsensical babbling as it revealed the subject's hidden anxieties and patterns of psychological defense. Having spent a good deal of her childhood in solitary pursuits, Jerrie felt fairly confident about this test, but she also knew that exploring the Oklahoma countryside alone was a far cry from complete sensory isolation. She ate a quick breakfast of milk and a mayonnaise sandwich, then set out for the hospital.

When she arrived, Jay and his assistant, Cathy Walters, made sure she was ready. She assured them that she'd slept enough and had eaten breakfast, though she didn't admit to having any feelings of nervousness. Keeping a brave face, she confirmed she was ready to get started.

They led her into an underground room where the air was heated to match her body temperature. Dressed in a bathing suit, Jerrie entered the eight-and-a-half-foot-deep, ten-foot-diameter water tank. She wore a rubber collar to keep her head safely above water and a piece of foam rubber under the small of her back to help her float. Sensitive microphones hung over the tank to pick up her every word and record it on tape. She was given an identification number for the test—52—and was told she could speak as much as she wanted to Cathy in the adjacent observation room so long as she used her number to protect her anonymity on the record. It would, however, be a one-way conversation. There were no cameras in the test room; once she was alone, she could remove her bathing suit, taking away the final sensation of fabric on skin. The water in the tank, heated to match her body temperature, circulated so she could urinate knowing it would be filtered out. When she wanted to end the test, she just had to tell Cathy, and the lights would come up slowly. With that, Jay and Cathy left Jerrie alone in the room.

The eight-inch-thick door closed, and the insulated room became silent as a tomb. Then the lights went down. Jerrie quickly realized that this was a different kind of solitude. Flying for hours alone at night was dark, but she had instruments to watch, weather to think about, radio signals to hear, and stars to see. There was none of that now. There wasn't so much as a sliver of light coming in from under the door, the room was so well insulated. If she lay perfectly still, she couldn't feel any physical sensation on her skin whatsoever.

Jerrie floated. She tried to relax but, knowing it was still a test, she gave continuous status updates. "I'm very comfortable lying on my back, my feet sort of dangling, my arms behind me." She made sure to speak normally since the microphones were so sensitive. She wondered if this complete blackness was what it was like to be blind. She yawned, and in the absolute silence, noticed her teeth clicked when she did. She was curious whether her teeth always clicked when she yawned, and whether the sound had been picked up on the microphone. She thought of the pile of fan mail on her desk and wondered if three secretaries would be enough to help her manage all the replies. Every so often, she caught herself speculating whether she was awake or dreaming, but the peacefulness of the tank eventually relaxed her into a state of forgetfulness.

"Reporting in again that everything's fine. I'll try to tell you how I feel. That's something you said you wanted to know," she said out loud. "Still no feeling in the water whatsoever, because there's no motion. The minute I move, of course, I can feel the water. But I dislike moving. I'd rather be real still."

Jerrie pushed herself around the tank. She felt the wall. She saw what she thought looked like a faint spot of light on the wall. She stretched to relieve a crick in her back. She imagined she was helping her older sister Carolyn take her kids to the doctor's and then out shopping—and that they'd left their black dachshund Schatzi in the car, but then the dog appeared in the doctor's office where no one seemed to mind watching him. Realizing she'd fallen asleep, Jerrie decided not to report the dream. It was a ridiculous one, anyway, and certainly not something she thought needed to be on the record.

After what felt like a few hours, Jerrie spoke directly to Cathy. "This is subject 52 reporting that everything's fine in here. I think I'll get out of the tank unless you want me to stay in longer. I don't have any strong feeling either way."

"Can you hear me?" The sudden booming voice made Jerrie jump.

"Loud and clear."

"Why do you want to come out at this particular time?" Cathy's voice asked.

"I don't have any particular reasons for coming out or staying in,"

Jerrie replied, still floating in the dark. "I don't think my feelings are going to change by staying in here any longer, and I don't see any need in doing so."

The lights in the room came up slowly. Jerrie made her way to the side of the tank. She climbed down the ladder, toweled off, dressed, and opened the heavy door to see Jay and Cathy there to greet her, both smiling.

"What time do you think it is?" Dr. Shurley asked.

Jerrie couldn't say. Two o'clock maybe? He invited her to remove the cover on the wall clock. It was seven o'clock. She'd been in the tank for nine hours and forty minutes, shattering previous records.

Jay Shurley was pleased with Jerrie's results. Like Randy Lovelace, he felt she possessed some mental qualities that could be an asset in spaceflight, namely how calm she had been in the face of extreme boredom and isolation. But something about her calmness felt off, almost as though she was hiding how she really felt. He was struck by her comments about immobility and the way she'd described her physical sensations in the tank. To Jay, she seemed strangely self-constrained, like she was forcing herself to be the stoic pilot she assumed would be best suited to spaceflight, almost as if she was unable or perhaps unwilling to allow herself the freedom to let her mind wander.

Considering how seriously Jerrie seemed to be about proving her merits as a pilot, Jay was somewhat taken aback when she returned days later with her friend and writer for *LIFE* magazine Jane Rieker in tow. Jerrie returned to the tank—this time in a bathing suit—while Jane sat in the dark taking infrared photographs and notes. Apparently, Jay thought, getting into space for Jerrie was less about her abilities and more about making sure the world knew how capable she really was.

Toward the end of October 1960, Mary Wallace Funk, who preferred to be called Wally, picked up the latest issue of *LIFE* magazine. Flipping through its pages, she stopped at a photo spread under the headline "Damp Prelude to Space." It was the infrared pictures of Jerrie in the isolation tank accompanied by an article identifying her as an accomplished pilot and the first woman to undergo tests for space travel. Wally barely knew anything about what an astronaut did, but she knew how to fly and, like so many pilots, was drawn to the idea of exploring a new frontier. Space seemed like as good a frontier as any, and she thought she might like to be the second woman to undergo tests for space travel. So she wrote to the Dr. Shurley mentioned in the article asking about the requirements, because she was "most interested in these tests to become an astronaut." She included her credentials—a university degree from Oklahoma State, a handful of collegiate flight team awards, a flight instructor rating, a single-engine airplane rating, and 3,000 hours in the air. For good measure, she sent a similar letter to Jerrie. She didn't want to overlook any possible route into this apparent astronaut testing program.

Jay Shurley was impressed when he read Wally's letter. Not only did this woman have an excellent flying background, but she was also just twenty-one years old, which suggested incredible initiative and drive. He replied right away explaining that he had no formal role in any female astronaut program but suggested she write to both Jerrie and Randy Lovelace. "May I congratulate you on your attainments thus far," he ended his letter, "and encourage you generally in your astronautic aspirations. I firmly believe that women, as well as men, have a vital and personal role to play in the coming exploration of space."

Wally wasted no time in writing to Randy, who was similarly impressed by her qualifications and drive. Her apparent career focus made her the exception to his minimum age requirement. He wrote back inclosing a card listing further information he needed, the same details he wanted from every interested female pilot. She was asked to provide basic background and personal things like her birthplace, height, and weight, as well as her Federal Aviation Administration medical certification, any further degrees, research interests, publications, membership in organizations and societies, military experience, and aviation instruction experience. To get a sense of her personality, Randy asked for a list of hobbies, sports, personal references, and a small recent photograph. A section at the bottom asked for particulars about her aviation experience as well as less immediately pertinent information about her lineage, church affiliation, marital status, number of children, and foreign languages spoken.

Less than a month after she'd first seen the article about Jerrie's sensory deprivation tests, Wally sent all her information to Randy. Soon after, she got a reply from Jerrie herself. "I was glad to hear that you are interested in the research program for women in space and you did the right thing by writing to Dr. Randolph Lovelace II in Albuquerque," Jerrie wrote, availing herself should Wally need anything else in her quest to join the research program. Now Wally, like Jerri Sloan and Jan and Marion Dietrich, had nothing to do but wait for the next letter to come.

LBJ had nothing to do but wait on the night of November 8, 1960. He and Lady Bird were holed up in the Driskill Hotel in Austin to watch the results of the presidential election on TV. Jack Kennedy and his team were elsewhere.

The campaign had been excruciating for Lyndon. Animosity from the Kennedy camp hadn't abated, mainly from Jack's brother Bobby, who just didn't like the vice presidential candidate; he still held a grudge against the Texan for a trip to the ranch during which Jack had been forced to go hunting and wear a ten-gallon hat. Internal friction was one thing, but the press had picked up on the tension. Rumors of a rift on the Democratic ticket were strengthened when people noticed Jack and Lyndon rarely appeared at events together. It turned out that the Kennedy team thought the contrast in ages between JFK and LBJ made the presidential hopeful look like a kid who was wholly unfit to run the country. Not until Lyndon pitched a fit demanding they present a unified front had the two appeared on stage together.

LBJ and JFK touched base over the phone as the election results trickled in. "I see you are losing Ohio," Lyndon needled his running mate. "I am carrying Texas and we are doing pretty well in Pennsylvania." It wasn't until seven o'clock the following morning that news anchors called the election for Kennedy-Johnson. They'd won by a razor-thin margin. LBJ similarly won the Texas Senate race by a hair, but enough to keep his seat.

For all the work he'd done campaigning, for all the stress of the previous months, the victory left Lyndon with a sense of foreboding. A lifetime in politics had led to this. He would have to give up his position as Senate majority leader, surrendering one of the most powerful positions in Washington for perhaps the least powerful office in the world. For the first time in his life, he felt no joy for winning an election, no jubilation. To his gathered staff, the vice president-elect looked worse than miserable. He looked as though he was in mourning.

Around the time her friend Lyndon was elected vice president, Jackie got a letter from Randy about his interest in putting women through the astronaut medical tests. As she was the former director of the Women's Auxiliary Service Pilots during the war, a staple in the flying world, and a trusted friend, he wanted her help and advice in setting up this research program. She assumed that he would eventually want her to take the tests, too. He had, after all, been using her flights to gather data for decades, and she was the only woman in the country to have worked as a test pilot, giving her a sense of the skill and experience the Mercury astronauts had under their belts. Not a moment too soon, she thought. She knew how important a test program like this could be for the future of women flying in space, even if it was years down the line, and she already had some concerns. From what he had told her, Jackie knew letters were on their way to potential female test subjects, but the selection looked haphazard. Of the twenty women he had contacted, seven didn't meet his own basic requirements. That left thirteen candidates, and Jackie believed some of those would not pass the physical tests, and still others would drop out for one reason or another. The result, she worried, would be a group too small for him to reach any conclusions about how women could perform as astronauts compared to men, which would limit potential next steps. Her own digging into the FAA databases in Washington and Oklahoma City revealed some 9,870 female pilots in the United States. There had to be more than twenty who qualified as candidates for these medical tests.

Assuming the role of Randy's special consultant, Jackie arranged her thoughts in a letter. Suspecting a future government program would benefit from his own preliminary test data, she urged him to liberalize the entrance requirements and relax the age requirements as well. She pointed out that he might be hard-pressed to find young women with more than 1,000 hours in the air. Older women would not only have the required hours, but they would also have more experience. Besides, testing older women could generate valuable data on women's long-term potential in an astronaut program. As part of this older dataset, Jackie offered to take the tests herself in January; she could fly out to Albuquerque before a speaking engagement in Dallas. The foundation had most of her medical information on file, so she would only need to take the specialized tests like the bicycle stress test.

Jackie urged Randy to open the testing to married women, too, and make it clear that marital status had nothing to do with fitness for testing. As she'd done with the WASPs, she didn't want any girl's decision not to marry to open her to criticism, and she felt married women showed emotional stability. Finally, though Jackie recognized it would be difficult for the women to make it to Albuquerque for the tests—they weren't funded as a group, so each woman would have to arrive when her schedule lined up with a free moment at the clinic—she advised that there be a final phase with everyone tested together. What this program needed, she wrote, was for all the women to have equal standing, to avoid any jealousies or criticisms. No one should have any preferential treatment, real or perceived, in something as serious as a potential woman-in-space program. The women should know this, and so should the press. Jackie also urged Randy to approach the women with a sense of reality about their testing. "At this stage at least," she wrote, "you should bear in mind, that it is very likely to be a long time before any one or more of the candidates is put into space flight."

Through more letters and in-person meetings, Jackie and Randy continued to discuss their plans for a more in-depth women's testing program as 1960 drew to a close. But Jackie was starting to have some concerns over Jerrie's press coverage, much of which she was hearing about secondhand when friends sent her the magazine clippings. "What is the 'isolation chamber' that Jerry Cobb talks about?" she asked Randy just before the new year. She didn't know whether the *LIFE* article meant Jerrie had been advanced to a second phase of testing at Wright Field or whether it was part of some routine medical tests in New Mexico. In any case, she felt it was unacceptable that she, Randy's consultant and old friend, should be finding out about these program developments at the same time as the rest of the world. To avoid any further surprises, Jackie decided she and Jerrie should meet and get the conversation about leadership out of the way. She invited Jerrie to the Ranch, her preferred meeting spot as it gave her the home-court advantage, but Jerrie turned down one invitation after another with flimsy excuses ranging from urgent travel plans to unforeseen work commitments.

In the beginning of January 1961, Jackie landed her Lodestar at the Dallas airport to find a small welcome party gathered on the runway. They were women from the local Zonta Club and the Women's Group of the Dallas Council on World Affairs, the two groups she was in town to address. Jerri Sloan was there, too, representing the local chapter of the Ninety-Nines, as was Jerrie Cobb, in town for the same events as Jackie.

Bringing her plane to a stop, Jackie leaned out the window and waved to the smiling crowd. Photographers scrambled for pictures while journalists took furious notes for articles that would appear in the next day's paper. Seeking a better angle for a shot, one photographer asked if Jackie could wave from the other side of the plane. She obliged but made sure everyone knew the implications of the seemingly simple request. "You're photographing me from the copilot's seat. I'm the pilot," she yelled down to him. There were more pictures taken as Jackie mingled with the crowd on the runway. One photographer managed to get local Zonta Club president, Noreen Nicol, on Jackie's right and Jerrie Cobb on her left. Their bodies were angled toward each other, but the two women kept a considerable distance from each other as they strained smiles for the photographer.

When the picture appeared in the *Dallas Morning Herald* the following day, their unease was palpable. The caption called both Noreen and Jerrie "fans of Miss Cochran."

Days later, Jerri Sloan sat on the stage while Jackie addressed a gathered group of business pilots at another Dallas event. Speaking to a mixed crowd about her time with the WASPs, Jackie praised the courage her girls had demonstrated in flying planes like the B-17 that the men were wary of, painting them as the best and bravest of the war.

Jerri Sloan was shocked. She couldn't believe that Jackie would stand in front of a group of pilots, many of them combat veterans, and insinuate that she had led the most significant contribution to the war. She couldn't let it lie and refused to be intimidated.

"Those were good men," she said when she approached Jackie after the speech was finished, "and they were shocked by what they just heard." Then she let loose. "I will tell you something. I will never sit on a dais with you again. Never. You embarrassed the hell out of me and I am ashamed to be a woman pilot."

"Damn you!" Jackie wasn't used to anyone speaking to her like that, least of all a woman junior to her. "And you want to be in the Mercury program?"

"Jerrie Cobb has already told me I've been accepted," she shot back, happy to know something about the program that Jackie apparently didn't. Which only served to make her angrier.

"Jerrie Cobb isn't running this program," Jackie shouted back. "I am!"

CHAPTER 18

Lovelace Clinic, January 1961

JACKIE WAS A LITTLE MIFFED AS SHE WATCHED JAN WALK ON A TREADMILL, a clamp on her nose and a tube in her mouth so doctors could measure her lung function under stress. She had the unhappy feeling that Randy either didn't want her there or that he was trying to keep her at arm's length from the women's testing program. He hadn't seen fit to answer one of her letters himself, and he hadn't been overly communicative on the issue during his latest visit to the Ranch. Jackie pushed the thought from her mind as Jan walked, taking care to position herself, clipboard in hand, in view of the photographer capturing the scene for *Parade* magazine. The media was still running with the story of Jerrie's testing, erroneously referencing a "women in space" program and calling her the first female astronaut trainee. Jackie was determined to use the Dietrich twins' testing to set the record straight; she'd picked them because they espoused things Jackie herself valued—both were hardworking and beautiful. As such, she'd not only paid for Jan's visit to the clinic, but she'd also hosted the pilot at the Ranch for a week as preparation. Jackie had invited Jerrie along, too, but she had declined, citing an agreement with *LIFE* magazine that precluded her from any association with another publication. This was the first Jackie had heard about any of the girls having any exclusive publicity "contract" for a test program that was supposed to be kept quiet until a formal press release, and Randy wasn't forthcoming with any explanation.

On Jackie's dime, Jan spent the third week of 1961 submitting to the Lovelace doctors' poking and prodding. Every night Jan wrote a letter to her sister Marion, not divulging any details, but offering sound advice. She warned her twin she'd be up at five-thirty or six every day and suggested she arrive with some extra weight on her since the days would be long, exhausting, and without time for three square meals. She also advised Marion to try not to have her color portrait taken on the day the doctors rubbed clay on her head for the electroencephalogram and not to take certain exercise tests on the day that would start with three enemas in two hours, though she knew her twin would have little control over the schedule.

At the end of the week, the clinic doctors told Jan she was in the upper 10 percent of all pilots who had gone through this medical testing program and the second woman to pass. A delighted Jan wrote to tell Floyd, one of the few people with whom she could share her good news since the testing still wasn't being publicly announced.

As few details as Randy had given Jackie on the women's testing program, he did agree with her that a sample size of two was hardly enough. Even the women Jerrie had helped him pick—all of whom would be getting their formal invitations soon—were too small a group. Jackie had the list and knew that even if half of them passed, the group would still be too small, so she started looking for more candidates. Floyd, meanwhile, took over the financial side of the program. Estimating that each girl would cost the Lovelace Foundation about $600 to $800 apiece,[17] he sent Randy 2,500 shares of the Federal Resources Corporation that, at market value, amounted to $7,400.[18] The money was earmarked to reimburse the Lovelace Foundation for the cost of the tests with the understanding that Floyd would advise Randy on where any excess funds should go.

In the first week of February, Wally received her formal invitation from the Lovelace Clinic at her home in Lawton, Oklahoma. "Examination of potential women astronauts is continuing," the letter began. "We have reviewed the credentials you have sent in and find that you are acceptable for these examinations." Pleased to finally hear back after so many months, Wally read on to find she was only getting three weeks' notice; the letter asked her to arrive in Albuquerque on Sunday, February 26, to begin testing the following morning. Randy also reminded her that the program was, for the moment, secret. There would be no announcement or press release until all the women had gone through the exams. Only then would the names of those who passed be released, with their consent, as a group. She was in it alone for the moment, but Randy's mention of a potential meeting of all the candidates "late this spring" hinted at some future sisterhood. Randy also made it clear that she could thank Jackie for being "kind enough to review the entire program."

Jim Webb arrived in Lyndon Johnson's office in the Senate Office Building on January 30. A lawyer turned lobbyist and businessman who had served as head of the Bureau of the Budget and as undersecretary of state during the Truman administration, Jim had been

17 Between about $5,150 and $6,865 in 2019.
18 About $63,500 in 2019.

handpicked by the vice president to serve as NASA's new administrator. He was, however, wary of taking the job. In its two years of existence, NASA had grown to absorb Wernher von Braun's team at the Army Ballistic Missile Agency in Huntsville, Alabama, as well as Caltech's Jet Propulsion Laboratory in Pasadena, California. Still, the agency had nevertheless gained a reputation as something of a boondoggle. The Mercury program was going through early tests, and while there was some talk about space stations and missions to the Moon, everyone agreed nothing so exciting would happen before 1970 at the very earliest. In the meantime, whoever ran the agency would have to stand up to the army and the air force trying to wrest control of space activities, all while also handling the responsibility of his agency potentially very publicly killing an astronaut. NASA scientists were excitedly pushing the boundaries of exploration, but the word in Washington was that as soon as America launched the first man into space, the country would lose interest in its real-life Buck Rogers. Space was glamorous now, but when the appeal faded, it would take its administrator down with it. It was, in essence, a career killer that Jim wanted no part of. To Jim's great relief, he found NASA's deputy administrator, Hugh Latimer Dryden, sitting on the sofa in the front room.

"I don't think I'm the right person for this job," Jim told Hugh as he joined him in the waiting room. "I'm not an engineer and I've never seen a rocket fly."

"I agree," Hugh replied. "I don't think you are, either."

"Well, can you tell the vice president?" Jim was desperate to find some endorsement for him not taking the job.

"I don't believe he wants to listen to me on that." Hugh had a hunch Lyndon's mind was made up, and he wouldn't be dissuaded. His sense of unease growing, Jim went in to meet with LBJ.

Hugh was right; Lyndon's mind was indeed set on having Jim run NASA. His lingering post-election depression was exacerbated by the powerlessness of an office that left him feeling useless and ridiculed as

Kennedy staffers increasingly "forgot" to include him in key meetings. The space program was proving to be the only area where the vice president had any power, so Lyndon threw his efforts into the job. He needed Jim Webb to run NASA because Jack Kennedy wanted Jim to run NASA.

Jim made it clear to Lyndon that he wouldn't accept the appointment secondhand; if the president wanted him to run the space agency, he'd have to ask him himself. Furthermore, the whole discussion would be moot unless the president also consented to keep Hugh on as deputy administrator. So LBJ got on the phone to JFK and set up another meeting, this time at the White House.

"You need a scientist or an engineer," Jim repeated his objections to the president later that day.

"No," Jack replied. "I need somebody who understands policy. You've been undersecretary of state, and director of the budget. This program involves great issues of national and international policy, and that's why I want you to do it." Jack assured Jim that he wouldn't be a figurehead, that he'd actually have a chance to direct the agency and shape its future. In the end, Jim couldn't turn down the chance to do something that might be important for his country, and he certainly couldn't say no to a president. He was sworn in as administrator on February 14 with Hugh Dryden as his deputy and Robert Seamans as associate administrator. With Hugh as his technical sounding board and Bob managing the employees, the triad running the space agency was a powerful one. Now they just needed to know what to do with that agency.

On March 6, 1961, Floyd sent Jackie an interoffice memo. "The girl who passed the tests last week is Mary Funk," he wrote, adding that "she is a flight instructor for Army pilots but I don't know just what that means nor do I know about her air hours, etc." Mrs. Virginia Holmes, meanwhile, was a heavy smoker and her heart hadn't stood up to the bicycle test. Wally Funk was officially the third woman to make it through the medical tests. "The twin," Floyd's memo told her, referencing the Dietrich girls, "starts her tests this coming Wednesday."

On the morning of Wednesday, March 8, Marion Dietrich crossed the road from the Bird of Paradise Motel to the Lovelace Clinic. It had been a bit of a whirlwind. She'd had just two days' notice that the clinic had a window to put her through the medical tests, but she wasn't worried. Jerrie had told her, Joan Merriam, and Patricia Jetton—the other women testing that week—that it was really just a series of medical exams. She also took strength from Jan, who reminded her that no matter the pain or discomfort, the doctors weren't going to do anything to hurt her. Besides, Marion reasoned, if Jan had passed the tests, she could, too.

What Marion did feel that morning in Albuquerque was enormous pressure. She believed that these tests were an opportunity she couldn't let slip away, that these medical tests were her first step in qualifying to become an astronaut candidate. Filled with pride and honor, she checked in with Vivian at the front desk.

Her first stop was the lab. Watching five vials of blood flow out of her arm, Marion wondered whether her physical performance might be affected. She didn't have time to dwell on the idea. She immediately moved in to the circulation tests, stress tests, and thorough x-rays. That night she ate like a small horse and exhaustedly went to bed on fresh sheets.

Thursday started with a gastric exam, for which she followed Jan's advice and removed her dress before donning a smock. She soon found out why—it was with considerable gagging that the nurse helped her swallow a length of the tube. The rest of the week brought more doctors who checked every possible nook and cranny on her body. All the while, Marion managed to maintain a chipper disposition. She knew taking each test cheerfully and without hesitation or complaint mattered. Every time things felt difficult, she reminded herself why she was doing these tests, and whatever pain she was experiencing melted away. Her last morning at the clinic, doctors gave Marion the results. Both Joan and Patricia were unfit, but she had made it through the medical tests.

Rhea Hurrle hadn't told anyone, not even her parents, that she was taking a series of medical tests at the Lovelace Clinic. They knew she was an ace pilot and had a handful of races under her belt, but her week at the Lovelace Clinic was a private affair. The only people who knew she was there were the clinic staff; Jerrie, who had told her about the testing in the first place; Jackie, who had paid for her week in Albuquerque; and Betty Miller, who checked into the Bird of Paradise Motel the same night. Both girls arrived right after Marion, Joan, and Patricia left. At the end of their week, Betty proved medically unfit owing to a sinus problem that impacted her breathing. Rhea, meanwhile, became the fifth woman to pass the Lovelace medical exams.

By the end of March 1961, John Glenn was growing increasingly frustrated.

Two months earlier, in January, Bob Gilruth had called the seven astronauts into his office and asked each of them to vote for the man other than himself who he thought should make the first flight. John had been horrified. He'd been giving everything he had to his astronaut training, and now the first flight assignment had come down to a peer vote. He knew he'd been rubbing his peers the wrong way for months, taking it upon himself to chastise their extramarital indiscretions and overall bad behavior for the sake of their public image, and sure enough, he lost the peer vote. Al Shepard would make the first flight with both John and Gus Grissom serving as his backups. John had been upset enough to write Bob a letter urging him to reconsider, saying a peer vote was an unfair system and that if NASA wanted to put its best foot forward, it should send him up first. The letter hadn't garnered any reply.

Now, he, Gus, and Al were presented as a trio in an attempt to keep the media pressure off Al. They appeared on the cover of *LIFE* magazine together and were jointly described as the three front-runners for America's first spaceflight. But because articles listed them in alphabetical order, many gave the impression that John was the favorite. He himself recognized that he was the one most often quoted in newspaper articles and knew that the media's love affair with the verbose family man persisted.

But there was a bigger problem than the flight order. After launching a handful of successful unmanned missions and chimpanzee flights, NASA was still nervous about launching a human, and the astronauts were united in their frustration over the agency's inaction. Especially Al. He wanted to fly, he was ready, and the other six astronauts backed him up as a group. They knew what they were doing was dangerous, but they also knew they were prepared, as individuals and as a program. The agency's final cautious move came from Wernher von Braun. The man behind the Redstone rocket wanted one more test in order to be completely sure the rocket was functioning perfectly. On March 24, 1961, an unmanned mission classified as a booster development flight launched flawlessly. It could have been Al's flight, and the astronauts worried that that unmanned mission had cost America its chance to beat the Soviet Union into space.

The day of the Mercury booster development test, the Lovelace Clinic sent a letter to Bernice Steadman, who preferred to go by her initial B. "Examinations of potential women astronauts is continuing," the letter began before saying that, after reviewing her credentials, it deemed her "acceptable for these examinations." Miss Jacqueline Cochran, it went on to say, would be reviewing the program as well as offering financial assistance.

In Flint, Michigan, where most women went to college in pursuit of their "Mrs." degree then worked for the Bell Telephone Company or a local bank until they got pregnant and started to "show," B stood out. She was the successful owner and operator of Trimble Aviation, the pilot school she'd founded before she got married. She'd read about Dr. Lovelace and the Mercury astronauts, and she knew Jackie by reputation. Because of the two of them, there wasn't a doubt in B's mind that this program was not only legitimate, it would surely become a real female astronaut program. She would, of course, be accepting the invitation.

Less than two weeks later, on April 2, B checked into the Bird of Paradise Motel and was delighted to find she wasn't alone. Jerri Sloan was checking in at the same time. Both women, who were married with children, were so happy to have a companion for the testing that, although they had separate rooms, they immediately took to calling each other "roomie." Having someone else in Albuquerque gave B the feeling that the program was a little more legitimate. Everything had felt so distant and strange to that point, but B, who believed Randy was testing women on behalf of the US government, figured disorder was a product of bureaucracy.

B also had the support of her husband, Bob, back home in Michigan. Bob had no qualms about women taking on something as audacious as spaceflight; he was thrilled that his wife had the opportunity to do these medical tests. During her week in Albuquerque, Bob called B every evening, teasing her to make her laugh so she could release a little stress. She, in turn, confided everything to him. "I really don't know if I can do this test. I haven't ridden a bike in so long," she confessed the night before the bicycle stress test.

"Just remember," he told her, "when you get really exhausted, you have a second wind."

She kept his words in mind when her turn came for the stress test. As the doctors increased resistance on the bike's back wheel, she took a deep breath and found her second wind.

Jerri Sloan, however, didn't have the same support from home. Though unsure about a space program that seemed to consist of a bunch of rockets with a tendency to explode, she thought the whole thing was exciting enough to go through the medical tests. Her stepfather felt differently. He had warned her that by taking the tests, she would be competing with her husband, Lou, a pilot himself, and that could only mean trouble. Jerri had ultimately ignored her stepfather's advice and arranged for her mother to babysit the kids for the week she was in New Mexico. It turned out her stepfather was right. Though Lou had supported his wife as a flyer, talk of "astronaut testing" proved to be too much. He called her every night, but far from the uplifting calls B got from Bob, these spousal check-ins left Jerri despondent.

On their last day in Albuquerque, Randy brought B and Jerri into his office. He told them they had both performed as well as some men in a number of tests. B was delighted, convinced that their performance would determine when—not if—they went into space and sure that they were on the cusp of something big. She returned home to share the news with her proud and loving husband. Jerri, meanwhile, was met at the airport by an attorney with divorce papers. Deciding between supporting his wife and distancing himself from a woman who left him feeling emasculated, Lou Sloan had chosen the latter.

Shorty Powers was sleeping fitfully on a small cot in his office. He and a handful of other NASA staffers had been working sixteen-hour days leading up to Al's flight, making sure every angle was covered, the press informed, and the country was along for the ride. A ringing phone cut through the blissful predawn silence. Barely awake, Shorty answered to find someone on the line asking something about a man in space, but all he registered was the time.

"It's three a.m. in the morning, you jerk," he snapped into the phone. "If you're wanting something from us the answer is we're all asleep!"

A few hours later, America woke up to newspaper headlines proclaiming that the Soviet Union had put a man in space and recovered him safely. Subheadings on nationally syndicated wire service stories featured some variation of an American spokesman saying that the whole country was sleeping.

Without any of the fanfare afforded the Mercury astronauts, twenty-seven-year-old pilot Yuri Gagarin had been launched into space on April 12, made one full orbit around the planet, then landed safely by parachute near a small village in the Saratov Oblast. Assuming the Soviet flight had knocked John out for contention as history's first man in space, the press sought him out for a comment. "They just beat the pants off us, that's all, and there's no use kidding ourselves about that," he told reporters. It was his honest reaction, and he couldn't say anything more since no one knew Al was in line to fly first. "But now that the space age has begun there's going to be plenty of work for everybody." John remained the media darling through and through.

Jack Kennedy knew the Soviets had been working on a manned mission but hadn't seriously thought they would be the first into orbit. A month earlier he'd approved Jim Webb's proposal of building up America's space program by authorizing funding for a larger rocket called Saturn, a rocket everyone at NASA believed would be the key to the country's future in space, but he had denied the space agency money to build a three-man spacecraft. Now he had to reconsider his next move.

That afternoon, Jack held a press conference, and for the first time since taking office, America saw their usually confident leader looking unsure. After his prepared announcements, the press began demanding answers about space.

"Mr. President," came one question, "a member of Congress said today that he was tired of seeing the United States second to Russia in the space field. What is the prospect that we will catch up with Russia and perhaps surpass Russia in this field?"

"However tired anybody may be," the president began, "and no one is more tired than I am, it is a fact that it is going to take some time, and I think we have to recognize it. They secured these large boosters which have led to their being first in *Sputnik*, and led to their first putting their man in space." He said that NASA was going to be putting more emphasis on its boosters and long-range planning, but had to admit that "the news will be worse before it is better, and it will be some time before we catch up. We are, I hope, going to go in other areas where we can be first, and which will bring perhaps more long-range benefits to mankind. But here we are behind."

While Jack was busy with the press conference, Speaker Sam Rayburn addressed the House of Representatives Committee on Science and Astronautics. Speaking as the president's proxy, Sam stated Jack's belief that the vice president could contribute importantly to the space program while at the same time providing valuable counsel to the office of the president. Jack wanted to leverage Lyndon's background and experience in the fields of space and defense in guiding the council, giving him the most responsibility in shaping the space program.

Sam then presented House Resolution 6169, an amendment to the National Aeronautics and Space Act nominating Lyndon as chairman of the Space Council, which the committee quickly approved. Now it would be up to Lyndon to select committee members, develop space goals and budgets, and facilitate making the space program jump from ideas in meetings to real launches. The president would be the one announcing everything, but the vice president would be making the decisions. That meant Jack could take the credit if things went well and pass the blame onto Lyndon if things went poorly.

Three days later, the Kennedy administration was dealt another blow with the failed Cuban invasion at the Bay of Pigs. The Central Intelligence Agency program, begun during the Eisenhower administration, had trained Cuban exiles to storm the beach at the Bay of Pigs and retake their homeland from Fidel Castro, the communist leader with close ties to Soviet premier Nikita Khrushchev. The plan hinged on the belief that the Cuban people would allow the United States to establish a friendly, non-communist government, but as soon as the attack began on April 15, things went wrong. Bombers missed their targets, photographs of American planes revealed US involvement and forced JFK to cancel the second airstrike, and the 1,200 troops that landed on the beach on April 17 were met with heavy fire from the Cuban army and bad weather. America's return with air cover two days afterward was late owing to a misunderstanding of time zones.

Almost the whole Cuban exile force was captured, and nearly one hundred were killed.

On April 18, Irene Leverton arrived alone at the Lovelace Clinic for her medical testing. Irene had learned to fly through the Civil Air Patrol as a teenager, and almost from the moment she was airborne knew she'd found her calling. Finding an outlet for that calling had been another matter altogether. Just seventeen years old with nine hours in her logbook when the WASPs started recruiting during the Second World War, she'd borrowed an older friend's birth certificate and forged her logbook in an attempt to fly for her country, but the ruse hadn't fooled anyone. Denied entry to the WASPs, she'd taken on a host of part-time flying jobs. In her brief time as a crop duster, she'd flown one infamous prank flight: she'd sprayed attendees at the dedication of the new Chicago lakefront airport, Meigs Field, with Jacqueline Cochran Cosmetics Tailspin perfume. She quickly gained a reputation as a serious pilot and clever prankster.

Her series of odd jobs had eventually landed Irene a position making charter flights and instructing pilots out of an airport in Santa Monica. But when she told her boss that she'd been invited to participate in a research program at the Lovelace Clinic, she was denied the time off for her week in Albuquerque. Unsure what might come from taking the medical tests but sure they were worth doing, Irene circumvented the issue. She told her boss's wife that she would be gone for a week, and without any further explanation left for Albuquerque.

A week later, Irene returned to Santa Monica as the eighth woman to pass the Lovelace medical tests to find she had been demoted to instructing primary students. Her boss wasn't willing to fire her over her transgression but found a way to punish her all the same.

On April 20, JFK sent a memo to LBJ in his capacity as chairman of the Space Council. "Do we have a chance of beating the Soviets by putting a laboratory in space, or by a trip to the Moon, or by a rocket to land on the moon, or by a rocket to go to the moon and back with a man? Is there any other space program which promises dramatic results in which we could win? How much additional money would it cost? Are we making maximum efforts? Are we achieving necessary results?"

Between Gagarin's flight and the Bay of Pigs fiasco, the president needed a win against the Soviet Union, and just as Lyndon had done in the wake of *Sputnik*, he thought space might be his way forward.

Lyndon moved quickly. He huddled with congressmen, talked to Jim Webb, consulted with presidential science advisor Jerome Wiesner, Secretary of Defense Robert McNamara, and NASA's Wernher von Braun. Every conversation confirmed what Lyndon already knew: the Soviets' early emphasis on big rockets had put them ahead of America when it came to launch power. But what the US had that the Soviets didn't, he learned, were resources and better leadership potential. So, rather than try to catch up with Russia, it made sense for America to embark on its own path and take on a long-range goal that would depend more on management and problem-solving than might. Of all the potential long-range goals, NASA had done the most research into a Moon landing mission, a program internally called Apollo. The caveat from Jim was that a mission this big would take ongoing support over a period of at least a decade with a price tag on the order of thirty or forty billion dollars,[19] meaning Jack would have to be ready to start something he might not be in office to see finished. Originally a distant goal, the Moon was suddenly something to fast-track, a so-called "crash program" that would take the easier path to this goal for the sake of beating the Soviets. Gradually, Lyndon came to see that there wasn't any other option: America's best bet was to focus on a manned lunar landing. The political and propaganda implications were as important, if not more important, as the technological coup. Control of space was the modern-day equivalent of controlling the oceans, and first in space would mean first in the world.

After just eight days, Lyndon passed his formal recommendation to Jack Kennedy. "As for a manned trip around the moon or a safe landing and return by a man on the moon, neither the US nor the

---

19 About $340 billion in 2019.

USSR has such capability at this time, so far as we know," his memo read. "The Russians have had more experience with large boosters and with flights of dogs and men. Hence they might be conceded a time advantage in circumnavigating of the Moon and also a manned trip to the Moon. However, with a strong effort, the United States could conceivably be first in those two accomplishments by 1966 or 1967."

He had just recommended America put a man on the Moon.

The April 30, 1961, edition of *Parade* magazine introduced America to a pair of smiling brunettes dressed in matching red flight suits and holding matching helmets. The caption identified them as "Jan and Marion Dietrich: the First Astronaut Twins." Under this baiting heading was Jackie's attempt to bring truth to the story. She'd run her manuscript by Randy, so the article that ran under her byline served as his first approved publicity around the women's testing, too. "Women will fly in space just as certainly as men will—only not so soon," the piece began, accompanied by the picture of Jackie overseeing Jan walking on a treadmill. Jackie made it clear that the women's testing, enabled through private donations she didn't say were her own, was unofficial and the idea of a female astronaut program was little more than a gleam in the eyes of the doctors interested in aerospace medicine. "I myself," she wrote, "expect to fly into space before I hang up my flying gear." The short article ended with a call for other women to send a letter expressing their interest in the program and credentials to Jackie at the Ranch. If the right woman was out there and wrote to her, Jackie teased, she might become the first woman astronaut to really earn that designation.

By her own admission Myrtle "K" Thompson Cagle, who felt her given name sounded too much like "turtle" and so went by the initial

K, was Jackie's most ardent fan. She'd made a trip to a Robinson's department store to have her copy of *The Stars at Noon* autographed. She'd been thrilled but too shy to say hello when she'd seen Jackie out one night in Palm Springs. She even had a Jackie doll that she'd dressed in flight clothes. K was also a pilot with multiple licenses and qualifications currently teaching at the Air Force Aero Club in Georgia. Having been too young and too short to fly with the WASPs during the Second World War, K read the *Parade* article and saw it as her way to make up for the lost experience of flying with her hero. She immediately answered Jackie's call with a letter. "I want to participate in your space program to make up for having missed the WASPs. Then, I can name drop...Jacqueline this, and Jacqueline that." K felt certain that if anyone could make a female astronaut program go from talk to reality, it was the woman who made the WASPs the envy of every female pilot during the war.

K's was among a barrage of letters that reached Jackie's desk in the wake of the *Parade* article, but hers stood out. As a pilot who had founded her own aviation school, K's was among the few names Jackie passed on to Randy at the end of the month.

Box 136
Warner Robins, Ga.
May 1, 1961

Dear Jackie:

I missed your WASPS, because I was too young and too short. Perhaps I can participate in your space program

I have an airline transport license, singel and muliti-engine land, a flight instructor's certificate for airplanes and instruments.

We have talked briefly to each other. I first saw you at the 99 convent in New York at the Waldorf in 1948 or 1949. You gave a party. for us.

Then I saw you at Robinson's and you signed your book for me. I was visiting Zaddie Bunker in Palm Springs. I saw you again that evening at Mrs. Bennetts at a reception for you. I am one of your most ardent fans.

I still have the Jacquline doll, the one I dressed like you in flying clothes. She's the only one of my baby dolls I kept.

At present, I am teaching for the Air Force Aero Club here, and helpin my handsome bridegroom of one year build us a house out from Macon. He's an industrial engineer working for Air Material.

Again, may I say, I want to participate in your space program to mak up for having missed the Wasps. Then, I can name drop...Jacquline this, and Jacquline that.

I flew a jet to 35,000 feet in 1953 but that wasn't high enough for m I want to go out and life among the stars for awhile....to see "the stars at n

Sincerely,

Myrtle

Myrtle Thompson Gagle
Box 136, Warner Robins, Ga.

Eisenhower Presidential Library

A week after the Dietrichs' issue of *Parade* came out, Jerrie spoke as part of the missile and space panel at the twenty-third annual meeting of the Aviation/Space Writers' Association.

"There are at this time sixteen highly qualified and experienced women pilots undergoing the astronaut examinations at the Lovelace Foundation in Albuquerque," she began, emphasizing that they were the exact same initial medical tests the male candidates had taken in 1959. She spoke to the scientific benefits of launching smaller and lighter women who consume fewer resources, women whom science has proved are less susceptible to boredom than their male peers. This wasn't to say she thought women were *better* than men, she said, just that women should be considered alongside men. What she didn't mention was that the women were testing alone or in pairs, not as a group. She also didn't discuss the non-medical qualifications that went into NASA's astronaut selection. Instead, she determinedly used the phrase "astronaut testing," letting the image of an organized program sit in the minds of her audience.

She closed saying the Lovelace tests would ultimately produce a group of seven to nine women who would form a core group for further testing. "This program has no official government sponsorship and the NASA does not include women astronauts in their program at this time," she said, then made her case for why she thought America ought to launch a woman sooner rather than later. "Russia may have put the first man in orbit but the United States can now put the first woman in space—and here's one who'd like to be riding that Redstone tomorrow."

Four days later, on May 5, after two cancellations due to bad weather, it was Al Shepard sealed into his *Freedom 7* Mercury spacecraft ready to ride a Redstone. The world watched as the Redstone's engine fired at 9:34 in the morning local time. Jackie, serving as a timer on the mission, watched from the control room as the rocket lifted off the ground. Eight hundred miles away in Washington, Jack, Lyndon, and a handful of others stood huddled around a television, their focus unwavering. If the flight failed, the country would blame LBJ as head of the Space Council for leading NASA to the death of an astronaut.

In just fifteen minutes and twenty-eight seconds, Al reached his peak altitude of 116.5 miles on a parabolic flight that afforded him just five minutes of weightlessness before he landed gently in the Atlantic Ocean. When it was all said and done, he'd traveled 303 miles, a far cry from Yuri Gagarin's full orbit around the globe, but it was a start.

Al's flight was the tonic the country needed: evidence of the competence and the courage America needed to secure a position of international leadership in space. Though he hadn't gone into orbit—and analysts were quick to point out that his flight paled in comparison to Gagarin's—Americans everywhere were thrilled that the first manned spaceflight had been a success. Ticker-tape parades in New York, Washington, and Los Angeles saw thousands upon thousands of people lining the streets for a glimpse of the first astronaut.

On a muggy day two weeks later, Jerrie awoke from a night of poor sleep in a noisy military barracks. But excitement took over as she reported to the Naval Air Station in Pensacola, Florida. She was there for advanced astronaut simulation tests, the third phase of her testing, which she was convinced would determine whether she could handle the predicted physical demands of spaceflight.

The schedule was daunting. She began with a two-day investigation into her physical condition, which included chin-ups, endurance runs, and sit-ups. Then it was on to a full pressure-suit run. After a full hour and a half to get laced and strapped into the smallest suit on hand, she climbed into an altitude chamber that took her to the equivalent of 60,000 feet without leaving the ground. In the simulated low atmospheric pressure, the suit expanded, engulfing her like a cocoon. Technicians asked her to move her arms and legs as though she were flying a plane and report whether she felt hampered in her movements; she didn't. Then they "dropped" her to 10,000 feet, then 5,000 feet, then sea level, stopping each time to see if she had any problem clearing her ears or if her head felt stopped up. Again, she reported feeling completely fine.

Then came the Dilbert Dunker, a wholly different simulated environment. Fully dressed with a lifejacket and a parachute pack strapped on, she settled into the cockpit of a round capsule sitting on skids and tightened her harness. She took a deep breath as the capsule was plunged into a pool at a forty-five-degree angle, then tried to stay oriented as the capsule flipped upside-down underwater. Confident in the rescue divers on hand, Jerrie watched the water gurgle up all around her and tried to keep a clear head as she undid the harness and got herself safely above water.

Another test had Jerrie alone in a windowless room furnished like an apartment sitting on a forty-two-ton steel gyroscope. Spinning ten times a minute, it was fast enough for her to notice the movement, but without windows, she didn't have the visual cues to make the motion disorienting. It did, however, make simple movements difficult. Walking to the chair, she stumbled as though she were drunk. When she was asked to flip switches, she had to focus hard on the dials to stave off motion sickness. Still another test had her riding in a jet, wires measuring her brain activity while the pilot flew a series of acrobatic maneuvers.

While Jerrie was at Pensacola, the eight women who'd completed the Lovelace Clinic's medical tests got another letter from Randy about their own upcoming trips to Florida. "By sometime in June we hope that all of you that passed the examinations here will be able to go in a group to a service laboratory where further tests procedures will be carried out. Just as soon as a definite date is picked, I will let you know immediately. Meanwhile," Randy's letter continued, "I would like for you to achieve the best possible physical condition that you can as the forthcoming tests are going to require considerable physical stamina. I would recommend walking, swimming, and bicycle riding as well as calisthenics."

At the end of the week, Jerrie learned that she had successfully completed the Pensacola exams. As far as she could tell, that meant she had passed the same series of tests that the male astronauts had, and she felt as ready to train for spaceflight as they were. The best part, though, was that the government had allowed her to test at a military facility. She was at the top of the bureaucratic ladder and couldn't see any reason the other women going through the medical tests at the Lovelace Clinic shouldn't follow in her footsteps. She felt she was truly on her way to space.

Newly encouraged, Jerrie decided to reach out to the man she viewed as the decision-maker. She wrote to Jim Webb, detailing her experiences in Pensacola and inclosing a newspaper clipping reporting Russia's interest in launching a woman. She impressed upon Jim her willingness to begin training alongside the men and join the fast track for a spaceflight.

Jackie, meanwhile, received letters of thanks from all the women whose Lovelace testing she had financially facilitated. B, anxious about what the next step might be, was dying to tell the world about their testing in the wake of Al Shepard's flight. Irene admitted she couldn't have made it to Albuquerque without Jackie's financial help. Jerri Sloan, midway through divorce proceedings, made a point of telling Jackie her generosity was both great and appreciated.

6-23-61

Dear Miss Cochran,

Thank you very much for paying my expences during my Albuquerque stay. I couldn't have done it any other way.

Sincerely,
Irene Leverton
(4171 Arch Dr.
No. Hollywood, Cal.)

Eisenhower Presidential Library

Jack Kennedy had to take a decisive step in confronting the Soviet Union. He'd been mulling over Lyndon's recommendation of a lunar landing mission for weeks, and he knew America was going to the Moon. Now, with Al Shepard's flight a success, it was time he tell the country what was next. Standing before a joint session of Congress on May 25, 1961, after addressing military and intelligence issues, he spoke to the mounting issue of space.

"If we are to win the battle that is now going on around the world between freedom and tyranny," he said, "the dramatic achievements in space which occurred in recent weeks should have made clear to us all, as did the *Sputnik* in 1957, the impact of this adventure on the minds of men everywhere. Since early in my term, our efforts in space have been under review. With the advice of the Vice President, who is Chairman of the National Space Council, we have examined where we are strong and where we are not, where we may succeed and where we may not. Now it is time to take longer strides—time for a great new American enterprise." Then came the decision he feared could prove to be the worst of his political career. "I believe that this nation should commit itself to achieving the goal, before this decade is out, of landing a man on the moon and returning him safely to the Earth."

NASA had never recommended a crash program like this to the president. It had recommended exploring space and expanding technical capabilities; Jack was mandating fast-tracking this goal as per LBJ's recommendation to gain support for the agency's longevity. But he wasn't declaring anything. He could only ask Congress whether or not it would approve a $549 million[20] supplemental appropriation

---

20 About $4,710,952,470 in 2019.

for NASA that would not only get America to the Moon but would also build the country's space presence through satellites, global operations, and communications. Congress voted in favor of the program with a price tag estimated as high as $40 billion.[21]

No one outside the upper echelons of NASA and LBJ's Space Council had expected a Moon landing to be the nation's next step in space. But space was now a matter of international importance, and the Apollo lunar landing program foremost among all America's space goals.

The next evening, Jerrie, the native Oklahoman, and Jim Webb sat at the head table at the first Committee on the Peaceful Uses of Space conference in Tulsa. The event was filled with NASA brass, aerospace contractors, politicians, and journalists, all of whom were still grappling with the idea that men were going to walk on the Moon in nine short years. Randy was there, too. He'd given a talk about the women testing at his clinic, research that was garnering useful data but as yet was not a formal woman-in-space program. His talk had, nevertheless, energized the crowd, and as the lone tested female pilot in attendance, Jerrie became the focus of attention. She'd played the part well, posing on a small flight of stairs next to a mockup of a Mercury capsule looking more like a model than a pilot desperate to fly it. Helena Cobb and Tom Harris were also there, and Jerrie was pleased that her mother and boss were seeing her rubbing elbows with the country's space elite.

Over dinner, Jim was so hotly sought after that Jerrie barely had a chance to speak with him, and certainly had no chance to discuss her recent letter detailing her Pensacola testing. They still hadn't had a chance to talk when, toward the end of the evening, Jim took to the podium to address the 2,000-person crowd. He talked about the president's call to land a man on the Moon, and about the conversations

---

21 About $343,238,795,990 in 2019.

he'd had about the space program with Oklahoma senators Robert Kerr and Mike Monroney.

"On one matter of interest to Oklahoma, and I have kept silent until now. This relates to the appointment of Jerrie Cobb as a consultant to the National Aeronautics and Space Administration."

Jerrie hadn't expected to come up in Jim's remarks. She held her breath as he continued.

"Recently, at one of our meetings with consultants in Washington, I asked Dr. Randolph Lovelace if he thought Jerrie could contribute to our program. He enthusiastically endorsed this idea. I understand that Jerrie has now finished all of the psychological tests which were basic to the selection of NASA's seven astronauts. So I expect to ask her to serve as a consultant on the role of women in the national space program."

She barely heard the rest of his remarks; her mind was swimming with his mention of her role as a possible NASA consultant. *This is it*, Jerrie thought to herself. *I finally have a chance to have a say in America's move into space.*

CHAPTER 19

Paris, End of May 1961

"DEAR FELLOW LADY ASTRONAUT TRAINEE," JERRIE'S LETTER BEGAN.

Just three days after Jim mentioned making her a NASA consultant, Jerrie was in Paris for the Fédération Aéronautique Internationale's Salon Aéronautique, but she could scarcely keep her mind on aviation. She could only think about space and couldn't resist writing to the other women to bring some order to their group. She'd already written her first official letter to Jim Webb reiterating her opinion that fast-tracking a female orbital mission could be the way for America to jump ahead of the Soviet Union. She was urging him to right the power balance between nations with a female flight, at the same time bringing admiration and respect to the American space program. Launching a woman could even serve as a display of confidence in American technology, a clear demonstration that its boosters were safer than the Soviets'. Jerrie couldn't see any negatives to the idea, and she felt sure that her God-given purpose was to be the first woman in space.

From her room at the Hôtel de Crillon in the Place de la Concorde in Paris, Jerrie wrote congratulatory letters to the seven women who'd passed the Lovelace tests. She warned them that things were only going to get more difficult from here, that the Pensacola testing was demanding and that they should begin preparing for a new physical challenge. She also sent each woman a waiver for their week in Pensacola. But far from an official government form, it was something she'd drawn up herself and asked they return directly to her.

Place:

Date:

Remember!

RELEASE

KNOW ALL MEN BY THESE PRESENTS: WHEREBY, I Geraldine H. Sloan am about to take certain physical tests at the U. S. Naval Aviation Medical Center, Pensacola, Florida, which tests include but are not limited to airborne electro-encephalograph and the low pressure chambers, and

WHEREAS, I am doing so entirely upon my own initiative, risk and responsibility;

NOW, THEREFORE, in consideration of the permission extended to me by the United States through its officers and agents to take said tests, I do hereby, for myself, my heirs, executors and administrators, remiss, release and forever discharge the government of the United States and all its officers, agents, and employees, acting officially or otherwise, from any and all claims, demands, actions, or causes of action, on account of my death or on account of any injury to me or my property which may occur from any causes during said tests or as a result thereof, as well as any other incident associated with such tests.

Geraldine H. Sloan
Signature

Witnesses:

Beatrice Brody
Name

10303 Linkwood Dr.
Address

W. B. Hamilton
Name of person to be notified in emergency

[illegible]
Name

1403 [illegible]
Address of person to be notified in emergency

10306 Linkwood Dr.
Address

International Women's Air and Space Museum

* * *

Two days later, a different letter ruined Jackie's day at the same Salon Aéronautique. It said that the successful Lovelace pilots would all be at the Naval Air Station in Pensacola for their testing around mid-June, just two weeks hence. The last she'd heard from Randy was that this testing phase wouldn't happen before fall. The letter also told her that Jerrie, apparently on her way to an appointment as a NASA consultant, had already passed the Pensacola tests. The worst part, though, was that the letter hadn't come from Randy, the Lovelace Clinic, or even Jerrie. It had come from one of the other Lovelace women.

Jackie felt like the whole program was slipping away from her. She was Randy's financial backer and consultant, but she hadn't been consulted on the *LIFE* magazine stories about Jerrie's testing, and she hadn't had any say in the shortlist of women he'd invited to the clinic earlier that year. The press kept coming to her with questions about the "women astronaut" program that didn't exist, and she knew so little about it that the best she could do was answer with vague generalities. It all made her look clueless and the test program disorganized. And now, according to this letter, Randy had put Jerrie forward as a NASA consultant instead of her. She couldn't shake the feeling that he was deliberately trying to force her out for some reason, and after nearly a quarter of a century of friendship she couldn't understand it. Hurt and confused, she called Floyd and told him everything.

After hanging up with her husband, Jackie sought out Jerrie to figure out what was going on. Since both women were staying in the same hotel, Jackie had her secretary leave Jerrie a note suggesting they catch up. Jerrie left her reply with the hotel's front desk, a handwritten note on hotel stationery suggesting they meet for a drink in the bar that evening. Jackie's secretary replied, inviting Jerrie to Jackie's room for cocktails that evening, but she never arrived.

The following morning, Jackie picked up Jerrie's reply from the concierge. "I accepted a dinner invitation," said another handwritten note, "and as is usually the case in Paris, when I returned to the Hotel it was too late to call you. Am on my way to NYC now." Still without any resolution, Jackie was left frustrated in Paris.

HÔTEL DE CRILLON
PLACE DE LA CONCORDE
PARIS
R.C. 55 B 14147

TÉL. ANJOU 24-10

ADR. TÉLÉGR. "CRILONOTEL"

Dear Miss Cochran –

I received your message last night and certainly sorry to miss seeing you again. When we talked yesterday morning I thought there was no chance of seeing you last night so I accepted a dinner invitation and, as is usually the case in Paris, when I returned to the Hotel it was too late to call you.

Am on my way to NYC now and will contact you there or DCA when you return.

Thanks again for the invitations and I trust your stay in Europe will be pleasant.

All the best –

Jerrie

Eisenhower Presidential Library

* * *

While Jackie was trying to arrange a face-to-face meeting with Jerrie, Floyd composed a long letter to Randy marked "personal and confidential." Dispensing with pleasantries, he launched right into a defense of his wife. How was Jackie supposed to talk to the press or serve as a consultant if he, Randy, insisted on keeping her in the dark, Floyd wanted to know. "Jackie does not want to be around if you don't want her," Floyd wrote, adding that if Randy did want her involved that she had to be consulted on all communications and group planning. Whatever Randy's issue, it was to stop now. "If you have personal or other problems dealing with the point of this personal letter you should lay them on the table because, as I said above, Jackie is rather unhappy." Then Floyd offered Randy the benefit of the doubt. "I don't want it to get to a distressing state and it may be due to confusion or misunderstanding on her part and her reluctance to bring her thoughts or impressions up for discussion." There might be a simple way to smooth things over.

Randy replied a week later with a letter marked PERSONAL. He began with flattery, congratulating Floyd on his election to the board of directors for the Air Force Association and the Space Education Foundation. Then he got to the matter of the women's testing. He assured Floyd that the latest group of women Jackie had suggested had all been invited to the clinic and that the navy had expressed its willingness to host them in July or August, though the date could certainly be pushed back if that was her preference.

On the matter of Jerrie, Randy promised he'd had nothing to do with her being in contact with NASA. From what he knew, she'd met Jim through their mutual acquaintance Senator Robert Kerr from Oklahoma. "I was as much surprised as anyone else when Mr. Webb stated that Miss Cobb was to be a consultant to him," Randy wrote. "As you know, the previous position of the NASA had been 100% against any examination procedures for girls. As far as the *LIFE* commitment, the only statement we have made publicly came out in *Parade* magazine."

Randy made it clear that each woman knew Jackie was her financial benefactor and pointed to the letters of thanks she'd gotten from the candidates as proof. He ended with a promise to send Jackie carbon copies of all the correspondence in connection with the program going forward and asked that she be the one to write to the candidates about smoking and exercise in advance of the Pensacola tests. At the bottom of the last page, he added a hand-scrawled postscript. "P.S.: I consider Jackie and you the couple I am closest too of all the couples I know." Awkward syntax and messy grammar didn't matter. He needed his old friend to understand his raw emotion.

But Randy had other matters on his plate. Almost as soon as Jack Kennedy announced the lunar mission, Randy was asked to help NASA figure out who should go to the Moon. It was clear the agency needed a group larger than seven astronauts who could be trained for this complex and dangerous mission as it developed. In his own role as a NASA consultant, Randy co-advised that recruitment of more astronauts should begin immediately, potentially widening the pool to include civilians with special scientific skills that would enhance lunar exploration and research. Whatever he said to Floyd and however much he valued Jackie as a personal friend, Randy knew his position with NASA held national importance and implications for his career longevity. Though he had a medical interest in women as potential astronauts, he wasn't going to jeopardize his relationship with NASA by fighting for something that had no bearing on NASA's lunar landing goal. He needed to help choose moonwalkers.

Randy's letter assuaged Jackie's hurt feelings and restored her confidence. She wasted no time in voicing her desires for the program as well as her concerns in her reply midway through June.

Pleased though she was to hear that the navy had consented to examine the female pilots as a group at Pensacola, there were some final details that needed sorting out. Logistically, she needed to know how long the group would be in Florida and how she would get each of them the money to cover their expenses. As far as the tests were concerned, she wondered if there shouldn't be a final physical exam beforehand to make sure they were in peak shape. She also wanted to know whether the group would be narrowed down further to weed out candidates who weren't serious about the program. To that end, she suggested Randy get them all to commit to subsequent tests—should there be further tests—before she spent a lot of money sending them to Pensacola.

She ended the letter with a thinly veiled slight toward Jerrie. "It is apparent that one of the girls has an 'in' and expects to lead the pack." She knew, albeit secondhand, that Jerrie had written to the other pilots establishing herself as their de facto leader and seemed to believe they were actual astronaut candidates. "Favoritism would make the project smell to high heaven," Jackie wrote. "Furthermore, I think to make it a publicity project for *LIFE* or any one else would be a mistake. There has been much unfavorable comment as to this concerning the astronauts. This should be a serious quiet project from now on. Naturally, there will be some publicity concerning it from time to time but it should be of the conservative type issued by a segment of the Armed Service involved. To have any one or more of the participants tied exclusively or at all to any particular publication would be a mistake. Also it would be a mistake for a participant to be on a special status with the NASA or any other branch of the Armed Services." She suggested the best course of action was for a non-competitor, someone not taking the tests, to be made the official leader of this research program. Someone like herself.

Oblivious to the arguments between Jackie and Randy, Sarah Gorelick arrived in Albuquerque alone on June 18. She was excited about the medical testing. Like so many Americans, Sarah had been inspired by Jack Kennedy's rousing call in his inauguration speech for citizens to step up for their country. She felt that being part of preliminary research into a female astronaut program was the best thing she could do, because the letter made her feel that this was a legitimate program. The problem was that she had already used up her vacation time from her job as an engineer with AT&T. She did, however, have some savings she could dip into; she lived frugally to make sure she always had enough money to keep a plane. Taking unpaid leave, she left for her week in New Mexico.

Without another pilot with whom to debrief at the end of a long, hard day, Sarah found companionship in the clinic nurses, who took her out drinking or on the town every night, giving her a chance to blow off some steam. At the end of her week, Sarah became the ninth woman to complete the medical tests. Her conversations with Randy left her with the impression that the program had a definite future.

K Cagle arrived a week later, on June 30. She, too, went through the tests alone and became the tenth woman to finish the medical tests.

As July began, the nine women preparing to follow in Jerrie's footsteps at Pensacola made their arrangements, which was easier said than done. Women who couldn't get time off from work had to decide whether it was worth quitting their jobs. Those with children had to arrange a week of babysitting. For some, the challenge was managing their expectations of where this testing could lead. Across the board, they put everything on the line for something that held no promise yet captured their imagination.

Each of the women tried to clear her schedule as the date for the Pensacola testing kept changing. On July 8, ten days before they were due to arrive in Florida, they each got another letter from Randy saying the Pensacola tests had been moved from July 18 to September 18. They had two months to rearrange their lives yet again, this time to arrive on September 17 for a two-week stay at the US Navy School of Aviation Medicine. "There is to be no publicity whatsoever about these tests or your trip to Pensacola," he wrote. "Any and all news releases will be made only after testing is completed and then only with the permission of the US Navy and the girls participating in the program."

He followed up with a second letter four days later. As with their New Mexico testing, Jackie would be covering transportation and lodging fees for any candidate who needed financial help, so they were asked to please let the clinic know, as it would be coordinating the aid. "As Miss Cochran will be in Pensacola for a few days during the tests, you can thank her in person," Randy told them. He then suggested that, like the male astronauts, the women consider acting as a group in matters of publicity. He expected things to get interesting after this next phase of testing.

Floyd, meanwhile, kept a close eye on the financial side, both as benefactor and as chairman of the Lovelace Foundation's board of directors. The cost per candidate had nearly doubled from $700[22] per girl to $1,200.[23] Looking through the financial statements, he noticed that "the Cobb woman" had donated $300.[24] He didn't know whether she was covering her own testing or covering some added cost since the testing had been expanded. All he knew was the Cochran-Odlum Foundation wasn't paying for her, so he didn't know where her funding came from or what her intention was with a donation.

Jackie wrote to the girls the same day as Randy. After urging them to take the Pensacola tests seriously, she reminded them that "the medical checks at Albuquerque and the further tests to be made at Pensacola are purely experimental and in the nature of research,

---

22 About $5,900 in 2019.
23 About $10,100 in 2019.
24 About $2,500 in 2019.

fostered by some doctors and their associates interested in aerospace medicine...you were under no commitment to carry forward as a result of successfully passing your test at Albuquerque and you will be under no commitment as to the future if you pass the tests at Pensacola. But I think a properly organized astronaut program for women would be a fine thing. I would like to help see it come about." But, she said, things were different now than during the war. Women pilots had flown as WASPs because of a manpower shortage. There was no manpower shortage in space, but it was nevertheless important, she assured them, to start gathering data about women astronauts in anticipation of the day when their sex would orbit the Earth. Jackie ended her letter with a recommendation that the girls refrain from any individual publicity. The interests of the group, she said, would be better served with the girls presenting a united front to the media.

The postscript alerted each woman that the same letter was being sent to everyone who had received an invitation to test in Pensacola. The postscript on the letter Jackie sent to Jerrie was more explicit: "Dear Miss Cobb: This letter in precisely the same form is being sent to each of the approximately dozen women involved."

Jerrie wasn't the only pilot whose media presence frustrated Jackie. Marion Dietrich, she learned, had sold a story to *McCall's* magazine about her Lovelace testing that would be published in the September issue, exactly when the girls would be at Pensacola. "I sincerely hope this problem gets straightened out," she told Marion in a separate letter. There was little Marion could do; her piece was already scheduled to be printed.

The Lovelace women made their arrangements. K excitedly dropped Jackie's name, using her letters to be excused from classes at Mercer University for the week of the tests. Gene Nora wasn't as lucky. When her boss at the University of Oklahoma flight school refused her request for two weeks off, she quit. Wally replied to the barrage of letters, telling Jackie she would be sure to arrive in Florida in peak physical condition and that she was looking forward to meeting her in person. Sarah thanked her for funding her trip but, sensing she didn't want to be indebted to Jackie, she never cashed the check.

B, for her part, was thrilled by the prospect of meeting Jackie and threw her support behind the famous aviatrix as the leader of their program. She also touched base with Jerri Sloan; the two had remained friends since their Lovelace testing and talked on the phone rather often, comparing notes and trying to get what they deemed to be a real program off the ground. But Jerri had other things to deal with. With her divorce pending, she ultimately decided that being a mother to her children was most important, especially since there was no promise this testing would lead to anything. She withdrew herself from the Pensacola tests. Now there were eight women planning trips to Florida.

On July 21, the world watched as Gus Grissom became the second American launched into space. It was another suborbital flight as Al Shepard's had been, textbook all the way through splashdown. As Gus bobbed in his *Liberty Bell 7* spacecraft in the Atlantic Ocean, the hatch opened prematurely. Water rushed in and Gus, recalling his training in simulators like the Dilbert Dunker, freed himself from the flooding spacecraft. The recovery helicopter pilot focused on the spacecraft rather than the astronaut, prioritizing the vehicle that was now too heavy to haul from the ocean. It sank, nearly taking Gus with it as he struggled to stay afloat amid the waves created by the low-flying helicopter's blades.

Gus was finally recovered, but the mission underscored the true dangers of spaceflight. Everything about the missions was hazardous, even the parts that happened on Earth. It raised questions about

NASA's competence and whether spaceflight was worth the potential cost in human lives.

Three days later, her freezer full of prepared meals for her family and an additional milk delivery scheduled, Janey Hart arrived at the Bird of Paradise Motel. She'd heard about the tests from her longtime friend B, who had passed her name to the clinic after finishing her own testing. The two pilots had met at a Ninety-Nines meeting years earlier and since had gone on plenty of adventures together from flying to sailing the Caribbean. Her nineteen years' experience, helicopter license, and service as a captain in the Civil Air Patrol was apparently good enough for the clinic. Her invitation to take the medical tests arrived in short order.

Upon checking into the Bird of Paradise, Janey was struck by how perfectly it fit the image of the no-tell motel. Pushing aside images of what had happened on those bedsheets before her arrival, she focused on the tests she hoped would lead to a real spaceflight program. She was a mother of eight and wife to Senator Philip Hart, whose liberal agenda she supported at every event he attended. She didn't have time to be a lab rat.

Janey was relieved to find that she wasn't checking in alone. Gene Nora Stumbough arrived the same day. A faculty member at the University of Oklahoma, she was taking advantage of summer vacation to take the medical exams without taking time off. Both women were thrilled to have a companion, but their schedules weren't the same. They went their separate ways in the morning and met up every night to compare notes and offer each other support for the next day's tests.

For her part, Gene Nora didn't think there was much future in the program beyond that week of testing, but she nevertheless thought it seemed interesting and exciting enough to spend a week seeing what it was all about. She was impressed with the thoroughness of the testing—she hadn't realized so many parts of the human body could be checked individually—but even when it was done she didn't feel there was anything more to the program than a bunch of medical tests. Janey, meanwhile, found the whole thing baffling. She couldn't figure out why cold water had to go in her ear and needles as thick as fire matches stuck into her hands, but she trusted the tests had some deeper significance. Both Janey and Gene Nora found the unwritten test of keeping their spirits up among the hardest, but both ended the week victorious. The total number of female pilots slated to take the tests at Pensacola was now ten.

Jim Webb was stuck on the biggest decision for Apollo. Before he could award any contracts to the companies that would build the mission hardware, he had to figure out *how* Apollo was going to get there.

The first thing he had to consider was the timeline. Jack Kennedy had said before the end of the decade, but NASA management figured that the Soviets would probably aim to do something big like a lunar mission in 1967 to mark the fiftieth anniversary of the Russian Revolution, so that took away two years. The question thus became how to get to the Moon in six years. Which meant that, on some level, the program would never get out of its experimental stage. Every mission leading up to the landing would be a test flight of some sort. The decision to recruit astronauts from the ranks of test pilots suddenly made a lot more sense.

It also made sense to build off everything NASA engineers were working on with the Mercury program; NASA could cut down on research and development time by retaining the same blunt body design and splashdown landing. But it wasn't as simple as that. It would take two weeks to get to the Moon and back, and the three-man crew—an odd number would prevent a deadlock in decision-making—would need to be able to take off their pressure suits, so the cabin would have to provide a "shirtsleeves" environment. Most importantly, the spacecraft that landed on the Moon would need enough propellant to leave the surface and make the return journey. All this meant the spacecraft had to be significantly heavier than Mercury.

To get that heavy spacecraft off the Earth, NASA would either need a single massive rocket called Nova or it would have to launch the spacecraft in two parts on smaller Saturn rockets, then assemble it in Earth orbit before sending it to the Moon. The so-called mode decision between direct ascent with Nova and Earth-orbit rendezvous with the Saturns amounted to two different approaches to the lunar mission: either a simpler mission with a complicated rocket or a complicated mission with a simpler rocket. Picking the rocket would determine what the mission looked like, and until that decision was made, the agency couldn't even start preliminary designs on the spacecraft. Nothing could happen until Jim picked a rocket.

Jim approved Earth-orbit rendezvous with direct ascent as a backup method. NASA sent out requests for proposals to industry contractors who realized their spacecraft could become obsolete in a matter of months if Jim changed his mind. Just months into the program and already Apollo was consuming a larger portion of the agency's overall budget.

Robert Pirie, the navy's vice admiral in charge of air operations, wove in and out of cars in a fruitless attempt to escape Washington traffic. Jackie, sitting in the passenger's seat, recognized it might not be the best time for a meeting, but figured since Robert couldn't go anywhere, she might as well put their time to good use. As he drove, she aired her concerns about the Pensacola testing.

Jackie told Robert that she was in favor of a woman-in-space program providing it was activated at the proper time and in the proper way, that is, in a way that didn't interfere with or delay the male astronauts' training. She also advised that the women be treated as a group, that they take their tests at the same time and under the same conditions without any discrimination or favoritism. She suggested that the navy make it clear to the women that the Pensacola tests didn't mean there was an official woman astronaut program. Similarly, the navy should be the body to make any public statements, not individual girls whose uncontrolled publicity put the whole program at risk.

As they parted company that afternoon, Jackie worried that Robert's mind had been too focused on the traffic to have really heard her opinion.

That same day, Jerrie welcomed Rhea Hurrle at the large, three-bedroom home in Oklahoma City she shared with Ivy Coffey.

Jerrie had gotten worried when the Pensacola tests had been delayed yet again, concerned that the program was losing momentum. So she'd taken it upon herself to organize some interim activities. Even though the psychological testing and isolation tank run she'd done the previous year with Dr. Jay Shurley wasn't part of the official Lovelace evaluation—Jerrie had done this of her own volition—she considered it a worthwhile step in her personal quest for space. And so she thought it might be useful to have some of the other female pilots go through the same tests. With Dr. Shurley on board, Jerrie wrote to the group suggesting they arrange to spend a few days at the Veterans Administration Hospital in Oklahoma City.

It had been much harder to organize these trips. Jackie wasn't involved in this stage, so there was no financial help. Jerrie certainly couldn't foot the bill herself, either; all she could do was help ease the women's financial burden by offering them a place to stay. She'd decorated the spare room with spaceship-themed bedspreads, celestial wallpaper on the ceiling, and maps of the solar system on the walls. Signs proclaiming "Have urge, will orbit" completed the motif.

Rhea Hurrle was the first to spend a night in Jerrie's space dorm at the end of July. She spent two days going through the psychological tests. Then on Wednesday, while Rhea was in the tank, Wally arrived at Jerrie's house.

The two women had already met through the air racing circuit and passed the day talking about Pensacola. Jerrie showed Wally what kind of physical tests she should prepare for; they tried to outdo each other in a display of physical fitness doing sit-ups before deciding it was a draw. They also talked about the other women. Both knew some of them already, but sitting together in the kitchen that evening, they both felt a sense of camaraderie they couldn't get from letters. It felt like a real program. After hours together, the phone finally rang. Rhea was out of the tank, and Jerrie could pick her up in fifteen minutes. She'd spent almost the whole day in isolation with no ill effects.

The next day, Wally began her psychological testing, Rhea returned home to Texas, and Jackie reiterated her thoughts to Robert Pirie in a personal memorandum she sent to his home rather than his office; it was neither an official communication nor something she wanted making rounds in the navy's offices.

Wally, like Rhea, proved fit enough for the isolation tank and managed to stay inside for hours. At the end of the week, Dr. Shurley concluded that, like Jerrie, both Rhea and Wally showed a remarkable tolerance to extreme monotony and isolation. He couldn't find any significant liabilities in either woman's personality and found instead that each possessed exceptional if not unique qualities that he believed could serve well on space missions. But the psychological fitness of women was based on a sample size of only three. None of the other women had the time or means to spend a week taking these non-vital tests.

Taking Jackie's worries to heart, Navy Admiral Robert Pirie decided to go to the source on the whole "woman in space" matter. He figured he ought to ask Jim Webb to verify that, yes, NASA was indeed interested in women astronauts before allowing the Lovelace women to test in a government facility. But he found NASA was noncommittal on the issue. For the moment, the agency said, it didn't have a requirement for any women's training program or even an investigation into the potential for women to serve as astronauts. It did, however, recognize that it might decide to undertake a similar research program at some point in the future.

The woman-in-space issue was just one among the dozens of NASA-related items crossing Lyndon Johnson's desk in late summer. He was trying to increase the national focus on education by leveraging the government's emphasis on space exploration. Scores of letters asked him for details about America's communications satellite program. There were also letters from the public—kids with suggestions of how to salvage spacecraft after splashdown, people wondering whether the Russians could seize control of a Mercury capsule in orbit. One woman even wrote LBJ to say she had seen electrical currents on the Moon.

Jim, too, had more than enough on his plate figuring out the specifics of Apollo and trying to surpass the Russians without adding women to the mix. On the matter of beating the Russians, it looked like NASA was falling even further behind; on August 6, 1961, cosmonaut Gherman Titov spent just over twenty-four hours in orbit.

Eleven days after Gherman's flight, Jean Hixson arrived at the Lovelace Clinic. The Akron schoolteacher, who mercifully had the support of her principal to take time off for her week in Albuquerque, was the lone woman on Lovelace's list to have served as a WASP; she joined in

December 1943 and served until the group's deactivation. Staying with the Air Force Reserves after the war, she developed a close working relationship with the Wright-Patterson Air Force Base, where she did some early testing on the effects of zero gravity and broke the sound barrier as a passenger. The supersonic schoolmarm married her love of aviation with her work, developing aviation curricula at the school where she taught, establishing a school planetarium, and organizing field trips to NASA's Lewis Research Center. She'd given all this background to the Lovelace Clinic but had shaved two years off her birthday; she reported herself as thirty-five rather than nearing thirty-eight. Her age had no bearing on her physical fitness. Jean became the thirteenth female pilot to successfully take the Lovelace medical exams.

Almost immediately, with the Pensacola tests just weeks away, Jean joined the other ten women in making their final arrangements for Florida. Sarah Gorelick knew she was going to have a hard time taking additional leave from her job with AT&T so sought Jerrie's help. She gave her boss a letter from "NASA consultant Cobb," but he wouldn't be swayed. So Sarah quit, and ten days before she was due in Pensacola, coworkers threw her a goodbye party in the Kansas City office, complete with a toy astronaut helmet. The secret of why she was leaving had gotten out among her friends, who were all thrilled for her. K Cagle managed to persuade the dean of Mercer University to allow her husband to register in her stead. Gene Nora Stumbough quit her job as a flight instructor, trusting that if the tests went nowhere, she'd be able to find another flying job. Wally Funk was so passionate about the Pensacola tests and convinced this was her path into space that she rode her bike the eight miles to and from work in preparation for the tests. Rhea Hurrle, Janey Hart, Jean Hixson, Irene Leverton, and Marion and Jan Dietrich all trained with bicycle rides, push-ups, and long hours studying aviation manuals and meteorology.

The final weeks were marked by more letters flying across the country, but notable by its absence was any word from Jim Webb. Jerrie couldn't understand it. He'd said he was appointing her a consultant, but she was never told to whom she was meant to report nor was there any record of her affiliation with the space agency.

Jackie, meanwhile, changed her mind on being in Pensacola to oversee the testing. She had the chance to try to break nine records in a Northrop T-38 jet around the same time and couldn't ask the company to halt ground organization and let the plane remain idle while she took a trip that had no bearing on Northrop's interests. Besides, flying a jet was a better use of her time than standing around in Florida as a spectator. She did, however, assure Randy that she still wanted to see the program succeed providing it was adequately controlled. "The screening process that brings these girls together as a group was much less refined than in the case of the Mercury Astronauts," she wrote Randy in the final lead-up to Pensacola. "None of these has had specialized training, probably none of them has had military control and the volume of important kind of air hours is not great in many cases. I hope these facts will be taken into account." She knew the Lovelace pilots, however physically fit, weren't the test pilots NASA needed in its astronauts. Moreover, as she prepared for these T-38 flights, she knew she was still the only woman with any relevant test pilot experience.

As the Pensacola testing drew nearer, Randy wrote to each advising them of their financial coverage and urging them to do some last-minute studying. True to his word, he also kept Jackie apprised of all relevant communications. He sent her the final list of Pensacola candidates at the end of the month: Myrtle Cagle, Jan Dietrich, Marion Dietrich, Mary W. Funk, Sarah L. Gorelick, Jane B. Hart, Jean F. Hixson, Rhea Hurrle, Irene Leverton, Bernice Steadman, Gene N. Stumbough. Jerrie had already done the Pensacola tests. Jerri Sloan had withdrawn her name. The group's total number going into this final test phase was twelve.

By fall, NASA decided that the time for extreme caution was over. Operations and planning shifted to preparing the larger Atlas rocket that could put the Mercury spacecraft into orbit for manned flight, and John Glenn, after serving as backup twice, was finally prime pilot. He felt like he'd drawn the first orbital mission by chance more than anything else, but he didn't care. John was going to be the first American to orbit the Earth.

NASA wanted to launch one final orbital mission with a primate on board before announcing John's new assignment. Without a new astronaut to write about, journalists returned to the story of the female astronaut hopefuls. Jerrie's was, of course, a familiar face, but after Marion Dietrich's article appeared in the September issue of *McCall's*, she and Jan were added to the public's consciousness. Alongside articles about decorating children's rooms and the dangers of subsidized marriages, the story of the twins learning to fly together and details about Marion's time in Albuquerque appeared under the headline "First Woman Into Space." Across the board, headlines and reports inflated the program, turning it from a medical endeavor into an actual female spaceflight program. Though the world didn't know the names of the Lovelace women, they were obliquely dubbed the "Astranettes" and described as a team of woman pilots training for spaceflight. Articles named Randy Lovelace among the men behind the program and speculated wildly that the first women to fly would be flat-chested since it would simplify finding a spacesuit that fit.

The real female pilots, meanwhile, eleven of them in all shapes and sizes, were quietly preparing to travel to Florida. Their expectations varied, some hoping their time yielded good data and others hoping to join the space program. Jerrie, the twelfth member of the group and their self-appointed spokeswoman, hoped their success would fast-track her own flight into space.

CHAPTER 20

September 12, 1961

FIVE DAYS BEFORE THE LOVELACE WOMEN WERE DUE IN FLORIDA, RANDY Lovelace was caught in a storm in Texas when someone called his office. His secretary, Jeanne Williams, answered but didn't know what she should do. Figuring inaction was worse than anything, she sent telegrams to each of the pilots on behalf of her boss.

"Regret to advise arrangements at Pensacola canceled probably will not be possible to carry out this part of program you may return expense advance allotment to Lovelace Foundation % me letter will advise on additional developments when matter cleared further."

Then Jeanne sent a telegram to Jackie.

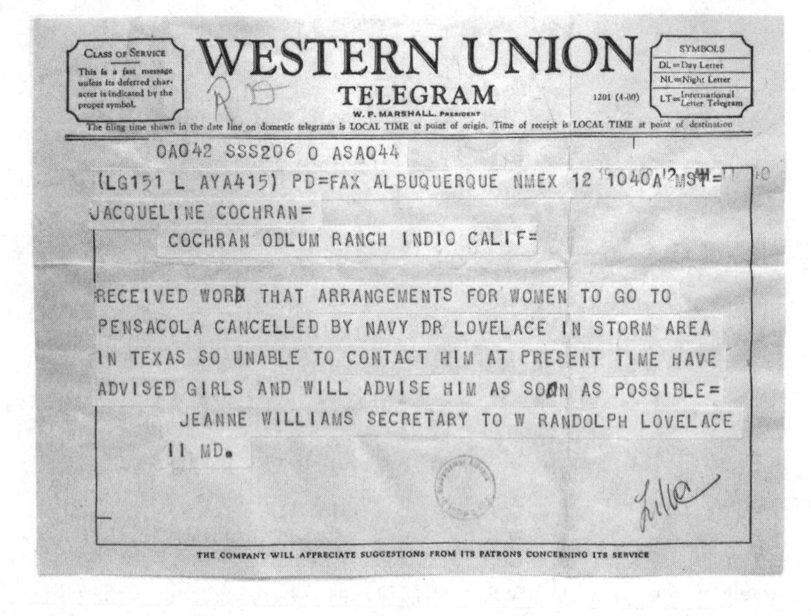
WESTERN UNION
TELEGRAM
W. P. MARSHALL, PRESIDENT

CLASS OF SERVICE
This is a fast message unless its deferred character is indicated by the proper symbol.

SYMBOLS
DL=Day Letter
NL=Night Letter
LT=International Letter Telegram

1201 (4-60)

The filing time shown in the date line on domestic telegrams is LOCAL TIME at point of origin. Time of receipt is LOCAL TIME at point of destination

OA042 SSS206 O ASA044
(LG151 L AYA415) PD=FAX ALBUQUERQUE NMEX 12 1040A MST=
JACQUELINE COCHRAN=
COCHRAN ODLUM RANCH INDIO CALIF=

RECEIVED WORD THAT ARRANGEMENTS FOR WOMEN TO GO TO PENSACOLA CANCELLED BY NAVY DR LOVELACE IN STORM AREA IN TEXAS SO UNABLE TO CONTACT HIM AT PRESENT TIME HAVE ADVISED GIRLS AND WILL ADVISE HIM AS SOON AS POSSIBLE=
JEANNE WILLIAMS SECRETARY TO W RANDOLPH LOVELACE II MD.

THE COMPANY WILL APPRECIATE SUGGESTIONS FROM ITS PATRONS CONCERNING ITS SERVICE

Eisenhower Presidential Library

* * *

Jerrie was in Florida preparing for the women's arrival when she heard the news. It had been two years since she met Randy Lovelace and Don Flickinger on that beach in Miami. She'd done all the tests and devoted time to raising public awareness on the value of women in space. She'd worked to get the group of women set for their trips to Pensacola. She had been so sure that she was on her way to space only to have someone pull the rug out from under her at the eleventh hour. She was livid, not with the Lovelace Clinic but with the unknown person or persons who had made the call. So she packed her bag and checked into a cheap Washington, DC, hotel with a shared bathroom down the hall.

The next day Jerrie started knocking on doors trying to determine who had cancelled the Pensacola tests and why. She went to NASA headquarters and to the Pentagon, cornering anyone she could get to listen to her, but the answers were always some variation on the same theme.

"Don't know."

"What testing?"

"Ask the navy."

"See NASA."

Wading through an unrelenting sea of red tape, Jerrie did learn one thing. The navy needed a "requirement" from NASA, a letter or memo indicating its interest in the program to justify the navy's time and money to host the women. That "requirement" was missing.

Back at the Ranch and without any legwork, Jackie succeeded where Jerrie failed and got to the heart of the matter. Her friend Admiral Robert Pirie sent her a copy of a letter from Hugh Dryden stating that NASA did not have a need for a women's program, though he recognized that some testing on women would likely happen in the future. Hugh's letter, in effect, explained the "missing requirement," and it left Robert's hands tied. "This doesn't leave much as to there being a requirement to perform these tests," he wrote to Jackie. "I do not see how we can proceed further with them." If NASA didn't need the test results, the navy couldn't justify letting a group of civilian women take up a week on expensive simulators.

THE DEPUTY CHIEF OF NAVAL OPERATIONS (AIR)
WASHINGTON

30 April 1962

Dear Jackie:

This will acknowledge your letter of 12 April, which enclosed the copy of your letter of 23 March to Miss Jerrie Cobb on the subject of women's participation in space.

As I have indicated to you before, there must be some NASA support of women in space, or women astronauts, programmed in order to get any support from any other agencies of the government. We have facilities which could be used in test and evaluation programs of this nature, but we cannot do it without a requirement of some kind being set up.

My congratulations on your recent additional record breaking feats in the T-38. You certainly have amassed a considerable number of them during this year.

Martin Decker is coming down this week, and we are going to discuss arrangements for the trip. I hope to be able to provide transportation from Paris to Athens for our group similar to the service we were able to render in the last two years.

The battle of the Pentagon wages more furiously than ever. I doubt very much that we will get out there to see you any time soon. We shall hope to see you here sometime in the spring.

With warm regards to you and Floyd, I am

Sincerely,

Bob

R. B. PIRIE
Vice Admiral, USN

Miss Jacqueline Cochran
Cochran-Odlum Ranches
Indio, California

Eisenhower Presidential Library

Unaware of the upheaval unfolding around the Pensacola tests, interested women continued sending Jackie letters asking for details about the program, hoping to get involved. Even girls as young as ten asked how they could grow up to qualify for the women-in-space program. Across the board, Jackie's replies echoed what she'd been saying for months: there is no women-in-space program, but the best way to be ready should NASA accept women in the future was to attend college, keep flying, and stay in good physical condition. Two pilots had heard that the Pensacola tests had been cancelled and wrote to Jackie hoping to join the program when the testing resumed.

For Floyd, his interest in the cancellation was financial. Randy had billed him $10,601.33[25] in connection with the Pensacola testing. The Cochran-Odlum foundation had covered the cost with a donation of 3,200 market shares of the Federal Resources Corporation, which were valued at about $18,700,[26] with the understanding that he would donate more if the cost of the testing rose. Now, he needed to know where his money was going.

Jerrie persisted in her efforts to track down a single party to hold accountable for killing her dream. She drew enough attention to the issue of the women's testing that Randy was forced to clarify his stance on the program for NASA. "The purpose of these comprehensive examinations here was not to try to put a woman into space at an early date but to collect detailed information on them and to endeavor to determine where they would best serve in the future," he wrote to Jim Webb, confirming that his interest in testing women was purely a medical curiosity. He had never promised any of the women that their testing would lead to a mission. "There seems to be no reason why there should be any urgency about putting a woman into space

---

25 About $90,970 in 2019.
26 About $160,465 in 2019.

at the moment. No program for women should be activated, in our opinion, that would slow up or interfere in any way with the men's program now underway or even give that appearance publicly." What he wanted was the same thing he'd done with Jerrie's original testing: to keep the results confidential until they could be collated and presented in a scientific paper. Randy knew—and he needed NASA to know that he understood—that the agency was facing far bigger challenges and he in no way wanted to interfere or compete with the lunar landing goal.

Letters and memos about the cancelled Pensacola tests got little notice as they crossed over Jim Webb's desk. In spite of the public image that NASA was a well-oiled, cohesive machine, it was, in reality, a group of disparate centers geographically far from one another working toward a goal not one person knew how to achieve. And Jim, from his desk in Washington, had to guide it. He had yet to make a decision on the mission mode for Apollo. The longer he waited, the more urgent the issue became, and now he had a third mode to consider.

Lunar-orbit rendezvous complicated the mission but solved the problem of launching a large spacecraft filled with heavy fuel off the Moon. This method hinged on the idea of keeping the heaviest element, the propellant, on board a mothership in orbit around the Moon while the crew landed in a dedicated small lander that would double as its own launchpad. Because the lunar landing payload was lighter, the whole mission would be light enough to launch on a single smaller Saturn rocket, which in turn promised to shorten the program's development timeline and give NASA a real shot at beating the Soviets to the Moon. It was an elegant solution with just two major hurdles. First, NASA engineers would have to figure out how to separate and reconnect the spacecraft in the unknown gravity field of lunar orbit—perform the actual lunar-orbit rendezvous. Second, the astronauts would have to learn how to fly these difficult maneuvers. It was clear that test pilots would be as vital for Apollo as they had been for the Mercury program.

The mode decision now more complicated, NASA received five bids from contractors hoping to build America's lunar spacecraft in early October. The space agency weighed the measures of each and couldn't deny that North American Aviation—the company behind the P-51 Mustang, the B-25 bomber, and the X-15 rocket-powered plane—had the most relevant experience. On November 28, NASA announced that North American would be building the Apollo spacecraft. The winning contractor just didn't know whether that spacecraft would be a full ship, a lander, or the mothership.

"This is ridiculous. I need to take some pictures, because people are going to want to see what it looks like to be an astronaut." John Glenn knew he was about to make history, and this seemingly small detail was hugely important to him. Engineers and mission planners argued that a camera would be little more than a distraction and that people already knew what the Earth looked like from space thanks to satellite imagery, but John wasn't budging. He knew the value that could come from a simple picture, that it could in effect bring every American along for the ride. He couldn't let it go. One day, during a trip into Cocoa Beach for a haircut, he popped into a drugstore and saw a little Minolta Hi-Matic camera boasting automatic exposure for forty-five dollars. He bought it on the spot, had it rigged to use while wearing bulky pressure gloves, and the matter was sorted.

As 1962 dawned and his launch date approached, John watched America get increasingly wrapped up in the excitement of the launch and knew he had been right to insist on a camera. The drama surrounding the mission increased even more as the initial launch date of January 16 slipped to January 23, then to January 27 when the weather failed to cooperate; for both launch safety and mission recovery, flight rules said no mission could launch in adverse weather or through cloud cover. Headlines focused on the "will he launch or won't he?" angle until yet another delay forced journalists to find a new take on the story. Articles speculated whether John should be taken off the flight since the mounting delays were surely taking a toll on his mental state.

In Washington, Lyndon Johnson felt the tension of the continuous launch delays. Jack Kennedy had decided that, as head of the Space Council, it would be most fitting for LBJ to congratulate John on his courage and his dedication after the flight. It would also fall to him to tell the world if any of the thousands of things that could go wrong did. So, though both the president and vice president had statements ready in case John died, the blame would fall firmly on Lyndon's shoulders.

The launch was finally set for February 20. ABC, NBC, and CBS were all planning uninterrupted coverage of the launch. TV executives knew that John's life would be hanging in the balance for the duration of the flight. Viewers wouldn't be watching a quiz show or soap opera when the real drama would be playing out in orbit.

The day before the launch, Jackie could feel Cocoa Beach buzzing with excitement, but she had work to do. She knew Jerrie was also in town for the launch, so she invited her to dinner. To Jackie's dismay, Jerrie accepted the invitation but wanted her friend Jane Rieker, the *LIFE* writer who'd photographed her in the isolation tank more than a year earlier, to join them as well. Irritated by the uninvited guest, Jackie nevertheless extended a polite invitation to Jane.

Jerrie had certainly gained confidence in the months she'd been giving public talks about women as astronauts, but facing Jackie was different. She felt she needed some moral support, and Jane, who was more socially at ease and had a stronger voice than she did, promised to make her feel that much more comfortable.

The three women sat down to dinner, and almost immediately the conversation turned to women astronauts. Jerrie asked Jackie a dozen times how she felt about the woman-in-space program and what she saw happening with it going forward. Jackie's answer was the same the first time as it was the twelfth: she didn't believe there was any real merit in fast-tracking a female spaceflight. The best thing for America and American women, she repeated throughout the meal, was to take a slow and steady approach to gathering data on female pilots as potential astronauts. That way, when NASA decided to fly women, the legwork would be done. The agency would have all the information it needed and maybe even some viable candidates on hand to make women's inevitable entrance into space that much smoother and quicker.

That answer didn't sit well with Jerrie. She wanted to fly now, and she pushed Jackie to commit to helping her build a public groundswell that would force NASA's hand. She reasoned that with enough pressure, the agency would cave and let her train alongside the male astronauts.

Growing increasingly exasperated, Jackie reiterated her unwillingness to force a fast-tracked women astronaut program, but her answer never seemed to sink in. When the women parted ways that evening, they hadn't come to any agreement on the status of Randy Lovelace's medical research program or the best next steps for possible future testing.

On the morning of February 20, 1962, John Glenn woke around one-thirty in the morning. It was the eleventh scheduled launch day, and he tried not to think about it ending with an eleventh launch delay. A little after two, flight surgeon Bill Douglas came in. The weather was fifty-fifty, he said, and John's backup, Scott, had already been up to check out the capsule. Everything looked ready to go, just so long as the weather held.

He showered, shaved, and dressed in a bathrobe, then after a quick breakfast, Bill gave him a final medical check and he was declared fit to fly. John donned his pressure suit, and under a cloudy sky, he took the elevator up the gantry to the white room that sat flush against his spacecraft christened *Friendship 7* by his children, Dave and Lyn. He wriggled in feet-first and settled in for the countdown as technicians bolted the hatch in place. The original launch time came and went as the clouds refused to break. Using a mirror to see through the small porthole window, he could just see the blockhouse and a hint of blue sky over the Atlantic.

While he waited, Scott patched a call from his wife, Annie, through to the cockpit so husband and wife could speak privately.

"How are you doing?" John asked. He knew there was no training course for being the wife of the first American in orbit. Annie told him they were all fine, watching all three networks' coverage at once, that the preparations looked exciting. But he knew she was worried.

"Hey, honey, don't be scared," he told her. "I'm just going down to the corner store for a pack of gum." It was the same thing he had told her when he'd gone off to fight in the Second World War and the Korean War.

"Don't be long," she whispered back, her well-practiced reply caught in her throat.

The countdown resumed, and at 9:47 in the morning local time, the clock reached zero. The Atlas's engine roared to life, thrust built up, and the rocket lifted off the launch pad. "The clock is operating. We're under way," John called from the cockpit. From their separate vantage points, Jerrie and Jackie cheered along with the gathered crowd as the Atlas and John disappeared into the sky, each wishing it were her inside that cramped spacecraft.

John's flight lasted three orbits. An erroneous light indicating that his landing bag was deployed forced mission controllers to bring him home four orbits earlier than planned. Nevertheless, when John splashed down not far from the Dominican Republic after four hours, fifty-five minutes, and twenty-three seconds in space, he was a hero. The mission was lauded as a stunning success. Celebratory letters poured into both NASA and the White House, including a note from Soviet premier Nikita Khrushchev. "If our countries pool their efforts—scientific, technical and material—to explore outer space," the Soviet premier wrote, "this could be very beneficial to the advance of science and would be acclaimed by all people who would like to see scientific achievements benefit man and not be used for 'cold war' purpose and the arms race."

Lyndon had long been a proponent for space being a common goal for all mankind. Jack, too, fostered the goal of having space be something peaceful rather than an element of competition. Now, in light of Khrushchev's call for a cooperative program, there was some question about sending the vice president to Moscow for a meeting to see how true that sentiment really was.

Two days after John became the first American in orbit, Jerrie was in Los Angeles giving a keynote speech at the First Women's Space Symposium conference. The organizers were thrilled to have a purported NASA consultant as their featured speaker, especially since Jackie couldn't make it owing to a scheduling conflict. NASA was less pleased. The space agency had warned the conference organizers that speaking in a commercial venue could jeopardize Jerrie's role with NASA, but neither the organizers nor Jerrie seemed to care. She stood proudly on stage, assuming as she spoke that she was the voice of all the Lovelace women.

"In this new and exciting space age in which we are now living, there *is* space for women," she began. "During the past eighteen months, twenty-five highly experienced women pilots have taken the astronaut physical examinations at the Lovelace Foundation in Albuquerque, New Mexico. The results have proven the fact—women are physically and emotionally capable of withstanding the stresses of spaceflight." She went on to say that both sexes have much to offer as humanity explores space, and that there should be no competition as the two work together hand in hand. The competition, she said, was with the Soviet Union.

"Just a couple of months ago I spoke with Russia's leading space medicine expert, Professor Yzadosky, who told me quite frankly that his country was training female cosmonauts. For several years Russian scientists have stated that they would use Mongolian women in space—these are a small, hardy race used to living at high altitudes in the Mongolian mountains with less oxygen." She used this as a rallying cry to try to drum up public support: America could score a propaganda victory in space by launching the first woman. "Let us go forward, then—there *is* space for women!"

Jerrie's speech had the crowd fired up and excited, but her words were at odds with the speaker who spoke immediately after her. Dr. Edward C. Welsh, NASA's executive secretary, painted a very different picture in his talk titled "Space: The Essential Business."

"For decades to come, none but a small percentage of the people of the Earth will become travelers in space," Edward began, qualifying this statement by saying that in fact only a fraction of the population had even flown in an airplane. "But, hundreds of millions of individuals have helped, directly or indirectly, to build the aircraft and airports and all the other essentials to air travel." It was the same with spaceflight. Though the astronauts were the most visible part of the space program, it took thousands of men and women working behind the scenes to put one astronaut into orbit. "It is disturbing to some that we do not have Astro-Eves as well as Astro-nauts," he said. "Be assured that we will have both in the near future. However, that is not the area in which we have a shortage—and thus it does not make much difference to our space leadership efforts whether a large number or only a small number of women go into training to become Astro-Eves, just so long as they are not drawn away from the educational, research, and other technical work where we need more competence, not less."

Edward's remarks echoed NASA's stance that women had an important role to play in space, just not the role in the cockpit. At least, not yet. But the media remained focused on Jerrie and the astronaut tests she said she'd passed. Journalists described her as "the attractive 31-year-old astronaut aspirant," and "America's first woman astronaut trainee." Articles erroneously placed her in the control tower when John Glenn was launched and quoted her as affirming that the United States "has the capability and the women to accomplish the first scientific feat of putting a woman in space—now." Reports failed to mention that the medical testing was an experimental and private program, and Jerrie did nothing to correct the stories that reached the American public.

Two weeks later, Jim Webb sat under bright studio lights with Senator Kenneth Keating. The nation's interest in space remained strong, and the senator wanted to bring Jim's own words to television and radio viewers throughout the country.

"We are going to have to institute training programs for others," Jim said of the need for more astronauts in advance of any Apollo missions. "You know we have a very active flight program. We'll be flying every sixty or ninety days now for some time. We have a two-man spaceship that we'll be flying in a year or so. Then we have the three-man Apollo which will come after that. So we will be training more astronauts. But these seven that we have already trained are going to make the flights right ahead of us."

"When will these other astronauts be called in for training?" Senator Keating wanted to know.

"It's a little too early for me to say. In about three weeks, we'll make some announcements as to the program."

Then Kenneth brought up what had crossed just about every American's mind at least once in the last month. "You know, I want to go up into space. Are the physical requirements for space flight going to be reduced so that a fellow like you and me can go up?"

"I think the best way I could answer that is that no man ought to go in space before he is well trained for it," Jim began. "You see, you've got to subject yourself to large g's and centrifuge operations, and you have to go through an extensive program. So I say there's no actual physical limitation that training can't overcome, but you wouldn't go through this expensive training unless you had a real reason to go there." For Jim, the reason to train anyone for the g-loads and stresses of space was Apollo, and the official call for more astronauts went out in a press release a week later on March 11, 1962.

For its second group, NASA wanted approximately ten additional men with the same background and qualifications, augmenting rather than diversifying its astronaut corps. The call that went out was for experienced American jet test pilots with engineering training who were younger than thirty-five and stood shorter than six feet, owing to the dimension of the Apollo spacecraft. The only change was that applicants didn't have to be actively flying with the military. NASA would now accept civilians, though the requirement of jet flight experience meant even civilian applicants would have to have military flying experience. NASA also relaxed its medical requirements. Having successfully launched four missions without any astronauts suffering ill effects, the agency was now looking for outstanding piloting skill more than physical fitness.

Adding to its astronaut corps was just the most visible element of NASA's growth, but the agency, on the whole, was growing. To keep Apollo on track, Jim asked the Space Council to grant it priority as soon as the Mercury program ended. It did, then passed the recommendation on to President Jack Kennedy.

Apollo's priority was just one of the space issues LBJ was juggling. He was still investigating the idea of a joint mission with the Soviet Union on behalf of the president; Jack had pushed his Soviet counterpart to discuss their countries working together in space, saying that "the tasks are so challenging, the costs so great, and the risks to the brave men who engage in space exploration so grave, that we must in all good conscience try every possibility of sharing these tasks and costs and of minimizing these risks." A less exciting but no less important goal, Lyndon was helping plan America's satellite programs. He was also dealing with letters and memos about Mercury astronaut Deke Slayton's flight status in light of a recently uncovered atrial fibrillation. For the man who had earmarked space as the way to boost his career, it was becoming little more than a headache. Through his lingering foul mood, Lyndon increasingly caught up on issues through brief memos from George Reedy and often signed letters without reading them in full.

Back in Oklahoma City with Aero Design in the spring, Jerrie redoubled her efforts on her Washington campaign as best she could from a distance. Unable to duplicate herself, she did the next best thing: she recruited Janey Hart.

Janey had been outraged when she learned of the Pensacola tests' cancellation. She saw it as discrimination on the basis of sex, plain and simple, part of a larger systemic bias that was keeping women out of so many aspects of American life. Janey saw no reason why the navy or NASA or whoever was behind this cancellation should want to deny them the jet time and simulator training they wanted. And the enemy, as far as Janey could tell, was Jim Webb. Jim held the whole of NASA in his hands. She knew that if they could get him to say yes, women would join the astronaut corps straight away. But as it was, he seemed unwilling to change his stance on women, preferring to pat them on the head in a patronizing "we'll take care of you little ladies later" sort of way to just get rid of them. In light of such gender bias, Janey relished the chance to take on Washington.

Together, Jerrie and Janey developed a strategy. Through a two-pronged attack, they would surround Jim and force him to give a definitive answer: would NASA fast-track a female astronaut program, yes or no? Jerrie would continue working the public angle, keeping their story in the press in the hope that taxpayers would put pressure on NASA and the Space Council to fly a woman. Janey would work from the inside, soliciting support from congressmen and senators by leveraging her secret weapon: her husband. Senator Philip Hart from Michigan was a liberal democrat, a longtime friend and supporter of Lyndon's, and an early recruit to Jack Kennedy's inner circle after he'd begun his presidential campaign. Being a senator's wife, Janey also knew how Washington worked and who really got things done. Instead of reaching out to the vice president directly for a meeting, she got in touch with Elizabeth Carpenter.

Liz was a former newswoman whom Lady Bird had hired as her assistant back when Lyndon was a senator. She had taken on increasingly important roles in the Johnson camp over the years and was now one of the few women working in the vice president's inner circle. Lyndon felt women were inherently trustworthy, but at the same time believed the highest office a woman could hold was that of a secretary, making Liz one of the highest-ranking women close to the vice president. Though Janey had never met Liz, they'd operated in the same Washington circles for years, and the senator's wife had an innate sense she could trust the vice president's employee. She'd heard Liz speak a number of times at meetings and on the radio as the press secretary for Lady Bird and Lyndon Johnson, and she knew they shared a similar political agenda. Most importantly, Janey knew Liz was passionate about extending opportunities for women.

Janey told Liz Jerrie's story. She told her how Jerrie had gone through three phases of testing—medical, psychological, and simulations at Pensacola. She told her how Jerrie had then helped more women take the same medical tests, how the successful dozen had been scheduled to go to Pensacola, but the rug had been whipped out from under them. Liz was sympathetic. The two women discussed how chimps were flying in space without engineering backgrounds or jet experience, and that surely the space agency could therefore see fit to send a woman to score a first over the Soviet Union. Liz couldn't promise Janey that LBJ would consent to meet with her and Jerrie, but she agreed to speak with him about them.

"Mr. Vice President: Bill Welsh, Senator Hart's Administrative Assistant, called me and asked whether it would be possible for you to see Mrs. Hart and Miss Cobb on March 15th or 16th. Bill sent me a memorandum which is attached and self-explanatory." The request, another of George Reedy's summary notes, landed on the vice president's desk on March 8.

Leafing through the attached memorandum, Lyndon familiarized himself with the issue. "Background: Twenty-Five carefully screened women pilots tested at Lovelace Clinic... twelve passed the same tests as the astronaut of the Mercury Project... Lovelace Foundation bears the expense and is willing to finance per diem and transportation for the twelve on further research..." It went on to explain that the "further research" needed to happen at Pensacola under the Navy or at Wright-Patterson under the US Air Force, in either case with NASA approval. "At this time," Lyndon went on reading, "Mr. Webb's attitude is negative. We urge that the program be moved forward. It has merit from scientific and research points of view and would be of propaganda value at home and abroad. Last week the President said: 'The only area in which Russia excels is in the use of women.'"

Lyndon figured he ought to at least hear the women out. He consented to meet them.

The same day Lyndon got up to speed on Jerrie's campaign to go to space, Jerrie wrote Jackie another letter. She was still bothered by how they'd left things in Florida and couldn't shake the feeling that she hadn't made her intentions on the matter of women in space clear, so she reiterated her standpoint. "I still feel that it is important for the United States to continue the research testing on women's reactions to various stresses," she wrote, "and to put the first woman in space." For good measure, she sent a copy to Floyd.

Meanwhile, in Washington, the press had got wind that Janey was seeking an audience with the vice president in an attempt to fly in space, and she quickly made headlines. "Senate Wife Could Be First Woman in Space" proclaimed Washington's Sunday *Star* just days later. The article described Janey as one of twelve still-unnamed women who had "passed all the grueling physical tests which Col. John Glenn and the other six Mercury Astronauts completed as a first step toward their historic missions." She was labeled "dynamic" by the press and her campaign "aggressive" as reports said she was seeking a congressional hearing wherein she and Jerrie could present a scientific breakdown of their case. Adding to the furor herself, Janey wrote letters to several members of the US House and Senate space committees, inclosing a copy of Jerrie's "Space for Women" speech, and wrote a letter to Congress that she released to the press.

The renewed media attention inspired a slew of women to write to the vice president in defense of the female astronaut hopefuls. Mrs. Catherine Smith urged Lyndon to "let the women, who are willing, have a chance to help in the Progress of our Country." After hearing Janey speak at the New Century Club, Mrs. George B. Ward told the vice president that "the intelligence, patriotism, initiative and creative ability of women is our most wasted resource in this country." Miss Sue Ann Winkelman was very shocked to learn that "future space flights were planned but none concerning female astronauts. I disapprove of this plan very whole-heartedly; and I agree one-hundred percent with Jerrie Cobb and Jane Hart when they say that unless we do something very soon, Russia will be the one to put the first woman in space." Rachel D. Jones asked, "Why not a Spacewoman?" Mrs. Howard F. White urged Lyndon to "give serious consideration to Mrs. P. A. Hart's proposal to incorporate women into the Space Program." Thirteen-year-old Fritzi Mann asked what she should study to become an astronaut. Fourteen-year-old Carole Glad asked the vice president, simply, if there was "a future for women in space and when?"

LBJ—or at least his office—sent replies to every letter. The girls were encouraged to study math and science and to speak with their counselors about college classes, provided they understood that no one could predict when women would be chosen as astronauts. The replies maintained that astronaut selection was not based on "sex, color, creed or other extraneous factors" but rather on the special requirements of the job, the principal factor being technical training—high-performance jet and experimental aircraft experience—in addition to the physical, emotional, and mental standards. "When women who meet all the requirements apply," the form letter from LBJ's office told the women, "then they will get equal opportunities with men for these important jobs in the future."

For every woman imploring Lyndon to open space to women, there was a man unwilling to see this change. An anonymous "A. Bachelor" scrawled a quick note on hotel stationery that made its way to the vice president. "I have just seen in the newspaper that a sen. is going to blast his wife into outer space. This is a hell of a way to get rid of one's wife. It's getting bad enough with women being the way they are, it's getting so a man can't go anywhere without having a woman hanging around. Now don't get me wrong, I don't hate them I just want to put them back in their place in the home where she belongs. I know that your judgement on this will be sound. I had this feeling when I voted for you. I will be looking for your answer in the newspaper."

SCOVILLE HOTEL
225 MITCHELL STREET, S. W.
Half Block from Terminal Station
ATLANTA, GEORGIA

file
SCIENCE
Space

Dear Sir
I have just seen
in the newspaper that a
sen. is going to blast his
wife into outer space. this
is a hell of a way to get
rid of one,s wife. it,s getting
bad enught with woman
being the way thay are. it,s
getting so a man can,t go
eanny where with out haveing
a waman hanging around. (now
don,t get me wrong,! I don,t
hate tham I just won,t to
put tham back in thar place
in the home where she belong
I know that your judgement
on this will be sound. I glad
thes feling when I voited for you
I well be looking fore your

The Lyndon Baines Johnson Presidential Library

ancer in the news paper.

Yours

a. Bachelor

The Lyndon Baines Johnson Presidential Library

292 Rockwell Ave.
Pontiac 19, Mich.
March 15, 1962

Vice President Lyndon B. Johnson
Chairman: National Aeronautics & Space Council
Washington 25, D.C.

Dear Mr Johnson:

I think the United States should send a Woman in Space. I don't think this should be a, Man Only Program. Let the Women, whom are willing, have a chance to help in the Progress Of Our Country.

I would like to have my name on the list, as, one of the First American Women in Space Program. I am an American Negro Woman. I am in my 40s. I am in good health. I think there is a job I can do in the Program. Please give me a chance to help.

I truly hope that your interest in Space Program could help in getting the United States to launch a Woman in Space Program.

Respectfully Yours,
Catherine Smith
(Mrs Catherine Smith)

The Lyndon Baines Johnson Presidential Library

Sitting behind the large desk in his office sifting through papers, Lyndon picked up the dossier Liz had prepared for him and began reading.

"Mrs. Hart and Miss Cobb are spearheading an effort to get NASA to give the nod for women to qualify as astronauts," it said. The memorandum told him that the women could not get additional training from Lovelace, the navy, or air force without NASA's approval. Jerrie Cobb had seen Jim Webb, whom Mrs. Hart described as "sympathetic but not willing to say yes or no at this time." The women, particularly Mrs. Hart, thought a word from him might help.

NASA, he read on, felt differently. "Dr. Dryden states: 'Orbital flight is not yet a routine operation and still a matter of too great a hazard to give anyone who has not yet had 1) high speed military test flying, and 2) engineering background, so they could take over the controls in the event it became necessary. If there was a woman with this background, we should consider her as orbital flight becomes more routine we can relax these rules.'"

Then Lyndon saw a personal note to him from Liz. "I think you could get good press out of this if you can tell Mrs. Hart and Miss Cobb something affirmative. The story about women astronauts is getting a big play and I hate for them to come here and not go away with some encouragement. Based on Dr. Dryden's statement, do you think you could write the attached letter to Dr. Webb, and show it to them before they leave."

Lyndon flipped the page to find the letter. Like so many he signed, it was written on his behalf with the space left for his signature to make it official. This one had tomorrow's date at the top, March 15. "Dear Jim," the letter began.

"I have conferred with Mrs. Philip Hart and Miss Jerrie Cobb concerning their effort to get women utilized as astronauts. I'm sure you agree that sex should not be a reason for disqualifying a candidate for orbital flight.

"Could you advise me whether NASA has disqualified anyone because of being a woman?

"As I understand it, two principle requirements for orbital flight at this stage are: 1) that the individual be experienced at high speed military test flying; and 2) that the individual have an engineering background enabling him to take over controls in the event it became necessary.

"Would you advise me whether there are any women who meet these qualifications?

"If not, could you estimate for me the time when orbital flight will have become sufficiently safe that these two requirements are no longer necessary and a larger number of individuals may qualify?

"I know we are both grateful for the desire to serve on the part of these women, and look forward to the time when they can."

Lyndon knew all the letter needed was his signature to open the conversation with Jim about women in space and effect a change with the agency. As chairman of the Space Council, he could champion the goal of America launching the first woman in space, could lead the nation to a small propaganda victory while NASA worked on the vastly more complicated Moon landing mission. On the other hand, he knew the space agency was facing much bigger challenges. It hadn't yet launched another orbital mission and already it was planning the newly announced interim Project Gemini. This two-man program was more experimental than Mercury, with each mission working out some facet of Apollo, be it long-duration power supplies in space or the tricky business of docking in orbit; this was why the agency was recruiting more test pilots. Apollo, meanwhile, was barely a half step closer to the Moon than when Jack had laid down the lunar landing challenge. No, if NASA didn't need this group of women, he wasn't going to take a stand for them, either. He decided instead to get this issue off his plate once and for all.

He picked up the blue ink pen on his desk, and in the space for his signature wrote, in inch-high letters, "Let's stop this now!" Below this, he wrote the word "file," indicating his final instruction on the matter to Liz.

Lyndon would still meet with Mrs. Hart and Miss Cobb—he would never turn down an opportunity for press coverage—but his decision was already made.

SCIENCE
Space

THE VICE PRESIDENT
WASHINGTON
March 15, 1962

Dear Jim:

I have conferred with Mrs. Philip Hart and Miss Jerrie Cobb concerning their effort to get women utilized as astronauts. I'm sure you agree that sex should not be a reason for disqualifying a candidate for orbital flight.

Could you advise me whether NASA has disqualified anyone because of being a woman?

As I understand it, two principal requirements for orbital flight at this stage are: 1) that the individual be experienced at high speed military test flying; and 2) that the individual have an engineering background enabling him to take over controls in the event it became necessary.

Would you advise me whether there are any women who meet these qualifications?

If not, could you estimate for me the time when orbital flight will have become sufficiently safe that these two requirements are no longer necessary and a larger number of individuals may qualify?

I know we both are grateful for the desire to serve on the part of these women, and look forward to the time when they can.

Sincerely,

Lets stop this now!

Lyndon B. Johnson

File

Mr. James E. Webb
Administrator
National Aeronautics and Space Administration
Washington, D. C.

The Lyndon Baines Johnson Presidential Library

CHAPTER 21

Washington, DC, March 17, 1962

JERRIE AND JANEY WERE SHOWN INTO LYNDON JOHNSON'S OFFICE JUST before eleven o'clock in the morning, a spacious office located across from the Senate chambers with a view of the Supreme Court. A large crystal chandelier hung from the ceiling adorned with ornate frescoes, decorum matched by the vice president's stately desk whose trinkets included a gold pillbox. The only thing missing was Lyndon.

The women had spent the previous day preparing. Though they'd been working together for weeks, the two hadn't spent much time together so had rushed to come up with a strategy. They agreed they should impress upon the vice president that they believed there was real scientific value in testing women for spaceflight and that the publicity that would come from launching the first woman could be significant both nationally and internationally. What their preparations hadn't included was a survey of how the other women involved felt, but neither was bothered. Jerrie assumed that they were all of the same mind, so she continued operating as their spokeswoman.

After a moment, Lyndon rushed in. They stood to shake his hand as he hurriedly apologized, saying that a bill signing ceremony had gone long. He had to open the Senate at noon, he told them, and also needed to eat lunch and get some preparations done beforehand, so they didn't have much time, but they were welcome to start.

Jerrie jumped right in. She began with a detailed explanation of what she believed were the benefits of sending women into space—how women are lighter, eat less, and consume less oxygen, so are more cost-effective as payload. Then she told the vice president about the medical and psychological exams—how she, Wally, and Rhea had proved women were emotionally and physically fit for space. Janey added her own appeal. Women shouldn't be barred from flying in space on account of their gender, she said. It was antiquated to suggest that women be relegated to the kitchen. If women would explore space eventually, she said in reference to the evasive line so many men had been telling her for months, why not start now?

Lyndon leaned back in his chair as he listened, fidgeting with his gold cufflinks and gold watchband. He didn't have anything against women flyers, he told them. In fact, one of his oldest and dearest friends was their colleague Miss Jacqueline Cochran, who had not only made him a fan of female flyers but had convinced him that women had much to offer the nation in all fields. The problem as he saw it, LBJ told the ladies, was that if NASA admitted women to its astronaut corps, every man on the street would argue they too were fit for spaceflight. The qualifications were put in place by highly informed professionals for the astronauts' and the program's own safety.

THE VICE PRESIDENT
WASHINGTON

April 13, 1962

Dear Jackie:

Thank you so much for your nice letter and the enclosure. It is always fun to see you and I had a good time telling Jerrie Cobb and Mrs. Philip Hart of our friendship when they came in to see me about lady astronauts. I told them you had made me in favor of lady fliers a long time ago.

Sincerely,

Lyndon B. Johnson

Miss Jacqueline Cochran
Cochran-Odlum Ranches
Indio, California

Eisenhower Presidential Library

Jerrie might have believed the vice president was taking their meeting seriously, but Janey knew she was witnessing a performance. LBJ was humoring them, feigning distress at being the bearer of bad news.

When Lyndon felt the meeting had gone on long enough, he leaned toward Jerrie and Janey and put on an earnest face. He wanted to help them, he said, he really did. But it wasn't up to him. It was up to Jim Webb and NASA to make that decision, he told them. He didn't mention the letter Liz had drafted in his name. With that, the meeting was over. The vice president picked up the private telephone on his desk as the two women were ushered out the door into the waiting crowd of journalists.

Jerrie went straight from LBJ's office to NASA's headquarters to try to speak with Jim but was told he didn't have time to see her. She was referred instead to a Dr. Cox to whom Jim Webb had turned the matter of women in space, but it turned out he didn't have any authority to change their standing with NASA, either. It was another brick wall.

When Jackie heard what Jerrie had been up to, she could scarcely believe the audacity. Running around Washington cornering politicians, speaking publicly with a level of authority she didn't have, getting all the way to her friend Lyndon. Then there was the matter of the letter, the last one Jerrie had written that made it clear she hadn't listened to a word Jackie had said over dinner that night in Cocoa Beach. Jackie understood Jerrie's passion; she, too, dreamed of flying in space. But the way Jerrie was carrying on, badgering decision-makers into seeing things her way—she wasn't endearing herself to the people who could make that happen, she was making herself a pest. Jackie knew from her days with the WASPs that a sound and organized program was the only way to advance the case for women, and watching Jerrie from afar, she felt she had to step in. Sure Jerrie would keep finding ways to avoid another face-to-face meeting, Jackie decided to put her thoughts very clearly in a return letter on March 23. "Dear Jerrie," it began. "I also enjoyed the evening with you and Mrs. Rieker at Cocoa Beach...

"You asked me several times that evening how I felt about a woman in space program and because I tried generally to express my views when the question was first put to me, I guess I was not communicating very clearly. Perhaps it's best that I place my views on record.

"Women will travel in space just as surely as men. It's only a question of when... I am sure you know from our various conversations that I am in favor of a space program for women. That I put up the money to pay the cost of most of the twenty candidates who took the medical checks at the Lovelace Foundation and then put the money for most of the ones who passed these checks to enable them to go to Pensacola...

"A space program for women is unlike all previous advents of women into various phases of aviation in that spaceflight for the present is terribly expensive, is an urgent project from the standpoint of national defense, and there is no lack of qualified candidates for the role of astronaut from among our already highly trained flight personnel. In other words, there is no present real national need for women in such a role...

"A sound, fully acceptable women in space program must, in my opinion, involve a sizable group... No clear conclusions can be reached by checking one person or even a very few. Extremes of tolerance, reaction, etc. will be found in any group while arriving at an average of tolerable limits. The government did not pick one person to be the astronaut nor did it limit itself to seven at the start. A great number were screened. Seven was the final result...

"There is no real present national urgency about putting a woman into space. To attempt to do so in the near future might indeed interfere with the space program now underway which is urgent from a national standpoint. I believe you disagree with me about this on the ground that to put a woman into space before the Russians would be a victory... A hastily prepared flight by a less than completely trained woman could backfire...

"It's better to be sound than quick...

"I compliment all of the present candidates for their ambition and their apparent willingness to go through the needed stretch of specialized training... If you go along with the soundness of the group idea, as I hope you do, then you can be particularly helpful to a program by going out of your way to create the group image publicly."

Pleased with the clarity she'd managed, six days later Jackie sent a copy to each of the other women who had taken the Lovelace exams. "So that you will have an expression of my views about a women in space program," she wrote in her cover letter, "I am sending you herewith a copy of my letter of March 23, 1962, to Jerrie Cobb."

The same day, Floyd wrote to Randy. "At Cocoa Beach a day or two before the Glenn shot, Jerrie Cobb had dinner with Jackie and brought along some woman writer for *Life* Magazine," he wrote in his cover letter. "Because Miss Cobb asked Jackie at least a dozen times during the evening how she felt about the woman in space program—even though Jackie tried to explain her position when the question was first put to her—Jackie concluded to write Miss Cobb a letter." Inclosed was a copy of Jackie's letter to Jerrie.

But Randy and the women he'd tested weren't the only people wrapped up in the matter. More parties had been dragged into the fray—NASA management, navy brass, politicians—so Jackie sent each of them a copy of the letter to make her intentions and stance abundantly clear. She sent a copy to Lyndon. "I realize what an extremely busy person you are," she wrote to her old friend, "and I would not impose upon you to read the enclosed copy of a letter (reading time eight minutes) I have sent to Jerrie Cobb if I did not think it as of great national importance and if you were not the head of our entire space program." Jackie also sent copies of the letter to Hugh Dryden and to Olin Teague, a member of the Science and Astronautics Committee in the House of Representatives. She sent one to Walt Williams and Bob Gilruth at NASA's Manned Spacecraft Center. She sent one to Admiral Robert Pirie. She even, with considerable effort, tracked down Jane Rieker at a new address. "I tried my very best, that evening, to express my views concerning women in space," Jackie told Jane in her cover letter, "but obviously did not get over my point so I wrote Miss Cobb a letter and I am taking the liberty of sending you the enclosed copy as I think it will interest you." For good measure, she sent a copy to a second *LIFE* writer, Jane Nash.

It wasn't long before Jackie got confirmation on her views. Hugh was impressed with her "excellent analysis" of the role of women in space. "I am sure that the time will come when women will participate in space flights, but I agree with you that it does not seem wise to make this a national objective of the space program at the present time." Bob was pleased to see that she had so eloquently expressed very similar views to those he had tried to convey in his own recent letter to Jerrie. "Your advice to Miss Cobb that rushing into a woman astronaut program just to beat the Russians might prove disastrous is apparently based on much the same premise that has guided a lot of our own decision making." Walt Williams wished that he or "other people in NASA could state the situation in the same lucid manner. It would be of great help to our program if your views could be published in at least the trade journals in the form of either an article or guest editorial." Randy replied to Floyd saying he found Jackie's letter on the whole excellent.

The same praise wasn't forthcoming from the other women. Many were busy getting their professional lives back on track after the cancelled Pensacola tests and simply didn't reply. Gene Nora Stumbough, however, deep in preparations for a countrywide tour flying with Beech

Aircraft, threw her support behind Jackie. "I agree with your thoughts expressed to Jerrie wholeheartedly and 100%. Women will travel in space, but there is no need to train women right now. And to frantically send up a woman immediately just to beat the Russians, is, in my thinking, totally invalid. I am not under the delusion that one of us next week or even next year will be shot off to the Moon. I am only afraid that by nagging those who make the decisions, we are hurting ourselves."

Jerrie continued to nag the decision-makers. A crusader, her mother had called her, amazed that her formerly shy little girl was now taking on Washington.

Her patience for not getting her way, however, was wearing thin. Unable to secure a meeting with Jim Webb and worried she had not made her position clear, she sent him another series of letters. "If you do not think this proposal is important enough to warrant your attention and it is out of Dr. Cox's authority, would you please tell me what you would like for me to do? I repeat that I would like to work with NASA towards this goal. I cannot help but believe that it is of utmost importance to our country and I would willingly give my life for it. Please let me help."

Jerrie also followed up with LBJ, looking for loopholes to sway the vice president's opinion. "I am sure you realize the importance of the United States putting the First woman in space and the contributions women can make to space flight. The only objection which NASA has made is that there are no women military jet test pilots. If this experience is deemed so necessary, it should be proved scientifically by letting a qualified woman pilot fly the space flight simulators under stress conditions and compare the results." She included a copy of her "Space for Women" speech for emphasis.

When Lyndon's reply came, Jerrie was amazed. "The choice of training individuals who will make space flights is quite appropriately left to the operating agencies in the program. However, I see no reason why preparations should not be made for testing and training individuals who have the required physical and mental capabilities, regardless of sex." Lyndon was feeding her breadcrumbs without any promise of resolution, but Jerrie read it as a glimmer of hope. For the first time, she saw a crack in the wall barring women from space and felt sure that with enough pressure, she could break that wall wide open. She believed what the media was saying about her, believed that she was really qualified to fly in space and deserved a mission. "In all true conscience," she wrote in her reply, "I must continue to do everything I can for this important matter."

Getting Jackie on her side was a different matter. A month after receiving the scathing, four-page letter, Jerrie wrote her reply, her frustration palpable. "The qualification rules have been laid down for astronauts and although NASA says they have nothing against women, it just so happens that the requirements are such that no woman can meet them. I don't know much about politics but I do know that exceptions have been, and can be made without destroying the scientific basis of the program... The people who are saying that women should not go into space now, are the same type who were saying that man would never fly, or that women did not have the mechanical or scientific mind to pilot an airplane. All I'm saying is that I am not content to sit back and listen to their silly excuses while waiting for Russia to prove the scientific importance of putting women in space."

At the end of the day, Jerrie finished her letter, she didn't need to be the first woman in space. She just wanted that first woman to be an American. "My aim is not personal but to get a group of women included in the space program on a sound, orderly scientific basis. No less I can do, with a sound conscience."

* * *

Whether because of Jerrie's dogged persistence or because the right people were starting to get wind of her crusade, the conversation began to shift in Washington. There was some talk about whether women ought to be included in future astronaut classes. Congressman Ken Hechler on the House space committee was in favor of seriously considering women, so much so that he inserted a copy of Jerrie's "Space for Women" talk into the *Congressional Record*.

The tone of press coverage around women in space was also starting to change. With the story of Jerrie and the other "astronaut hopefuls" growing old, journalists began highlighting women who were actually working with NASA, like the astronauts' nurse Dee O'Hara and astronomer Nancy Grace Roman. But Jerrie's name was rarely absent from these pieces. She was still considered "America's first woman astronaut," often quoted as extolling the virtues of women's emotional and physical stability in space—"so brace yourselves, men!"

The more such pieces she read, the more Jackie realized that the media was the problem, that these stories about women astronauts were doing little more than feeding Jerrie's false impressions of her own importance in a program that did not exist. So she did what she could to set the record straight. Jackie wrote to the journalists whose stories she read explaining that there was no program for women, that the early medical tests were part of a research program, and warned that any false publicity risked delaying or even cancelling possible future steps for the women. She kept NASA abreast of these reports, too. When *Current Biography* magazine published an entry on Jerrie that said she had been "invited by the National Aeronautics and Space Administration to undergo a series of qualifying tests for the astronaut training program"—described as NASA's own and not Randy's—on account of her "brilliant flying record," Jackie sent the entry to Hugh Dryden. The clipping set off a flurry of internal memos at the space agency, all of which asserted that "Miss Jerrie Cobb is not 'the first woman to satisfy the criteria for space flight set by NASA,' since she does not meet the selection criteria for astronaut candidates."

Jackie landed her Lodestar at the Seattle-Tacoma International Airport on the afternoon of May 7, 1962, to find a pack of journalists waiting for her. They knew about her ambitions of flying higher than 100,000 feet and at twice the speed of sound; she'd said as much after securing nineteen records on a recent flight from Gander, Newfoundland, to Hannover, Germany. But as she was in town for the second Peaceful Uses of Space conference, they wanted to know if she hoped to add first woman in space to her growing list of accolades.

"First woman in space?" She repeated the question, then paused before answering. "I'm not even fifty. But they seem to want younger astronauts. Personally I do not think age should bar a trained person from space journeys. If I were permitted, I would give anything to get into the space program." Her drive was real, but her age was not; she was days from turning fifty-six and wholly unwilling to admit that to the press.

The next night she sat at the head table alongside John Glenn, Wernher von Braun, Senator Robert Kerr of Oklahoma, and two dozen other major players in the nation's space program as they listened to Jim Webb's after-dinner remarks. There were no surprise consultancy suggestions or discussions of women in space this year. Instead, the conference focused on how far NASA had come in the past twelve months and how far it had yet to go.

The conference was a busy two days for Jackie, so much so that she couldn't find time to have a private word with Jim about women in space.

* * *

Days later, back in his office at NASA headquarters in Washington, Jackie's name landed on Jim's desk in a handwritten note from Elvis Stahr, secretary of the army. "This strikes me as such a well-reasoned letter that I asked Jackie's permission to show it to you. It's not my business, of course, but Jackie is a good friend—whom you would enjoy meeting. Will arrange if you like." Attached was a copy of Jackie's March 23 letter to Jerrie.

Jim read the four-page correspondence and found Jackie's argument both reasonable and well expressed, but he didn't have time to meet with her. He was in the midst of seemingly endless appearances before Congress about agency funding and preparations for the second orbital Mercury mission while guiding the space agency through its second round of astronaut selection. And he still had to commit Apollo to a lunar mission mode. "When this lets up a bit," he wrote to Elvis Stahr in response, "perhaps the three of us can meet."

He did, however, make a point of following up with Jackie. "There have been many pressures to go into a program that would be aimed at putting the first woman into space," he wrote on May 24, the same day Scott Carpenter completed three orbits around the globe in his Mercury spacecraft named *Aurora 7*. "We have felt it imperative to use astronauts who had extensive flight experience with experimental jet high-performance aircraft. The ability to react from a background of long experience with high-performance aircraft is very important, particularly in these early flights, and it has seemed to us should continue to be overriding in the selection of the next group of astronauts. However, if there are women with this experience who wish to apply, we certainly will consider them along with all other applicants."

He ended the letter thanking Jackie for her support of NASA's programs.

* * *

Reading Jim's reply, Jackie worried he'd badly misinterpreted her intentions. To her, it read as though he believed she was one of the people pressing to get women training as astronauts right away, which she wasn't. Restating her position in still another letter, she added a warning. "I think you are likely to let yourself in for a lot of continuing harassment if you continue to take the position that even now women with such experience will, if they apply, be considered along with all other applicants. This position stated by you may be looked upon by some as a slightly cracked door to be opened with a push." She had a plan, she said, to eliminate this pushing from women.

On June 13, Jerrie's yearlong unrealized consultancy with NASA came to an end. There was neither a formal letter about her firing nor any reference to her unpaid stipend of fifty dollars a day; she was simply not retained. Janey, meanwhile, was receiving a growing number of letters from college coeds who wrote asking how they could plan for a future in science. Janey could almost feel these students' discouragement. The nation needed scientists, but these young women could barely see beyond the obstacles stacked up in their way.

But both women felt the lawmakers in Washington were starting to take them seriously. By sheer luck, Jerrie managed to meet with George P. Miller, the representative from California who was serving as chairman of the House space committee. He supported the women's cause and believed both Jerrie and Janey when they said it was all a matter of discrimination against women. George threw his weight behind them, eventually generating enough support to set up a congressional hearing. Things moved quickly, and by the middle of June the hearing was set for a month hence, a hearing wherein the House Committee on Science and Astronautics would investigate the alleged discrimination in the space agency against women.

Janey arrived at Jackie's Manhattan apartment for a lunch meeting on the first Wednesday in June. She knew Jackie had heard about the House hearing and suspected she wanted to have her say on the record.

Stepping inside, Janey was stunned; Jackie's foyer alone was a display of power. The front hallway featured a large compass inlay on the floor. The sky blue walls were painted with scenes of planes flying through the air. Shelves against both walls held Jackie's dozens of trophies and awards, so many that they spilled onto the floor. It all made Janey more nervous than she already was; beyond their differing stances on female spaceflight, Janey and Philip were wild-eyed liberals compared to Jackie and Floyd's conservative politics. She wondered whether they would get along at all.

Janey told Jackie what she knew about the pending hearing, which wasn't much. She knew that the subcommittee was headed by Democratic Congressman George Miller of California and that both she and Jerrie would be testifying, but couldn't say if Jackie would be able to appear as a witness as well. In any case, Janey asked that Jackie submit a statement so her views could at least be on the record. Then she listened as Jackie told her the same thing she'd been saying all along: there was no national need to rush a women-in-space program. Throughout the meal, Floyd stayed quiet.

To Janey's ear, it was almost as if Jackie was adamant that if she couldn't be the first woman in space, no woman should fly. She couldn't say she was surprised by Jackie's attitude, but she had hoped the aviatrix might be more sympathetic. By the end of their lunch, Janey knew that without Jackie on their side the cards were stacked against her and Jerrie.

A little over a week later, stories about the upcoming hearing popped up all over the country. Reports said Democratic Congressman Victor Anfuso of New York had agreed to serve as the subcommittee's chairman. He had a long history of interest in space and a desire to serve on the committees that would shape the nation's space program. He was also a longtime friend of Lyndon Johnson's, someone from whom the vice president had drawn inspiration throughout his career. The two had joked that Victor might leverage his political position to fly in space himself. Victor was such a fan of the space program and the astronauts that he'd even made moves to have Cape Canaveral renamed in John Glenn's honor after his orbital mission.

Jackie read about the hearing in the newspaper, and though she was going to Europe that afternoon, she didn't want to leave the country without making sure her opinion was on the record. "I have drafted such a statement and I leave for Europe within the hour to be back early July," she told Janey, adding that she would send a final copy to her before the hearing.

She also wrote to Jim Webb, including a copy of her position in eight bullet points. "You will note," she said in her cover letter, "that I am taking the position that these things should be left to the decision of NASA." Her formal statement reiterated what she'd been saying for months. She didn't believe there had been any intentional or actual discrimination against women and that the determination for astronauts "should not depend on the question of sex but on whether such inclusion will speed up, slow down, make more expensive or complicate the schedule of exploratory space flights our country has undertaken." What she wanted, more than anything, was to see a soundly organized women's testing program akin to the WASPs.

Copies made their way to all the important players—Hugh Dryden, Bob Gilruth, chief of staff of the US Air Force General Curtis LeMay, Admiral Robert Pirie, and of course Randy. She also sent her statement to Janey and to Gene Nora, the one pilot who had shown her the most support. Across the board, her statement earned favorable responses from all the NASA brass.

Only Randy offered Jackie any feedback; since he was unwilling to jeopardize his position with NASA by taking a stance, he told her his opinion so she might speak for him: "The main reason we went into the examination of potential women astronauts was not from the standpoint that they would actually pilot the spacecraft at all, or at least for any appreciable time, but that they would fill a definite need for the experimental program that is anticipated. We thought it was essential that the first women in space should be highly trained pilots as they have proved they have courage, determination and are highly motivated or else they wouldn't be where they are in their flying careers. There has never been any time when any of us felt we could determine woman's future in space on the basis of examining one individual or even a few. We do not recommend injecting women into the middle of the present program but we do need to subject them on the ground to the stresses that they would be exposed to there. We feel that much more would be gained if they were pilots and that is the reason we would like to limit these tests to women pilots at the moment."

Again, Randy used a handwritten postscript to make his personal feelings known. "We have missed seeing Floyd and you. Congratulations again on all the records."

Janey was disappointed when she learned that Victor Anfuso had been appointed chairman of the subcommittee. She felt George Miller was more supportive of the women and favored establishing a long-term research program rather than a crash program. He had, after all, called for the hearing in the first place. Now she worried the whole thing would become little more than an attention play by Victor. She wrote to Jackie, relaying her thoughts on the hearing. She figured she might as well give Jackie all the facts in case she was called upon to testify.

On July 9, Jackie was formally asked to testify at the hearing in a letter from Richard P. Hines, a staff consultant to the Committee on Science and Astronautics. "Hearings have therefore been scheduled before the subcommittee to begin July 17," the letter told her. "Mr. Anfuso would like to have you appear at that time, if convenient. He is also inviting Miss Jerrie Cobb and Mrs. Philip Hart to appear. I am sure you realize all our Committee hearings are held for the purpose of determining facts in the most objective manner, and we try to remain as non-partisan as possible and express judgements and conclusions based upon testimony presented by advocates both pro and con." It came as no surprise to Jackie that Jerrie and Janey would be testifying; it had, after all, been their waves in Washington that had gotten this hearing scheduled in the first place. Jackie could actually look at this lineup as a positive. Having followed Jerrie's every move for more than a year, she was sure there wouldn't be any surprises in the proceedings.

Jerrie, meanwhile, wrote the other women another "Dear FLATS" letter alerting them to the coming hearing. "It is probable that names and a brief biography of each of you will be released officially," she wrote. In the event the press contacted them, she urged them to all "work together and represent this as the sound and serious program it is." She didn't stop to think that each woman's experience with her singular week of medical testing had been vastly different than her own months-long fight.

Dear FLATS:

Next Tuesday, July 17th in Washington, D.C. a hearing on "the practicability of training and using women as astronauts: will begin before a special subcommittee of the House Committee on Science and Astronautics.

Three days have been established for these hearings and both sides of the question will be discussed. It is probable that names and a brief biography of each of you will be released officially for the first time, so the press may contact you. Let's all work together and represent this as the sound, serious program it is. You may want to review the points made in the enclosed reprint from the Congressional Record.

I will be the first to testify Tuesday morning at 10 a.m. and enclosed is a copy of my prepared opening statement. Also to testify will be Mrs. Hart and Miss Cochran. From NASA, Dr. Seamans and Dr. Roadman have been invited to testify as well as John Glenn, Scott Carpenter and Joe Walker, the Committee Counsel says.

I knew you all would want to know this latest development in our project, and hopefully it will help get us underway. Wish us well on Capitol Hill and I'll be in touch.

Best regards,

Jerrie

Jerrie Cobb

JC:bd

Encls.

International Women's Air and Space Museum

* * *

On July 11, six days before the hearing, Jackie sent her testimony to Victor Anfuso. That same day, Jim sat before the gathered media at a NASA press conference. "Ladies and gentlemen," he opened the afternoon's proceedings. "It seems to me that we might look back very briefly to a little more than a year ago when President Kennedy made his original decision to put forward an increased augmented space program." In that time, NASA had awarded contracts to build the essential hardware and spent months considering the intricacies of the mission, but "in all of this program there remained the final decision as to exactly how the first effort to make an exploration of the moon with men would be achieved." But they finally had an answer. Lunar-orbit rendezvous, exhaustive studies confirmed, was the only way forward. NASA was thus inviting industry contractors to submit their bids to build the lunar module that would actually land on the Moon. With just over seven years to the end of the decade, the agency was not only short on time, it was also making this its top priority. "We will have a period of perhaps three months within which to get the proposals from industry, to evaluate them carefully, and to reach a final decision." On the question of astronaut selection, Jim announced that the preliminary group had been narrowed to thirty-two, all qualified test pilots whose skill was up to the challenge of this demanding flight. Of those selected, he expected one of them would be on the first lunar mission.

Five days before the hearing, while the Space Council met to discuss the ramifications of a lunar-orbit rendezvous on NASA's growth over the rest of the decade, Jackie got another note from Richard P. Hines's office. Victor Anfuso knew Jackie was chairing a meeting of the National Aeronautical Association at the same time the hearing was scheduled to begin but wanted her to present her testimony in person. He asked that she arrive around 11:15; this would put her arrival after Jerrie and Janey had both testified and should hopefully give her enough time for her NAA meeting beforehand. The note asked, finally, that she call Victor directly to touch base about her testimony.

Two days before the hearing, Jerrie arrived in Washington alone. While some of the other women tested lived close to her, and though she had plenty of room in her plane, she hadn't picked anyone up, hadn't asked them to sit in as witnesses, or even invited them to submit a statement or opinion for inclusion in the official record. She and Janey had decided they would speak on behalf of the group. Neither gave any thought to the fact that they might be ignoring the other women's perspectives.

The day before the hearing, Jackie arrived in Washington. From her hotel suite, she called Victor Anfuso to confirm her testimony. She also called Randy one last time to make sure that he hadn't changed his feeling on women astronauts. He hadn't. Her due diligence done, Jackie prepared to present her unchanged testimony the following morning.

CHAPTER 22

Washington, DC, Tuesday, July 17, 1962

IT WAS ALREADY WARM AND HUMID WHEN JERRIE ARRIVED AT THE NEW House Office Building a little before ten o'clock. She was dressed in a short-sleeved, button-down black dress cinched at the waist with a black-and-white belt and modest black pumps, fitting for both the event and the weather. Her hair was styled in her favored ponytail with some softness in the front. She went inside and found a mix of congressional representatives, journalists, stenographers, NASA brass, and curious onlookers milling around in the hallway outside Committee Hearing Room B-214. They were all there for the subcommittee hearing on the qualifications of astronauts. Jerrie looked around and saw some familiar faces. Her friend Jane Rieker from *LIFE* magazine was there, as was Cathryn Walters from Dr. Jay Shurley's office. But it was Janey whom Jerrie was happiest to find, dressed similarly conservatively with her pilot's lapsed-time chronometer on one wrist and a delicate gold bracelet on the other.

Janey was nervous. She and Jerrie hadn't had a chance to compare testimonies, and Janey wondered whether if they overlapped at all in their statements it could hurt their cause. As the two pilots quickly conferred in the hallway, Congressman Victor Anfuso joined them and explained how the hearing would go. Jerrie would testify first, then Janey, then the floor would be open to questions. In a quick, whispered conversation, Jerrie admitted to Victor that she was scared to death. She'd landed an unpowered plane blind, survived internment in Ecuador, endured heartache, and proven her physical and psychological fitness, but part of her was still the tongue-tied little girl who hated public speaking.

The crowd drifted into the hearing room. Victor showed the other ten subcommittee members to their seats along one wall. All told, nine of the group were men, eight of them war veterans, and the two women were both a generation older than Jerrie and Janey. The press and other observers filed into the benches along the other three walls. Surrounded by spectators, the witness table faced the subcommittee and was the unquestioned center of attention. Looking around as she took her seat at the small table, Jerrie realized there was standing room only. Behind them, John Glenn's portrait hung on the wall.

"This meeting will come to order." Victor spoke loudly enough to quell the chattering crowd. "Ladies and gentlemen, we meet this morning to consider the very important problem of determining to the satisfaction of this committee what are the basic qualifications required for the selection and training of astronauts." He continued, "We are particularly concerned that the talents required should not be prejudged or prequalified by the fact that these talents happened to be possessed by men or women. Rather, we are deeply concerned that all human resources be utilized."

These opening words quieted some of Janey's concerns. It sounded to her like Victor might be willing to consider the women as pilots, not female pilots, looking objectively at the merits they could bring to the space program. Victor reminded the subcommittee that NASA had a number of women working behind the scenes, laid out the day's schedule, and then called on Jerrie.

Jerrie looked down at the annotated copy of her testimony. Slashes indicated a pause. She'd noted which words to emphasize, where to smile, and worked in material from previous speeches she knew could get a laugh from her audience. She took a steadying breath.

"Thank you, Mr. Chairman, honorable Members of Congress." Almost as though she were watching from the outside, Jerrie noticed there was no waver in her voice. She sounded calm, measured, and articulate. "Our purpose in appearing before you is single and simple: We hope that you ladies and gentlemen will, after these hearings and due consideration, help implement the inclusion of qualified women in the US manned space program." She was the lead witness, she explained, because she had been the first woman Dr. Randy Lovelace had invited to take the astronaut medical tests, and she alone had completed the advanced testing phases. "Early last summer I was sworn in as a NASA consultant," she added to her list of credentials, unbothered that a curious journalist could make a simple phone call and uncover that her swearing-in ceremony had never happened. "After a group of twelve other women had passed the Mercury astronaut tests, I was sort of drafted to be spokesman for all thirteen of us." The whole group, she said, couldn't be in Washington because no funds had been available to cover their travel costs, so she introduced them all, including Jerri Sloan, who'd withdrawn her name from contention, through short biographies. Pens scratched furiously in the viewing gallery. The elusive women the press had obliquely been mentioning for months now had names.

Jerrie summarized her story for the subcommittee. She emphasized women's resilience to radiation, their tolerance for monotony, and the overall ease of launching a smaller astronaut into space. Then there was the matter of national pride. "No nation has yet sent a human female into space," she said. "We offer you thirteen woman pilot volunteers." With that, Jerrie ended her prepared remarks. She couldn't fathom what the subcommittee was thinking, whether they were interested or just being courteous.

"Miss Cobb," Victor addressed her, "that was an excellent statement. I think that we can safely say at this time that the whole purpose of space exploration is to someday colonize these other planets and I don't see how we can do that without women." The room laughed as he turned his focus to Janey. "Mrs. Hart, it is a real pleasure to welcome you here. Besides being the wife of a very distinguished senator, you are also the mother of eight children—four boys and four girls—is that correct?"

"Yes, sir," Janey replied. "I couldn't help but notice that you call upon me immediately after you referred to colonizing space."

"That's why I did it." Victor chuckled.

Unfazed, Janey began her testimony. "The subject at hand is certainly one of my favorite ones." She left no doubt as to her passion. "I am not arguing that women be admitted to space merely so that they won't feel discriminated against, I am arguing that they be admitted because they have a very real contribution to make." Janey knew it was these kinds of statements that earned women condescending smiles and humoring winks, but she also knew that progress only came when women had a chance to prove themselves.

"A hundred years ago, it was quite inconceivable that women should serve as hospital attendants...it was somehow indecent for a woman to be among all those soldiers, wounded or not. Well, the rest of the story is altogether familiar to you. The women were insistent. There was a shortage of men to do the job. And finally it was agreed to allow some women to try it, provided they were middle-aged and ugly—ugly women presumably having more strength of character." A woman in space, Janey said, was no more preposterous than a woman in a field hospital during the Civil War. Waiting for a manpower shortage before tapping into the contribution women had to offer was ridiculous, and it dissuaded young women from pursuing an education. "Why must we handicap ourselves with the idea that every woman's place is in the kitchen despite what her talents and capabilities might be?" But at the same time, as a mother of eight, Janey certainly understood the importance of raising children. If a woman wanted to marry and become a homemaker, that was her choice, but to force women into marriage as a career when their aptitudes might lie elsewhere was simply wrong. "Our affluent society has provided so many household aids that the intelligent, energetic housewife can find many hours to devote to other useful purposes. Let's face it: for many women the PTA just is not enough." She went on to recount the same story Jerrie had told about their testing, urging the subcommittee to consider reinstating their research program.

"I don't want to be the Susan B. Anthony of the space age," Janey said as she finished her statement. "I just think we would be making a serious mistake if we assumed that women just have no contribution to make to space exploration."

The floor now open to questions, Victor asked Janey whether she felt flying in space would inspire other women to follow in her footsteps. She replied that it would. "Would you go so far, Mrs. Hart, to say that anything man can do, woman can do better?"

"No, sir, I would not," Janey replied, taken aback by the sudden shift in tone.

Jerrie stepped in to walk the subcommittee through the tests the women had passed, using pictures to illustrate exactly what they'd done. But Victor had done his homework on NASA's astronaut requirements and knew the agency demanded more qualifications than just passing a series of medical checks. He knew there was a reason they were grounded, but he wanted to see what Jerrie had to say.

"Miss Cobb, what do you think are the minimum qualifications for an astronaut?"

"I could not answer the minimum qualifications for an astronaut because I am not qualified," she answered.

"Is it because one of the requirements is that the astronauts be also engineers?" he asked.

"No. I don't think that is it at all. There are many women engineers. It is the jet test pilot experience that makes it impossible for a woman to meet the qualifications," Jerrie answered, pleased to have reached the heart of the issue.

"Are any of your women test pilots?" Victor pressed.

"Some of us have worked as test pilots but it is impossible for a woman in this country to be a jet test pilot."

"Do you feel it is essential to have been a test pilot before you could qualify as an astronaut?" Congressman Joseph Karth from Minnesota asked.

"I personally do not feel it is essential at all," Jerrie said. "An astronaut must pilot a spacecraft—not test jet fighters. If you total the flying hours of this group of women pilots you will find the women averaged four thousand five hundred hours each, which is more than the men astronauts have."

"There is a considerable difference between straight flying—commercial or private—and test piloting, isn't there?" Joseph continued his line of questioning.

"I suggest there is an 'equivalent experience' in flying that may be even more important in piloting spacecraft," Jerrie replied. "What counts is flawless judgment, fast reaction, and the ability to transmit that to the proper control of the craft." She went on to explain how, though the Mercury astronauts had 1,500 hours apiece in jets, the women had upward of 10,000 hours in propeller planes. "Pilots with thousands of hours of flying time would not have lived so long without coping with emergencies calling for microsecond reaction—" Jerrie hadn't finished making her point when the door opened.

"May I interrupt to welcome Miss Jacqueline Cochran who has just arrived?" Victor cut across her. "We will proceed with her testimony after Miss Cobb and Mrs. Hart."

Jackie wasn't the least bit sorry about her late arrival. It was the same as making a judge wait after a race while she touched up her lipstick—a quiet way to make an impact. As she moved to take her seat, she took in the scene: the subcommittee in one long row facing the witness table where Jerrie, in her seat next to Janey, had kicked off her shoes and now sat with her stockinged feet on the floor crossed at the ankles.

The questions resumed. It was stated for the record that Jerrie was neither an engineer nor a jet test pilot, but she implored the subcommittee to see that she had a lot to offer the space program regardless. She told the subcommittee that while women were barred from astronaut testing and training, there was a clinic called "Chimp College" in New Mexico where doctors were training chimpanzees for space missions, one of which was a female named Glenda. If a female chimp could train alongside males, why couldn't female humans train with men? This led the conversation to the issue of the cancelled Pensacola testing.

"I notice, Mrs. Hart," Congressman Ken Hechler addressed himself to Janey, "you used the phrase 'somehow the program was cancelled.' Could you explain this a little bit? What were you informed about the program?"

"As to why it was cancelled?" she began. "I have no idea, sir. That is one of the mysteries of the past year." It was the only way she could express it.

"We will get the answer to that question tomorrow from NASA," Victor said. "This committee has the assurances that NASA wished to cooperate and is cooperating. I indicated in my earlier statement that these hearings are helpful. And I know that you don't want to criticize any branch of the government. You just want answers."

"I would like very much to work with NASA," Jerrie confirmed.

It was then that Congressman James Corman from California asked Jerrie if she knew where Irene Leverton was. "I have an address in Los Angeles. I believe it is Santa Monica."

"Is she one of your constituents?" Victor turned to ask his colleague.

"Yes," Congressman Corman answered. "I thought she might be back here for these hearings and I wanted to meet her."

"I wish she could have been," Jerrie answered, without offering any explanation as to why neither Irene nor any of the other pilots was present.

Congressman James G. Fulton of Pennsylvania, a Republican and longtime friend of Lyndon Johnson's who shared in the vice president's love of space, spoke next.

"Did it strike the women that the reason the tests were cancelled was because the men thought the women were too successful?" James's congressional counterparts didn't try to hide their disdain. He wasn't highly respected among his colleagues, so much so that it wasn't uncommon for his peers to vote against an issue because James spoke out in favor of it. Now, his question asked in favor of the Lovelace women was answered only with laughter.

"Miss Cobb, you showed a little bit of resentment toward Glenda out in this test center," Congressman Joe Waggoner asked. "You do not feel any resentment about the female monkeys in the cancer clinic, do you?"

Again, the room laughed at the comparison, but Jerrie didn't give them the satisfaction of showing her anger. "I think there is place for both," she said.

The questions went on, and Congressman Joe Waggoner put the key question to Jerrie. "Do you think that we ought to sacrifice anything in the way of accomplishment in time with regard to our lunar landings and other space activities, or to go into this program to the extent that we would put a woman in space at the expense of slowing down another program?" Operating under the assumption that a women's program would have to be distinct from the regular astronauts' training, the congressman was putting Jerrie in the very position she didn't want to be. He was forcing her to answer which she valued more: her desire to be the first woman in space or her desire to see the United States reach the Moon before the Soviets.

"No, sir," she replied. "I don't think you have to start a whole new program for women," she said. She just wanted the women to have what she viewed as a fair chance at proving themselves with the opportunity of training for a real mission if they were deemed ready.

There was no resolution when the hearing took a brief recess.

"The committee will come to order," Victor reconvened the hearing. "I have the honor and privilege of welcoming Miss Jacqueline Cochran who, without a question, is the foremost woman pilot in the world, and who holds more national and international speed, distance, and altitude records than any other living person." He turned to Jackie. "Do you have a prepared statement?"

"I do, Mr. Chairman," Jackie replied. She held a copy of her statement in a large, double-spaced font, just as she had done during her congressional campaign. "I only heard, Mr. Chairman, on Thursday, when I was out west, that I was going to be requested to come before your committee, and I had no opportunity to prepare very much of anything." Victor didn't mention he'd already read her statement, nor did she say she'd sent it to NASA, Randy Lovelace, and a handful of other key decision-makers.

"I do not believe that there has been any intentional or actual discrimination against women in the astronaut program to date," she began, adding that she, an accomplished test pilot with ample jet flight experience and a strong desire to fly in space, had never once felt discriminated against. She stated her long-held belief that manned spaceflight was of such national urgency that it was natural that men should fly first; they were the ones in the country who were "experienced, competent and qualified to meet possible emergencies in a new environment." To that end, the decision on whether women should fly came down to the question of whether they would hamper NASA's ongoing programs. But either way, the existing pool of women was too small. A proper testing program should be large enough to establish norms for future generations.

When Jackie finished, Victor opened the questions. "Does any woman to your knowledge meet those specifications that you, yourself, have laid out?"

"I don't think anyone could make such a statement," Jackie answered before drawing a comparison with the WASPs as the only benchmark she had. "The sad part of the program was, our attrition rate was very high due to marriage. I don't know the exact figure. Somewhere in the neighborhood of 40 percent. They flew every type of aircraft the nation had, just as successfully. In fact, their fatality rate was only slightly under that for cadets." The WASPs were great pilots, but they didn't necessarily stay devoted to their careers. It was the same with commercial aviation, she explained. "Airlines spend $50,000[27] average to check a pilot out on a 707 or Convair 880," she offered as an example. "That is expensive if you lose them through marriage." No one interjected to point out that many of the Lovelace women were married, nor that Jackie had learned to fly at Floyd's suggestion. The subcommittee instead remained focused on NASA's astronaut criteria.

"I would like to ascertain whether you, in your opinion, feel it is a reasonable requirement that NASA has laid down that you be a test pilot before you be considered as one of the astronauts in the space program?" Congressman Karth asked.

"If you want my own honest personal opinion—"

"Your personal opinion."

"I don't think so, no," Jackie said. "Apart from technical training, you learn as you go along. I have learned a great deal. I know about the practical side of flying, shall we say, and the way planes are rigged."

---

27 About $424,760 in 2019.

"Would you say that a program to train selected American women as astronauts, apart from our present astronaut program activities, in order to launch a woman pilot into space before the Soviet Union, is a worthwhile national objective?" Victor inquired next.

"Well, sir, that is a very difficult question to answer," Jackie began. "Sure, it is nice to be first, but it is also nice to be sure. I don't think it would justify having a crash program. It would make the hard years of training these men took look a little silly, even if it succeeded. So, no, I can't quite say I think there should be."

"In other words, we should not try to launch a woman in space merely for propaganda purposes, we must be sure of the safety?" Victor continued.

"Yes, I believe that with all my heart," Jackie confirmed. She reiterated her stance that the best thing would be to undertake a large-scale program including women of various ages and levels of experience, perhaps even including women as old as herself. No such program should be done simply for the sake of beating the Soviets.

James Fulton pushed Jackie to elaborate on this hypothetical program, how many women she wanted tested and whether or not she would seek the same opportunities for women with the air force. He spoke so forcefully that Victor interrupted twice asking him to yield, but Congressman Fulton wasn't finished defending the Lovelace women's continued testing. He invoked notable women of history as proof of female strength—Malinche serving as a guide for Cortés, Queen Elizabeth defeating the Spanish Armada, Queen Isabella financing Columbus's expeditions, and Pocahontas saving John Smith's life.

"I think that was done for love, sir," Jackie answered simply. "Women will do an awful lot for that."

"Ladies can be courageous for various reasons in space," James retorted.

"I think there is no doubt women can go into space and be as successful as men, but I say I don't want to see it done in a haphazard manner." Jackie was unwavering in her conviction on this point. "In 1938," she recalled, discussing her first Bendix win, "there were fourteen pilots took off. I won the race across the board from the boys. So women can fly as well as men. But we are in a new environment. We are in a new era. Even if we are second in getting a woman into the new environment, it's better than to take a chance on having women fall flat on their faces." It was Jackie's final word on the matter.

"Thank you very much," Victor said as he ended the hearing for the day.

Janey couldn't say she was shocked by Jackie's testimony; she'd read her statement and knew how Jackie thought a woman-in-space program should be run. As she left the hearing room with Jerrie, they were met by a swarm of reporters pelting them with questions: *Why do you want to compete with the men? Are you married? Aren't you scared?* Janey hung back and watched as Jerrie faced the press yet again.

"The committee will come to order."

At ten o'clock the following morning, Victor Anfuso called the subcommittee to order for the second day. "There is no question that the witnesses who appear before us today have demonstrated by their backgrounds as engineering test pilots, and as products of NASA's astronaut training, that the criteria for choosing space pilots at this point in our national space program were wisely selected." John Glenn and Scott Carpenter, the only two Americans who had orbited the Earth, sat at the witness table. Janey and Jerrie looked on from the viewing gallery surrounded by people eager for a glimpse of their national heroes. Jerrie couldn't help but notice that the two star witnesses weren't scientists, they were jet test pilots, and though she'd never met either of them she couldn't imagine they would be willing to share their prized status with women. She wished a scientist would testify, but the third man appearing on NASA's behalf was George Low, director of spacecraft and flight missions in the Office of Manned Space Flight. Jackie wasn't there at all; she didn't think sitting through the second day was worth her time when she could just read about it in the official *Congressional Record* later. After his opening remarks, Victor called on George to testify first.

"My statement this morning covers the qualifications that we have set in our current program for the selection of astronauts," he began. The qualifications for astronauts had been determined after lengthy consideration, he explained, adding that there was ample motivation for NASA to make the right decision because the success of America's manned space program hinged on the capabilities of these astronauts. "In many ways, manned spacecraft can be considered as a next generation of very high performance jet aircraft." Both jets and spacecraft have a similar number of onboard systems, he explained, as well as life-support, power, and fuel systems, and both fly at significant altitudes and velocities. "Thus, there is a logical reason for selecting jet test pilots," he explained. "In our manned space flight program, we are in a similar situation as in the early development flights on a new aircraft. Each spacecraft differs slightly from previous ones. Procedures are modified and improved from flight to flight. Test pilots are trained and experienced in just this type of work." NASA was still learning, and it needed knowledgeable eyes and ears in the spacecraft to give feedback to the engineers.

"Is the decision to take candidates from civilian employment something new?" Victor asked, referencing the agency's announcement that it would be taking nonmilitary pilots.

"This is a new provision," George said, adding that the need for test flight experience was still part of the basic qualifications.

"In your experience," Victor continued, "and from the experience of Mr. Shepard and Mr. Carpenter and Colonel Glenn, and other astronauts, is it necessary that he be a test pilot?"

"I have tried to answer that, Mr. Chairman, in my statement," George said. "Perhaps I should let one of the gentlemen with me clarify this."

"Colonel Glenn,"—Victor turned to address John—"what do you say? Do you think an astronaut or crewmember necessarily has to be a test pilot?"

"First," John began, "let me preface my remarks by one statement. I am not 'anti' any particular group. I am just pro space." He wasn't leaving anything up for interpretation. "Anything I say is towards the purpose of getting the best qualified people, of whatever sex, color, creed, or anything else they might happen to be."

"You are not against women. You are a married man, you have children." Victor wasn't asking a question, he was affirming John's status as an all-American hero.

John went on to echo George's sentiment that flying in space was an experimental job that required a certain background of technical skill and understanding, and it was only going to get more complicated and dangerous leading up to the Moon landing. Astronauts wouldn't be passengers on these missions, they would be an integral part of the machine. They would monitor systems, analyze the spacecraft's performance in flight, and be ready to take full control with a cool head if something went wrong. "This is not to say that no one else could be trained to do it," he allowed. "However, the test pilot program is built around people who continually demonstrate the emotional, physical, and mental stability to do this." Here, spelled out plainly, was the issue that Jerrie kept hoping would go away. Women pilots simply did not have the required years of jet test experience that the men did, experiences John explained were vital to the job.

Pursuing the question of background experience, Congressman

George Miller asked John how long it took to become a test pilot. John hazarded a guess: about two years.

"I had in mind," Congressman Miller turned to Victor, "that if we have to take people as astronauts who are not test pilots, and if it then takes two years to train them as test pilots before they can become astronauts, this shortens the span in which they can serve as astronauts." Maybe, he ventured, it made sense to wait until women could qualify as test pilots before allowing them into the astronaut corps so they could serve a full term.

"I would like to add one thought." Scott Carpenter spoke up for the first time. "I believe that there is nothing magic about a test pilot, although they have had the benefit of training and experience. The best reason for selecting test pilots for this job, I believe, is that they have had the opportunity to demonstrate that they have the capabilities required of the job by reason of the fact that they have been employed in the past in the profession that most nearly approximates spaceflight. Our training as astronauts really began when we began flying." In this way, he said, NASA was taking advantage of as much as fifteen years of experience with each astronaut.

Watching from the viewing gallery, Jerrie noted that the hearing was finally getting to the issue of "equivalent experience."

Victor turned back to John. "Colonel Glenn, from your experience—by the way, are you an engineer?"

"Not a graduate engineer, no," John admitted. "I was taking engineering in college."

"You have engineering experience?"

"Yes, sir."

Here was the loophole. Jerrie scribbled a quick note and slipped it to Janey. *Our group, average flying hours, 4,500. Male astronauts, 2,500. How's that for jet test equivalent?* Janey smiled as they wordlessly agreed that the case for women in space looked like it was in good shape. Neither acknowledged that their flying time was in propeller planes, not jets. Compared to the astronauts' jet test experience, it was like comparing their daily commute with a Formula One driver in a race.

"These qualifications that NASA has set out apply to women as well as men, do they not?" Victor addressed this question to George. "I want to get that point clear." They did, George explained, though no woman had yet met them. Victor pressed on. If civilians could now qualify as astronauts and women could be trained as test pilots, could a woman theoretically be selected as an astronaut? "Miss Cochran, for example, is a test pilot?" he asked.

"Yes, Miss Cochran is an outstanding example," George confirmed, adding that he didn't know another woman with her same outstanding qualifications. But if a woman met the qualifications, NASA would certainly consider her for the astronaut corps. "We are certainly not opposed to anything like that in the future," George said.

The future seemed to be where Jerrie, Janey, and the other women pilots belonged.

"One more point," George resumed. "The equipment available for training pilots for our flights, the centrifuges, the vacuum chambers, all of this equipment is very loaded up at the present time."

"That is the best point you have made," Victor said, pleased that George had brought it up himself. "In other words, you are not objecting to women, but at the present time, to let them use the things that you are using now for the astronauts would be interfering with that program."

"We would be interfering with the current program," George affirmed.

At this point, Congressman Joseph Karth returned to the idea of relevant experience as it related to jet flying. "Yesterday two of the witnesses spoke very strongly about this qualification. They felt, quite frankly, that an extensive number of logged hours in actual flight compensated for all of the variables or invariables and the emergencies that one might meet as a test pilot. Therefore the test pilot requirement was not a fair one, because it ruled out too many people who normally get this training if they had logged a great number of flying hours." Then he turned to Scott. "Commander Carpenter, would you care to remark on that aspect of the testimony that we received yesterday?"

"I feel that this analogy might be valid," Scott offered. "A person can't enter a backstroke swimming race and by swimming twice the distance in a crawl qualify as a backstroker. I believe there is the same difficulty in the type of aviation experience that thirty-five thousand hours provides a civilian pilot and the experience a military test pilot receives." Thousands of hours as a commercial pilot didn't equate to hundreds of hours of high-speed test flying.

"I disagree basically on your approach," Congressman James Fulton jumped in, again rising to the women's defense. "I believe that space is not an experiment or adventure. I think it is a new area where everybody will operate. Under those circumstances, when women are paying the taxes here, as much or more than the men, I don't think they should be kept out of space because of rigid requirements." He returned to the one point that gave Jerrie hope. "On the basis of the requirements that Mr. Low has stated, obviously Colonel Glenn would have been eliminated. You wouldn't have passed, because you don't have an engineering degree, do you?"

"I have one now. I did not at the time of selection," John explained.

"So we can't look at these methods of selection and requirements as rigid. They must be variable, to get various characteristics. Wouldn't you agree with that?"

John had to acknowledge that he had been an exception to the rule, but not without cause. "My background at the time of the original selection, I believe, was gone into," he explained. "It was felt, with my in-service experience and the schools I had been to, while I did not have the actual hours at college, I had more than the equivalency of an engineering degree."

"If a woman, then, through her experience, and her flight experience, can give equivalent capabilities and characteristics for a good astronaut, she should not be rejected because of a requirement which she is unable to fulfill." James Fulton was not backing down. He urged the subcommittee to see what he thought was NASA's clear bias against women. "I believe that the United States should adopt a program of the first woman in space. We should set that as a national goal. I think for the world it would be a tremendous step forward," he finished. "I would hope President Kennedy would state such a program." He turned to John. "Would you agree?"

"I think this is a little out of my province, sir." The room chuckled at John's deflective answer, but James wasn't letting up. He asked the question again, this time urging John to answer not as an astronaut but as a taxpaying American.

"Will the gentleman yield," Victor called out to cool James's heated temperament. "I don't think it's fair to ask any of the astronauts that question."

"I don't want them to shy away from any question," James shot back.

"I think you might ask that question to NASA," Victor suggested. "I think you should ask Mr. Low." James did, and George's reply was evasive.

"I feel, Mr. Chairman," George answered, "that you gentlemen of the committee are much better qualified than any of us here to advise us on what the national goal should be."

"You would not object to that as a national goal—first woman in space program?" James pressed.

"I don't believe, Mr. Fulton, that I am wise enough to state what our national goals should be."

"How about Commander Carpenter?" James was determined to get an answer.

"I think at this time it is definitely an experiment," Scott replied. "There are so many unknowns and it is important for us to eliminate as many of these unknowns before the flights take place as is possible."

"But, you see, doesn't that lead you into the old question of protecting women?" James asked.

"No, I believe it is protecting our program," Scott replied.

"Against women?" James shot back.

"No, sir; not against women."

"If the gentleman will yield!" Victor tried to nip the argument in the bud, but James refused to give up.

"Jacqueline Cochran," James began, "holds more aviation and speed records than any living human being, and everybody admits her qualities as a jet pilot. Secondly, in 1959 there were the Lovelace Foundation tests at Albuquerque, New Mexico; seventy-five physical tests were completed in February 1960 by Miss Jerrie Cobb. Miss Cobb and this twelve-woman group passed these tests. Then Miss Cobb underwent the two-week series of tests at the US Navy School of Aviation Medicine in Pensacola in April 1961, and passed. Then what happened?" He addressed the room at large now. "In May of 1961 Administrator James E. Webb names Miss Cobb a NASA consultant. So she holds a position and says what is to be done but is not allowed to do it." He was only getting more heated.

"Now my feeling is this," James continued. "Since this group of women has passed these tests successfully, NASA should outline a training program that does not interfere with the current programs but will let women participate." Victor moved to cut him off, but James wouldn't let him. "If I could finish with this, Mr. Chairman: it is the same old thing cropping up, where men want to protect women and keep them out of the field so that it is kept for men." Now he turned to John. "When you go to the Moon you would want a scientist or astronomer along. Why wouldn't a woman be good company on a trip to the Moon?"

"I'm not looking for company, Mr. Congressman. I am looking for the best qualified person to do the job at hand," John answered.

James went back and forth with the NASA representatives, who remained steadfast in their argument that training women would take resources away from astronauts training for Apollo missions, so he made his appeal to his fellow subcommittee members. "I will urge

NASA, and I am sure the committee will too,"—he looked at his colleagues on either side of him—"that they carry on some kind of a parallel program, without interfering in your present program, to give these women a chance to someday become test pilots. I think the military test pilots schools should be opened to them. They should be permitted to take those tests."

"I would like to add on this physical examination program—the program run out there for some of the women at Albuquerque," John jumped in. "I think sometimes in the papers and magazines the write-ups on this have been a little misleading," he offered. "I think the tests mainly are run to see if there is anything wrong with a person physically. It isn't that it qualifies anybody for anything. It just shows that they are a good healthy person... A real crude analogy might be: We have the Washington Redskins football team. My mother could probably pass the physical exam that they give preseason for the Redskins, but I doubt if she could play too many games for them."

"You picked a bad team," Congresswoman Jessica Weis replied playfully. "Maybe she could."

The whole room chuckled at the idea.

"I think this gets back to the way our social order is organized, really," John finished with a simple summary. "The men go off and fight the wars and fly the airplanes and come back and help design and build and test them. The fact that women are not in this field is a fact of our social order. It may be undesirable. It obviously is, but we are only looking, as I said before, to people with certain qualifications. If anybody can meet them I am all for them."

"I think Colonel Glenn hit in his statement on the exact differences of opinion which exist here," Congressman Joe Waggonner said, in praise of John's analysis. "This program is developed to this point because of the differences in our social order which time has laid down for us." The men then went back and forth passing the buck on who cancelled the Pensacola testing without reaching any conclusions.

Congressman Walter Moeller from Ohio jumped into the fray. "If today our priority program is getting a man on the Moon maybe we should ask the good ladies to be patient and let us get this thing accomplished first and then go after training women astronauts," he suggested.

George agreed. "I think that even Miss Cochran yesterday in her testimony had this very same point of view, that a training program for women should in no way interfere with our program."

James Fulton remained the lone voice in support of Jerrie and Janey's cause. "That means that if we are going to the Moon within this decade, it is telling these women they are going to wait ten years," he said. "That is the same thing that has been said to women when they were interested in suffrage, or when they were interested in planes." He tried one last time to argue that launching history's first woman into space was a coup against the Soviet Union, but the subcommittee remained unconvinced.

Victor finally made his recommendation to NASA on behalf of the subcommittee. "Go back and talk to Mr. Webb and Dr. Dryden and come up with some kind of a program so that you can continue to have the bipartisan support which you have always seen and enjoyed," he said. "I want to take this last opportunity of congratulating our two great American astronauts who have demonstrated not only a great ability in the field in which they are engaged but also that they have assumed a personal responsibility, whether they like it or not, of demonstrating leadership throughout the world." With that, he concluded the hearing.

In the viewing gallery, Jerrie sat stunned. There was no mention of why the third day had been cancelled, and Jerrie couldn't understand it. She'd been hoping for another day to repeat her views, for another rebuttal, to offer her scientific analysis of her merits and another look at her qualifications. The room filled with chatter as the men and women around her rose to leave. She turned to Janey and saw her own shocked expression looking back at her.

CHAPTER 23

After the Hearing, July 1962

"BREAK LOOSE NOW, THE DOOR IS OPEN," BEGAN THE TELEGRAM JERRIE sent the Lovelace women almost as soon as she left the hearing. "Now you can help further our cause. Suggest you contact everyone you can requesting they wire President Kennedy at the White House, urging immediate program for women in space." It was her call to action. Though the House subcommittee hearing had shown that her flying credentials weren't up to what NASA needed, she remained convinced that pushing even harder would somehow reverse the decision. With all of their names finally public, she hoped the pilots could all band together and present a united front to the president. Coupled with the expected press coverage, she believed they had a good chance of Victor's recommendation actually nudging NASA in their favor.

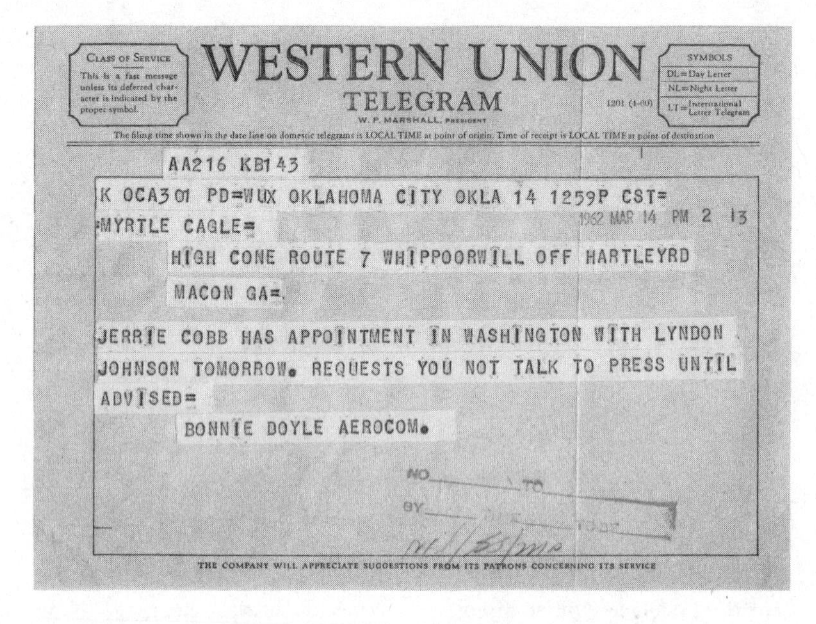

CLASS OF SERVICE
This is a fast message unless its deferred character is indicated by the proper symbol.

WESTERN UNION
TELEGRAM
W. P. MARSHALL, PRESIDENT

SYMBOLS
DL=Day Letter
NL=Night Letter
LT=International Letter Telegram

1201 (4-60)

The filing time shown in the date line on domestic telegrams is LOCAL TIME at point of origin. Time of receipt is LOCAL TIME at point of destination

AA216 KB143

K OCA301 PD=WUX OKLAHOMA CITY OKLA 14 1259P CST=
1962 MAR 14 PM 2 13
MYRTLE CAGLE=
HIGH CONE ROUTE 7 WHIPPOORWILL OFF HARTLEYRD
MACON GA=

JERRIE COBB HAS APPOINTMENT IN WASHINGTON WITH LYNDON JOHNSON TOMORROW. REQUESTS YOU NOT TALK TO PRESS UNTIL ADVISED=
BONNIE DOYLE AEROCOM.

THE COMPANY WILL APPRECIATE SUGGESTIONS FROM ITS PATRONS CONCERNING ITS SERVICE

International Women's Air and Space Museum

* * *

"Barring women unfair, unwise, and un-American; third aviatrix urges caution." Press coverage began the following day with short articles about the subcommittee hearing buried midway through national newspapers. They told the public the names of all the women who successfully took the medical tests at the Lovelace Clinic, that the experts' testimonies had disagreed on key points, and that all three female witnesses were blond.

Jerrie and Janey were lauded for fighting on behalf of women everywhere. Stories focused on Jerrie's statement that she didn't want to slow down the nation's space program, she just wanted women to start training so they would be ready when the time came. Many journalists agreed, editorializing that society ought to cease frowning on women who sought to combine family life with a career. The press also picked up Jerrie's assertion that she represented the whole group who had taken the Lovelace medical tests; she and Janey were described as leaders of an "intrepid band of would-be lady astronauts," courageous women who were taking on the space agency that had cancelled their navy tests two days before they were due to start. Other writers picked up Jerrie's comment about Glenda the chimp and suggested that trim and tailored Jerrie "had every reason to be satisfied with what she was" and shouldn't draw a comparison between herself and a chimpanzee. Jerrie was described as the lone invitee to a testing program that was only expanded after her successful completion of the medical phase. Robert C. Ruark wrote in the *Telegram* that the requirement for astronauts to be pilots was little more than splitting hairs. "I am inclined to swing along with Jerrie Cobb," he wrote, "who is practically rarin' to jostle a star." The *Daily Oklahoman* labeled its home state hero "the Space Suffragette."

They weren't the only "lady astronauts" mentioned. With their names now known, journalists tracked down the other Lovelace pilots. "I've been saying this for two years," Wally Funk told the *Los Angeles Times*. "Just let Russia send one woman into space and the United States will launch a crash program to do the same... and I understand Russia is training women for space missions."

Jackie, meanwhile, was portrayed as the villain. Though every article acknowledged that she held "more speed, distance, and altitude records than any other living person," she was criticized for her "go-slow approach in licensing women for spaceflight." Her comment about attrition due to marriage was cited in nearly every article alongside her statement in favor of a large research program to avoid the spectacle of women falling flat on their faces. Phyllis Battelle writing for the *San Francisco News Call Bulletin* went so far as to suggest that Jackie was trying to keep women out of space to keep her crown as the foremost flyer in the country.

The men's reviews were as mixed as the women's. George Low was described as "chivalry itself as he skated delicately around questions as to why women aren't in the program." James Fulton, on the other hand, was mocked for his inconsistency. "Just about every famous female except Cleopatra was brought into the act yesterday on behalf of women as astronauts," one reporter wrote, adding that the same congressman directed his "tongue in cheek barrage" at Jackie, the best-known woman pilot. John Glenn emerged every bit the national hero, the press asserting that the Mercury astronaut "would welcome a qualified woman aboard his spacecraft 'with open arms.'" Some articles opined that women were just trying to invade the last male frontier of space. Male and female writers were divided, some believing that there were enough men to fill out NASA's astronaut corps and others supporting seeing women in space. Others were just pleased to know that not a dime of taxpayer money had gone to testing the women.

None of these articles offered an honest discussion of the stark differences in background and experiences between the Mercury astronauts and the female pilots.

Jerrie wasn't willing to let the hearing be the last word on her dream of flying in space. On the morning of July 20, she submitted an additional statement for the hearing's official record; she felt compelled to do something in light of the hearing's final day being cancelled. Then she sent a telegram to Jack Kennedy's office: "Respectfully request appointment to discuss women in space program."

The telegram was routed from the White House to NASA with the request that Jim Webb acknowledge or otherwise suitably handle the request.

Two weeks later, Jim's reply told Jerrie that the president's schedule "is such that it is impossible for him to grant you the appointment which you requested," and furthermore the agency's views on women in space had been thoroughly examined during the subcommittee hearing. But, Jim wrote, he would consent to discuss the matter with her further "if you think it would serve any useful purpose, and a mutually convenient time can be arranged."

Jerrie jumped at the chance to continue her conversation with NASA. She wrote back immediately saying she was willing to travel to Washington on a moment's notice to "discuss, work, serve, assist, train" or do anything that would expedite getting a woman into space. In her fervor, she revealed her motivation wasn't to start a women's astronaut program. Her aim was purely personal; the only woman she cared about getting into space was herself. "If I went back to college and got an engineering degree and managed some way to get some jet test pilot experience, could you tell me if I'd be acceptable as an astronaut candidate then?" she asked before making another appeal. "If you would just give me a chance to work out on the simulators when they are not busy, then we would know if I needed to go back and get the degree and jet test time. I beg of you just for the opportunity to prove myself."

The press's tone in covering the hearing irked Jackie. All the articles read as if the writers didn't understand the dangers of spaceflight, the experimental nature of Randy's medical program, or that forcing a crash program would do the women more harm than good. She was particularly irritated with Phyllis Battelle for her unfair piece that overlooked the fact that she, Jackie, had funded the women's medical testing and was footing the bill for Pensacola, too.

Even Floyd was aggravated, not only with the media's unfavorable portrayal of his wife, but also with some of the things Jerrie had said in the hearing. Reading the transcript of her statement, he penciled a list of questions on scratch paper he'd like to ask her on the record. "Were you selected as an astronaut officially by NASA or by any other official body?

Who requested you to take the medical test at Albuquerque? Who paid for the tests? Did you sell a story to *LIFE* magazine based in large part on such tests? Do you have any present arrangement with *LIFE* magazine or any other magazine as to future publicity concerning yourself or the subject of women in space?"

Jackie was more concerned with making sure she hadn't misunderstood Jerrie's role in all this. She called Randy, who confirmed over the phone that not only were some of Jerrie's statements poorly worded, a number of them were not factual. At no time had he and Don Flickinger planned to test only one woman, and though he hadn't overseen it himself he knew the three girls who had done psychological testing hadn't done the correct tests. As far as he knew, the girls had neither met as a group nor agreed to put Jerrie forward as their spokeswoman as she implied in her testimony. Confirmation on this last point left Jackie dismayed. She couldn't understand Jerrie's audacity in presenting herself as the group's spokeswoman. She knew opinions among the women varied; Gene Nora's letter to Jackie saying "I'm on your team" was evidence enough that Jerrie didn't speak for everyone. She decided to see for herself what the women had to say.

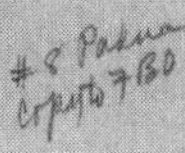

The Lovelace Foundation

For Medical Education and Research

4800 Gibson Boulevard, S.E.
Albuquerque, New Mexico

Director of Research
Clayton S. White, M.D.

Business Administrator
J. R. G. Fulcher

July 16, 1962

Jacqueline Cochran
630 Fifth Avenue
New York 20, New York

Dear Jackie:

Many thanks for sending me your testimony. There have been many favorable comments in Washington about your testimony in contrast to other testimony.

In answer to the letter of July 26, I have the following comments:

I At no time did Don Flickinger and I plan to just have one woman pilot checked. As you know, twenty girls were finally examined and of these thirteen passed. As far as I know, the other twelve girls have never met as a group and asked Miss Cobb to be their spokesman.

II There was no publicity because no scientific report of the results of the testing has been published as yet.

III Again, I question her spokesman idea.

IV We have not determined at all the ranking of the various thirteen candidates that have passed the tests here.

IX The girls have not been given the necessary mental and psychological tests as yet.

XI NASA is the agency that would have any decision making powers as to future tests.

Perhaps the best result of the testimony will be that a long range test program for women can be initiated.

Sincerely yours,

Randy

W. Randolph Lovelace II, M.D.

WRL/jw

Eisenhower Presidential Library

* * *

"As you probably know, there were some hearings before a House Committee of Congress last week on the subject of whether women have been or are being discriminated against in the astronaut program."

Each of the Lovelace women got the same letter from Jackie in the first week of August. She told them that she, Jerrie, and Janey had testified, and since she had sent them all her March 23 letter to Jerrie, her inclosure of a copy of her letter to the counsel for the committee didn't tell them anything they didn't know. "I spoke only for myself without claim to represent any of the women who passed the Lovelace tests," she told them. "Miss Cobb stated she was the spokeswoman for all of the ten not present. I am satisfied this is not true as to some who had already expressed their views to me. Was she authorized to act as spokeswoman to you? I'll be glad to hear from you."

B Steadman was shocked when she read Jackie's testimony. She knew full well that no one had promised them a flight in space, but she was nevertheless disappointed that Jackie hadn't taken advantage of a room full of congressmen to stand in defense of women's abilities. "For Heaven's sake, Miss Cochran," B wrote in her reply to Jackie, "if this Space Program has meaning, why don't you grab the reins and get it going. It needs one leader with wisdom, understanding and the ability to get people to work together. We are without this type of leadership now." She then asked for some resolution on the future; she had an aviation business to run, so if nothing was going to come from this testing she wanted to get back to her regular life.

Janey, of course, wasn't surprised; she had read Jackie's statement and heard her testimony. She'd long had a hunch that Jackie, too old to go into space herself, was attempting to control the program lest one of the other women eclipse her. But she also recognized that Jackie had funded their little experiment. Regardless of their opposing opinions, Janey also felt the press had misinterpreted their statements, taking their words out of context and focusing as much on their hair color as their flying skills. As such, her reply to Jackie was marked by the grace befitting a senator's wife. "Knowing how busy your schedule is, I am one who can appreciate the effort you made to appear in Washington at the hearing. Isn't it amazing what a flavor the press can put into a story? I am going to try to get the transcript into the mail as soon as they are corrected and printed. In the unlikely event you have not gotten yours I will be delighted to supply you."

Some of the other women were more upset with Jerrie than with Jackie. K Cagle, for one. "In your letter you asked if I had authorized Jerrie Cobb to be my spokesman," she wrote to Jackie. "No ma'am, I have not." K had assumed that Jerrie had been *appointed* their leader since they had all been asked to direct their questions and correspondence to her. In truth, K trusted Jackie's judgment. "You are asking for what we CAN get, Jerrie Cobb is asking for the impossible. I am on <u>your</u> team."

Route 8, Macon, Ga.

Aug. 8, 1962

Dear Miss Cochran:

Your letter of Aug. 1 reached me last night. We moved into our new half a house we built ourselves for the past two years on Sat. Aug. 4th. And we are so happy!

In your letter you asked if I had authorized Jerrie Cobb to be my spokesman. No ma'm, I have not. I assumed she had been appointed team captain because the replies about our Pensacola trip were directed to her. I talked to Doctor Lovelace, Jr. ("Randy") before leaving the clinic to find out what I could or couldn't say to the press. I have compleéd with his advice. The press (Atlanta Journal) has made up some remarks of their own, even assigned me an age which I don't feel is public information. The Macon news has been very nice. They, too, however, get things twisted to suit their selves. They called me the first day of the hearings and said they had a story from Washington containing my name and wanted a comment. I told her, Miss Gertrude Traywick, that I could not make a statement, that it would have to come thru the Lovelace Clinic or Miss Jacqueline Cochran. They printed the story, used my name but did not make up any comments.

This is the way I stand. Miss Jacqueline Cochran (you!) paid my expenses while at the Lovelace Clinic. I do not know who paid the Eleven hundred dollars plus for the testing. I thanked the Doctors Lovelace. I consider you my sponsor and spokesman. I believe in you, I trust your judgement in any situation. You have had a great deal of experience in dealing with government officials and important business people. I am sure I can't even visualize the forest of important people you have influenced to your side. I like your seasoned judgement, your soft, feminine approach. You know what we can get, so why ask for the impossible and make enimies? You are asking for what we CAN get, Jerrie Cobb is asking for the impossible, I think. I wish I could orbit tomorrw and I know you do, too.

Miss Cochran, I want to fit into your plans. I know you have plans. I know you want to go into space and I know you will. I am on <u>your</u> team. I know we can get from "the men" what we want, not by pushing, but by winning. You've got the know-how. So-o-o Take the reins, we are YOUR girls.

Yours very truly,

Mrs. Myrtle Thompson Cagle

Eisenhower Presidential Library

Jan Dietrich remained as devoted to Jackie as ever and told her that "Jerrie did not ask permission to represent me at the hearing." She was angry that Jerrie had appeared before Congress as her spokeswoman without permission, giving the press a false view of where she stood on the matter. She wanted to see the testing continued, but she was worried that the antagonism brought on by the hearings would prove more of a hindrance than a help. Marion echoed her sister. "I believe Jerrie thought it necessary for her to represent the whole group," she told Jackie. "She did not speak to me about this."

Irene Leverton didn't mince words. She was angry, angry that she only heard about the hearing after the fact and hadn't been asked to submit her views for the record. "Miss Cobb is not my spokesman, officially or unofficially. I disagree with many of her views although, I do want a continued program." She felt the overall media coverage was bad for their possible continued testing. "To beat Russia with a female astronaut was very poor thinking on Jerrie's part, I'm afraid the idea hurt us all a lot," she wrote. "I'm sorry that a statement by each of us wasn't read at the hearings. Perhaps sometime in the future we women will start pulling together." Irene was mad enough to wire Jerrie after the hearing to express her anger. She also wired George Low and George Miller. She wanted it on the record that Jerrie did not speak for her. She also granted interviews to journalists, as did Jan; they both felt compelled to set the record straight after Jerrie's misrepresentation. Both Jean Hixson and Rhea Hurrle—now going by her married name of Rhea Allison Woltman—were less reactionary. Both replied to Jackie that they simply wanted to see their testing continued.

At the end of the month, the Lovelace women got yet another "Dear FLATs" letter from Jerrie. After asking each of them to send Janey a "candid and glamorous" photograph of themselves "not in pants" for an upcoming article in *Town and Country* magazine, she addressed the issue of representation. "So that we can stick together as a group there should be a central contact for all publicity on a national level. I will handle it from here if you want me to," she offered. She was still holding on to the notion that this group of pilots who had never met were some kind of cohesive group, and that their very varied opinions could somehow be brought into line with her own. "Also, there should be one spokesman for the group and I would appreciate your views on that. As my prepared statement before the subcommittee pointed out, I assumed that role more or less by default, and because I have worked three years getting this program started." Her pleas and appeals for the other women to help promote her personal ambitions were becoming increasingly desperate.

On August 12, the Soviets scored another first. *Vostok 4* launched to join *Vostok 3* in orbit, marking the first time two missions with two cosmonauts flew simultaneously. The next day Jerrie sent another telegram to Jack Kennedy. "Hope in light of new Russian space achievement you will reconsider my request to discuss with you United States putting first women in space."

Again, the White House rerouted her request to NASA asking Jim Webb to deal with her. Jim reiterated that the president couldn't see her before explaining that "the Soviet space activities of the past week do not warrant, in my judgement, a reappraisal of this agency's policy with respect to the utilization of women astronauts at this juncture in our program."

Though she was getting the brush-off from NASA at every turn, Jerrie still had some media support. That month's issue of *LIFE* magazine named her among the 100 most important young movers and shakers in the country, labeling her an "astronaut in training." The international media, too, was still keen on her story, mailing requests for official press pictures to go along with their stories. Every so often she sent copies of these requests to Jim. "Just a short note to accompany enclosures which might interest and amuse you. I hope you don't mind if I keep trying 'cause this means more to me than life itself. All I can do is keep working, trying and praying."

But America was starting to turn its attention to NASA's newest group of astronauts. Introduced on September 17, 1962, were nine men the *New York Times* told readers were "candidates for the lunar flight." Like the group that preceded them, they were all pilots with military experience and extensive test flying backgrounds. Though two among their number were civilians—Elliot See and Neil Armstrong—they were both former naval aviators who had left the service to fly as test pilots for General Electric and NASA respectively. Neil had even flown near space as a test pilot for the X-15 program, reaching a peak altitude of 207,500 feet. The experimental rocket plane was launched from under the wing of a B-52 bomber to high enough altitudes that he had to rely on the same reaction controls the Mercury spacecraft used at the peak of his flight until the plane descended enough for traditional flight control surfaces to bite into air. This was a wholly different kind of flying than any of the Lovelace women had done, so though Elliot and Neil were civilian pilots, their skills as aviators set them apart.

"We all remember the pioneering flights of America's beloved Amelia Earhart. Like many others, she gave her life to the advancement of aviation."

Jerrie spoke from the podium in the Grand Ballroom of the Statler Hilton Hotel in Cleveland before 250 people, most of them women, for the Zonta Club luncheon on November 28, 1962. The theme of the afternoon was "women in the space age" in tribute to Amelia. A handful of the older women there had probably met Amelia socially or at an air race, but only Jackie, who sat at one of the many tables set up in the ballroom, had really known her. The days they'd spent planning flights and the nights lounging in front of the fire at the Ranch seemed like a lifetime ago.

"Jacqueline Cochran," Jerrie continued her address, "has set more world aviation records than any other person in the world. In space it will be no different." Spaceflight was no childish lark, Jerrie said. Women had long contributed to the advancement of science. This was no different, she said, imploring everyone there to contribute to the best of their God-given abilities. "Having been a professional pilot for many years, I feel that I could best contribute to space research in the area of astronaut testing and training. To prove that women have the capabilities of actively participating in space flight, I had the privilege of undergoing three cases of astronaut testing." She had been appointed a NASA consultant, she erroneously told the gathered crowd, but, she said, "I'm the most unconsulted consultant in any government agency today.

"I'm certainly not a feminist, nor do I want to be a space age Harriet Beecher Stowe. Certainly I do not wish to be a heroine or a martyr, but I would willingly give my life to this purpose and count it a blessing to have served my God and my country to the utmost. With God as my pilot," she finished, "I hope to make that space flight."

Applause filled the room. Jerrie regained her seat as Jackie stood and moved to the podium before the Zonta women.

"The Lovelace Foundation for Medical Research gave the first medical checks to the large group of astronaut candidates." Jackie didn't waste time on any preamble. "This unofficial and volunteer medical research project, through publicity, editorial license, and wishful thinking became parlayed into the fairly widespread belief that there is an astronaut program for women." It was a very different version of the same story the Zonta women had just heard from Jerrie. "There is much misinformation on the subject. I know my facts." She was friends with Randy Lovelace and had been a guinea pig for his experiments for years, she told them. Floyd was chairman of the Lovelace Foundation. She was friends with the top men in the armed services, at NASA, and even the vice president. She funded the women's testing to the tune of $18,000.[28] "The misinformation that has been floating around has actually slowed down rather than speeded up the possibility of a space program for women. There is no astronaut program for women. There has never been such a program for women. No woman in this country has passed the full battery of tests that would make her a candidate for training even if such a program were planned. NASA has never nominated, selected, or approved any woman for such role." In establishing the qualifications for astronauts, she told the luncheon crowd, NASA hadn't discriminated against women, it had discriminated against most of the world save a handful who happened to meet the nearly impossible standards it wanted to participate in the grand experiment of seeing whether humans could survive in space. She herself didn't meet the standards and didn't feel discriminated against. Besides, the Moon was just the beginning. Future missions would need trained scientists and technicians of both genders. "As to women," she said as her speech came to an abrupt end, "it can well start with a program of the type I have mentioned."

Jackie's speech was far less inspiring than Jerrie's, but it resonated with decision-makers. As 1963 dawned, a copy made the rounds just as her March 23 letter to Jerrie had. She sent a copy to Jim Webb, who enjoyed her straightforward approach. She sent one to Hugh Dryden, who, pleased with her statesmanlike view of the role of women in the national space program, was happy to see that she was continuing to set the record straight. It even made its way to George P. Miller in the House of Representatives, who promptly had it entered as an appendix in the *Congressional Record*.

---

28 About $148,000 in 2019.

By the spring of 1963, Jerrie was losing her will to keep fighting, to keep pushing her increasingly lonely agenda. She'd been given answers by the experts in government, in aerospace, and hadn't liked them. Arguing for a different answer was no longer a crusade; it was an obsession. Her public appearances hadn't coerced NASA into changing its stance, so she made yet another appeal to the president. "It is difficult to write this letter knowing it will be read by your secretaries and assistants..." She didn't know how else to start the letter. She felt she couldn't do anything anymore but pour her feeling out and hope that this time, somehow, she wouldn't be ignored.

"It is a fact that the American people want the United States to put the first women in space. While NASA refuses, the Soviet Union openly boasts that they will capture this next important scientific first in space by putting their lady cosmonaut up *this* year. We could have accomplished this scientific feat last year, and even now, could still beat the U.S.S.R. if you would make the decision. It need not even be a long orbital shot, or interfere with the current space program; on a rush basis a sub-orbital shot would suffice or an X-15 flight to a 50 mile altitude. Any aerospace doctor or scientist will tell you the scientific data obtained from such an experiment would be of lasting benefit. May the Lord guide you in your decision." She packaged her letter with a scrapbook of her correspondence with NASA and several news clippings about her fight to get into space and sent it off to the White House.

Jerrie's parcel was rerouted to Jim Webb's desk, and it pushed him over the edge. He was fed up with her persistent letters, her increasingly critical tone, and that she was still passing herself off as a NASA consultant. He decided to end the matter once and for all. "It is certainly fine with everyone in NASA for you to work towards any goal that you believe in. Your association with NASA has come to an end." He was leaving no room for interpretation this time. "I did not think it would help you work towards your goal to criticize this relationship under which you had been asked to serve as a NASA consultant, and that this relationship has not proved of the value I had hoped it would. Now that your appointment as a consultant has expired, I think you should certainly pursue your goals in any way that you feel you should."

In Jerrie's stead, Jim appointed Jackie the new NASA consultant on women in space. It was official this time, with a private swearing-in ceremony on June 11 in Jim's office. Jackie was keen to avoid publicity; she didn't think any good would come of her celebrating this new appointment. She did, however, ask for copies of the pictures for her own records.

Jerrie had one more trick up her sleeve: with her old writer-friend Jane Rieker, she was releasing her memoir. *Woman into Space: The Jerrie Cobb Story* went through her teenage years learning to fly, her tumultuous relationship with Jack Ford, her meeting Randy Lovelace, and offered readers an intimate look at her astronaut medical tests. It was her last-ditch effort to draw attention to the issue by sharing her whole story with the nation. She could only hope that renewed public interest might finally force NASA to reconsider its stance on women.

"At this writing, we are waiting still," the final pages of the memoir read. "The real movers of this country are its more than 180,000,000 inhabitants, and they're going to see that an American woman has an early place in space!" It was a final plea directly to the American people, but it came too late. On June 16, just days before Jerrie's memoir was due to hit shelves, Valentina Tereshkova flew as the pilot of *Vostok 6*. The first woman in space was a Soviet.

Reading about Jerrie and the other women pilots whom the American media described as "astronaut hopefuls" and "candidates" in early 1962, Soviet scientists were interested. Particularly Nikolai Kamanin, the head of recruitment for the cosmonaut program, though his interest was political rather than scientific. He didn't want the first woman in space to be American; that, he thought, would be unacceptable to patriotic Soviet women. Moreover, he knew that the nation that launched the first woman would score a significant propaganda victory, and this alone made a female mission worthwhile. He figured that if the right woman could be trained in about six months and fast-tracked into the next Vostok mission, the Soviet Union had a real chance of launching the first woman.

From a list of pilots, skydivers, and parachutists who met the age, height, and weight limitations imposed by the Vostok spacecraft, fifty-eight names landed on Nikolai's desk. Though disappointed in the poor caliber of candidates, he didn't bother widening his search. The Vostok spacecraft the woman would fly was entirely automated, so she didn't need to be overly qualified. The most challenging part of the flight would be the landing. Vostok couldn't land softly enough to protect the cosmonaut inside, so whoever became the female pilot would, like Gagarin, eject at about 10,000 feet and land by personal parachute. With this in mind, he'd handpicked twenty-three women to put through medical tests. After three rounds of eliminations, Nikolai selected five women, all experienced parachutists and only one with any piloting experience.

The female cosmonaut candidates joined their male counterparts in March of 1962, but the transition from a unisex to coed program wasn't seamless. As the women received flight instruction, experienced centrifuge runs, and were subjected to extended isolation tests, the men remained largely hostile. The pervading opinion among those still waiting for a flight assignment was that the stunt of a female flight was doing little more than taking a mission away from a man who had earned it. A few men, however, enjoyed the social element of having women in their midst.

A year after the women started training, Nikolai made his choice: Valentina Tereshkova, a mill worker and amateur parachutist, would make the female flight. Soviet premier Nikita Khrushchev endorsed the choice owing to her favorable propaganda profile—she was single, good-looking, hardworking, thoughtful, and the daughter of a farmer who had been killed in 1940 during the Soviet-Finnish War. She was, in short, a fantastic role model for Soviet women of all ages.

Valentina was told on May 21 that she had been chosen to make the first female flight. Less than a month later, she was in space.

America was suddenly interested in women astronauts again. Valentina's flight was called "an embarrassment" more than once. Journalists reported that Jerrie could have been up first in articles that contained ads for her memoirs. Reports rehashed the hearing, saying US lady pilots were "boiling mad that Jimmy Webb turned them down." But not all the press was positive. Some people questioned whether Valentina's flight was essential or whether it was little more than a publicity stunt that had no real bearing on the space race.

News of a Soviet woman in space didn't change Jackie's mind on the issue. In light of the importance of the Mercury and Apollo programs, she still felt NASA had made the best decision. She was so confident she offered to make a statement and issue a press release in her official role as a NASA consultant on the matter. She wasn't alone in thinking the Soviet feat had little bearing on American women pilots. "I don't think Valentina's orbit is going to help us one bit," K Cagle told a local Macon newspaper that identified her as a representative for the thirteen women under consideration for a spaceflight of their own.

By November 22, Lyndon Johnson had to admit that his term as vice president had been one of the unhappiest times of his political life. He was even losing his zeal for space. The public was still enamored with the astronaut program, but support for NASA as a whole was waning. There was increased talk of space being used for national defense, and with the cost of Apollo steadily rising, JFK was thinking of scaling things back. Two months earlier at the eighteenth United Nations General Assembly, the president had asked why the United States and Soviet Union were duplicating their space technologies when they could be working together on a joint lunar mission. Nothing in space was certain anymore, and Lyndon was deeply unhappy about it. He walked down the hall to Jack's suite in their Dallas hotel and told the president that if he was going to seek reelection, he would have to find a new running mate.

Members of both teams in the hallway couldn't hear the words; they could only hear the heated exchange between Jack and Lyndon.

The next day, Lyndon sat in an open car driving behind Jack's as their motorcade moved slowly through Dallas. Out of nowhere, he heard a bang—then before he knew it one of his Secret Service men was lying on top of him, pressing him into the seat of the car. He was vaguely aware that Lady Bird was being similarly forcibly protected. In the car ahead of them, Jack Kennedy had been shot. Within hours, Lyndon was sworn in as president, suddenly thrust from the least powerful office in the nation to perhaps the most powerful position in the world.

Jack Kennedy had promised Americans a nation where government was good and the American dream was accessible to everyone. But his assassination revealed the country was more fractured than it seemed. Civil rights groups, impatient with the slow pace of change in Washington, were increasingly finding their voice. Betty Friedan wrote *The Feminine Mystique*. Wives and mothers nationwide realized they weren't alone in feeling unfulfilled. Women began speaking out against the "family values" that had been keeping them confined to domesticity, inspired to want more from their lives than being wives and mothers.

But not all of JFK's hopes risked crumbling in the wake of his death. The president's assassination added an emotional element to the lunar landing goal. Now, the space agency wasn't just following a presidential order, it was honoring the dream of a fallen president. NASA had no choice but to get to the Moon by the end of the decade, and Lyndon couldn't have cancelled Apollo if he had wanted to. Amid everything else he inherited from JFK, LBJ redoubled his support behind the lunar landing goal.

Three months later, as Lyndon was settling into his new role, a letter from Jerrie landed on his desk. In it, she urged him to reconsider the question of women in space as it was now a matter of catching up to the Russians. "If it is absolutely impossible to phase women astronauts into the Gemini or Apollo projects at this time," the letter read, "would you please consider the possibility of letting a woman pilot work in the X-15 research project and, if competent, fly the X-15 to the fringes of space?" He could tell she was trying any means necessary to get into space. Inclosed was a copy of his own letter from two years earlier, the one wherein he said he had no problems with women in space. As Jack Kennedy had done so many times, Lyndon passed the responsibility of answering off to someone else.

The response, signed by the president but written on his behalf, left no room for interpretation. NASA's decision to select test pilots as astronauts was as valid now as it had been in 1959, and the first likely "modification of this policy will be made to permit properly qualified scientists to serve as crew members during Project Apollo missions." As for her request to fly the X-15, this was patently out of the question. "From your own experience, I believe you are aware of the exacting demands placed on the pilots of extremely high-performance jet aircraft. These demands, however, are at least an order of magnitude less than the demands placed on the X-15 pilots." There was no chance NASA would let a pilot of propeller planes join a highly specialized program to fly this rocket-powered experimental plane to the fringes of space.

The letter ended reminding Jerrie that she had special skills and qualifications that could be used in other ways to help the country through any number of worthwhile organizations and encouraged her to pursue these options.

Jerrie walked along the beach in Jamaica. She'd needed an escape after everything that had happened in Washington and the final rejection letter from the president, so she had returned to the island where she'd once been so happy, but she couldn't stop the thoughts running through her head. She'd lost her fight to get into space. She was nearly broke from traveling so much in pursuit of that mission. She had lost her first love when Jack Ford had died, and now she felt her second love—her passion for spaceflight—had been stolen from her. She did have "special skills and qualifications," as the president had told her, but none of that felt important now that she didn't have a purpose. She couldn't understand *why*. Why had her loves been taken from her? Why had she spent years devoting herself to spaceflight only to have it lead to nothing? She had always believed in God, but she couldn't understand why He would take everything from her when she felt like she was doing everything right.

Looking around the secluded beach, she briefly thought she could happily live out her days in a hut, a hermit hidden away from society on this beautiful island. She had loved this island since she and Jack had shared their first kiss in the surf all those years ago. Thinking of Jack, she realized that when she had wanted to slow down and he had told her she would miss flying too much, he had been right. She was a pilot, and she needed to fly.

Slowly, like it had happened during the international air race all those years ago, Jerrie realized she was starting to make plans. Long before taking the Lovelace medical tests, she'd nursed a secret ambition of working as a missionary pilot. She'd always been deeply religious, so combining her love of flying with her love of God felt natural. If she could get a plane, she thought, she could start learning to navigate through the Amazonian jungle, learn how to live in this new world, and create a new life for herself. She began to see a very happy new chapter of her life just over the horizon.

On the morning of May 11, 1964, Jackie sat in the cockpit of the F-104G Super Starfighter thirty-five thousand feet in the air. The Lockheed-built cutting-edge supersonic fighter plane had never been flown by a woman, so Jackie had gotten herself hired as a Lockheed test pilot so she could use it to break her own 15–25-kilometer straight course record; she was determined to protect her status as the fastest woman on Earth as long as she could. Chuck Yeager had reprised his role as her teacher, much to the chagrin of the young male pilots fresh from the military. At fifty-eight, they grumbled, she was little more than an "old dame" trying to do their jobs. She'd won them over with picnics of home-cooked fried chicken, and once she got in the air, they were forced to admit that she could outfly all of them.

From her perch high above the ground, Jackie stared down the barrel of an invisible rectangular tunnel 300 feet tall, a quarter of a mile wide, and twenty miles long, a bounding box set by radar and ground instruments monitoring her speed and position. Her goal was to fly through that box without touching a single invisible wall. If she did, her speed record would be invalidated. It was a kind of precision flying some of the young men couldn't do; her decades of pushing the limits of speed in the air gave her the experience she needed to respond to things happening in fractions of seconds. Eyes glued to her instruments, she followed the precise direction of the space-positioning officer watching from the ground, guiding her through the corridor. She flew straight, level, and fast through that tunnel on two subsequent passes. Below her on the ground, two sonic booms shook the earth.

A week later, her records had been checked and validated. She'd averaged 1,429.297 miles per hour, smashing her old record and reaching a new top speed for women of Mach 2.2, a little over twice the speed of sound. "Successful businesswoman, happy housewife, confidante of presidents, royalty, and military leaders," Lockheed's press bureau described her, "Miss Cochran continues as the reigning feminine flier of the world."

EPILOGUE

Somewhere in the Amazon, Christmas 1971

SEVENTEEN-YEAR-OLD JULIANE KOEPCKE LOOKED UP AT THE CANOPY OF THE Amazonian jungle overhead. It took a moment before she remembered the turbulence that had shaken LANSA Flight 508, remembered her mother panicking next to her, and the whistling sound of wind rushing past her as she fell to Earth, hanging upside-down still strapped to her seat. As the scenes came flooding back to her, Juliane realized she couldn't hear anything but the sounds of the jungle. The horrifying realization dawned on her: she was the only survivor of the plane crash. She also knew the jungle could be as dangerous as what she had just endured. Rather than wait for help to find her, Juliane grabbed a bag of candy from the wreckage and set off, ignoring the pain of a broken collarbone and severely injured knee.

Ten days later, Juliane woke to the sound of male voices. Woodcutters had stumbled on her sleeping spot. Their shock at finding a young woman alone in the jungle doubled when she told them what had happened. The men took her to their village, where they gave her rudimentary care, but Juliane was grateful all the same. At some point while she was resting, an American pilot named Jerrie Cobb arrived. Jerrie offered to take Juliane to a nearby missionary site where she could be treated by proper doctors and have a safer place to rest. Juliane consented to go with Jerrie; nervous though she was about flying again so soon, she knew she needed better medical attention.

Boarding Jerrie's plane that afternoon, Juliane's anxiety hit a peak. Jerrie turned to her and offered what she thought were calming words: she assured Juliane that she was "the first woman in the world to be trained as an astronaut," so flying with her was akin to "flying in the arms of an angel."

Juliane, scared and eager to get back on the ground, didn't give this oddly self-aggrandizing statement a second thought.

Like an international game of broken telephone, the true account of Juliane's rescue was distorted by the time the story reached North America. Reports said that Jerrie had located LANSA Flight 508 on her own, that she'd recovered Juliane directly from the wreckage before transporting her to a missionary clinic in Yarinacocha, Peru. It was the latest in a growing list of exploits reaching the United States about Jerrie's activities in South America. In the late 1960s, she'd gained a reputation as a pilot willing to fly through dangerous situations to bring necessary medical aid to those in trouble in the Amazonas. She'd even located tribes thought lost. Increasingly seen as a humanitarian and hero, Jerrie gained fans in the United States, among them US Air Force Lieutenant Edwin J. Kirschner, who put her name forward to the Harmon Trophy committee for the 1971 award.

Floyd was alone at the Indio Ranch when he received a pile of mail Jackie sent home from the hospital in Albuquerque; the Lovelace Clinic remained her preferred medical facility even after Randy and his wife died in a plane crash in 1965. Jackie hadn't asked Floyd to reply or otherwise do anything with the correspondence. Rather, she just wanted to have it at home for when she returned. Nevertheless, Floyd sifted through the pile to make sure there was nothing pressing and stopped when he came across a note from General Jimmy Doolittle. It was addressed to the Harmon Advisory Committee members, of which Jackie was one, and included the list of possible recipients for the 1971 award. Floyd was familiar with the Harmon Trophy since Jackie had won it seven times, and he also knew how his wife felt about Jerrie's nomination. Feeling compelled to give his wife a voice, he took it upon himself to write to his friend Jimmy.

"I know what Jackie had in mind about Jerrie Cobb and she did not wish to commit it to a paper that would be generally passed around," Floyd's letter began. He relayed that Jackie didn't really know what kind of work Jerrie had been up to in the jungle, and though it was certainly courageous of her to rescue a survivor, it wasn't remarkable. But her issues were more complex than that.

"Miss Cobb's attitude and actions several years ago in connection with her posing as a woman astronaut, accepted or about to be accepted as such, Jackie thought were thoroughly unjustified, thoroughly uncalled for and harmful in effect to women pilots." Jackie, Floyd continued, knew the real story, and writing on his wife's behalf, he finally relayed the truth about the Lovelace testing. "Miss Cobb on her own application went to the Lovelace Clinic and asked to be given the same medical tests." He explained that she failed, and even though medical tests were but a small part of the qualifications to fly in space, she was so determined she returned months later, at Randy's invitation and after undergoing a fitness regimen, to retake the tests. This time she passed. Months later, Randy mentioned her results as a matter of interest in his presentation in Sweden, and this subsequently "broke into the press and in the course of time was quite blown up to the point where a great number of people thought Miss Cobb had been accepted for the astronaut program." It was Jackie, in an attempt to bring order to this would-be program, who funded the other women's testing and was working on getting them all to Pensacola for the second set of tests as a group for the sake of determining whether women would be fit for spaceflight; she knew representation mattered even though America wasn't ready for women astronauts at the time. But somehow Jerrie got there first and did the Pensacola tests herself, after which point "whoever was in charge of this work in Washington called the whole thing off for all the women in view of the poor showing that Miss Cobb made of these further tests."

Noting Jackie's opinions on the matter, the Harmon Trophy committee members ultimately agreed that Jerrie's flying activities were more than enough to earn her the award. Furthermore, in reading the letters she sent home and reports about her work, her rescue of Juliane was just one example of her hazardous job of flying as a relief pilot through the Amazonian jungle. Jerrie traveled to Washington for the luncheon on September 20, 1973, at the International Club, where President Nixon presented her with the Harmon Trophy.

Jerrie was happily settled in her dual-country life by the time she won her first and only Harmon Trophy. From her small bungalow in Florida, she could fly from her own airstrip to South America where her work as a relief and missionary pilot continued funded by the not-for-profit Jerrie Cobb Foundation. Flitting between North and South America, she was never out of the news for long. In 1981, someone nominated her for the Nobel Peace Prize, bringing her a momentary surge of popularity. In 1983, after Sally Ride became the first American woman in space, Jerrie was in the news again, this time as the "foiled astronaut." In 1995, a *Dateline* segment on the Lovelace women brought the story to a new generation. For this retelling, producer Jim Cross gave the women the moniker "the Mercury 13" as a play on NASA's Mercury 7, but failed to mention that Jerri Sloan had dropped out of the Pensacola testing, and positioned Jackie as the villain who kept them grounded. The segment also explained that NASA had run the tests because it was curious about women's fitness, effectively cementing in the public's psyche the mistaken idea that they were indeed trained as part of an official program.

In 1998, when NASA announced that a now seventy-seven-year-old John Glenn had a spot on the space shuttle crew of STS-95 as a mission specialist, Jerrie reignited her thirty-five-year-old quest of fighting for a mission in space. She returned to the United States as fired up as she had been in 1962. At sixty-seven, she still wanted to fly in space and figured if NASA was sending John up to gather data on one senior citizen, surely two data points—one male and one female—would be even more valuable for the agency.

Jerrie was back making headlines. For months, the public bombarded NASA with letters and petitions urging it to let Jerrie fly. Letters came from the National Women's History Project, the American Association of University Women of California, and congressmen including Republicans James M. Inhofe and Don Nickles of Oklahoma and Democrats Barbara Boxer and Dianne Feinstein of California. Jerrie's campaign got all the way to the White House. First Lady Hillary Clinton took a stand in Jerrie's defense and even got her a meeting with NASA Administrator Dan Goldin about a possible flight. The media fell in love with her all over again, and the narrative this time was for NASA to right its decades-old wrong. But it didn't work. Whatever press coverage she got, Jerrie wasn't a national hero; no one flying in space was on the same level of John's return to orbit. Even when NASA briefly considered opening seats to passengers on shuttle flights, it never offered her the commanding role she'd coveted since the idea first got in her head in the fall of 1959. Her spaceflight dream, she felt, died a second time.

Jerrie did, however, see the first woman fill that role. On July 23, 1999, thirty-seven years and five days after the congressional hearing, Jerrie watched Eileen Collins, the second woman in the country to graduate from a test pilot school, launch as the first female mission commander.

Jerrie became increasingly reclusive in the years after Eileen's flight. Except for a handful of interviews for articles or books about her life as an "almost astronaut," she lived her days out in near-obscurity until her death in 2019.

Jackie never saw an American woman fly in space.

By the time Jerrie was awarded the Harmon Trophy, Jackie's own flying days were over. Two years earlier she'd decided it was time to add a new class of records to her impressive résumé. She wanted to fly a helicopter, specifically a new Lockheed design that promised to break records. No one expressed any dismay over this ambition; everyone in the flying world knew that if any senior citizen was going to break a flight record, it would be Jackie. To teach her to fly this new aircraft, a friend had brought a Lockheed model to the Ranch for private lessons. Everything was fine until the end of her first solo flight. She brought the helicopter down hard, bouncing up and crashing hard enough to break the landing gear right off. In her whole flying career, it was the first time she'd ever suffered a medical issue in flight. She'd started having seizures.

Seizures were the latest in an ongoing list of medical problems. She'd had ongoing issues with persistent abdominal adhesions after her teenage appendectomy, eye surgery to regain her depth perception, and foot surgery in England. At the Paris Air Show in 1971, she'd been struck with chest pains she assumed were symptomatic of pneumonia. In 1972, she'd suffered a severe heart attack. Over the years, she'd miscarried the two children that she and Floyd would have loved to raise. By and large, she kept the severity of her illnesses from Floyd; his arthritis had become so crippling over the years that he couldn't even shake hands to close a deal anymore, and she never wanted to add to his worries when she knew deep down that she was fine. But her own health was getting so bad she couldn't pretend everything was fine.

One night, after suffering another seizure, Jackie decided it was time to go to the Lovelace Clinic. Seeing his wife off into the car with their driver, Floyd pleaded with her in a whisper, "Oh, Jackie, don't die and leave me."

She suffered three more seizures on the trip to Albuquerque. When she got there, doctors found the source of the problem was her heart. Her pulse had been dipping so low the loss of blood to her brain was causing the seizures. The solution was a pacemaker.

Talking about flying and even looking at airplanes became painful. Adding to her anguish, the life she had grown accustomed to with Floyd was also coming to an end. Floyd had lost a fair amount of money to some bad business deals, and Bruce Odlum, his son from his marriage to Hortense, was taxing their already strained financial resources with bad business deals of his own. In the late 1960s, Floyd and Jackie were forced to sell their Manhattan home and retire full-time to the Ranch, where they kept up appearances and entertained as many famous friends as ever. Ike Eisenhower and Lyndon Johnson held a tête-à-tête there in 1968, and Ike even moved into one of the guesthouses to write his memoir. But now the Ranch was the only asset they had left, and they had no choice but to sell it. Bruce managed the sale so poorly he took his own life rather than face the debt he'd accrued. Floyd and Jackie were left with barely enough money to move into a small house across the street from their former compound.

When Floyd died in 1976, he'd named Jackie and Chuck Yeager co-executors of his estate. As he helped Jackie sort through all of Floyd's papers and financial records, Chuck found an envelope sealed with a quarter set in wax. It was the letter that Jackie had given Floyd before their marriage forty years earlier, the one she'd told him contained the true story of her past.

"Do you want to open this?" Chuck asked.

"No," Jackie replied.

"Is there any reason to keep this letter around?"

She didn't say anything. Chuck burned the still-sealed letter.

Jackie's health declined rapidly after that, and her mood took an irreversible turn. The woman who had always gotten her way couldn't keep aging at bay. Even when heart and kidney failure left her so swollen and uncomfortable she had to sleep upright in a chair, she drank heavily and made sure visitors knew what trophies she'd won.

In a final act, she organized her own burial down to the simple flat headstone in a small Indio cemetery. In the process, she had the headstones marking her family's graves in DeFuniak Springs replaced, except for Robert Jr.'s; that one she left black with age.

Right up until her death in 1980, Bessie Pittman remained Jacqueline Cochran's secret.

APPENDIX

AERO COMMANDER, INC.

BETHANY, OKLAHOMA

March 8, 1962

R7

Miss Jacquelin Cochran
435 E. 52nd Street
New York, N.Y.

Dear Jackie:

It was wonderful to have the opportunity to become better acquainted and I truly enjoyed the evening with you at Cocoa Beach. It was nice of you to invite Mrs. Rieker too, and I know she enjoyed the evening also.

After the Glenn shoot I flew to Los Angeles to be the luncheon speaker at the First International Women's Space Symposium. It was a fine meeting and, I think, did much to encourage high school students to pursue science careers.

I still feel that it is important for the United States to continue the research testing on womens' reactions to various stresses, and to put the first woman in space. Only during your WASP program, were any medical baseline studies made.

I trust you had (or are having) a good trip in Europe and I hope we'll have the chance to visit again soon. May the Lord bless and guide you, and with every good wish, I remain

Sincerely,

Jerrie

Miss Jerrie Cobb, Ass't. to
Vice President Marketing

(Sent copy to FBO at apt.)
7.W

SUBSIDIARY OF ROCKWELL-STANDARD CORPORATION
Coraopolis, Pennsylvania

Eisenhower Presidential Library

March 23, 1962

Miss Jerrie Cobb
Aero-Commander, Inc.
Bethany, Oklahoma

Dear Jerrie:

Thanks for your letter of March 8.

I also enjoyed the evening with you and Mrs. Rieker at Coca Beach.

You asked me several times that evening how I felt about a women in space program and because I tried generally to express my views when the question was first put to me, I guess I was not communicating very clearly. Perhaps it's best that I place my views on record.

Women will travel in space just as surely as men. It's only a question of when. Women have engaged in all phases of aviation to date, including ballooning, gliding, parachuting and powered flight. There is no reason to believe that women will be eliminated now that we are leaving the atmosphere and getting into space.

There are quite a lot of women who would like to be the first woman or the first American woman to go into space. I feel confident that this is true with respect to each of the twelve women who passed the medical checks at the Lovelace Foundation in Albuquerque. It's a laudable ambition. I, myself, would like to make that first flight by a woman into space. But after nearly thirty years of flying I decided that I could serve the others best by not becoming a competitor with them as an active participant. I crossed that bridge when I did not take the tests at Albuquerque that were taken by the selected group of volunteers. I would like, however, to be helpful to the active aspirants. Because of this detached attitude I can perhaps more objectively consider the factors involved in trying to help get a women's astronaut program by other participants off to a sound start.

Eisenhower Presidential Library

- 2 -

I am sure you know from our various conversations that I am in favor of a space program for women. That I put up the money to pay the cost of most of the twenty candidates who took the medical checks at the Lovelace Foundation and then put up the money for most of the ones who passed these checks to enable them to go to Pensacola for the second series of checks should be clear proof of this. But each time you asked me at Coca Beach how I felt about such a women's program I tied to my expressions of approval some such words as "if soundly organized." So what I mean by this must be what is in doubt in your mind.

A space program for women is unlike all previous advents of women into various phases of aviation in that space flight for the present is terribly expensive, is an urgent project from the standpoint of national defense, and there is no lack of qualified candidates for the role of astronaut from among our already highly trained flight personnal. In other words, there is no present real national need for women in such a role. When the women pilots (WASP) were organized in World War II there was an urgent need. We were short of manpower and pilot power. I was asked to organize for service the women pilots by General Hap Arnold, Chief of Staff of the Air Force. And the minute the pressure on manpower let up toward the end of the war and pilots were returning from abroad who were available at home for air duty, I recommended that the WASP be deactivated. Otherwise the WASP might have become the subject of resentment by the male pilots.

A sound, fully acceptable women in space program must, in my opinion, involve a sizeable group. It seems to me the reasons for this are apparent. No clear conclusions can be reached by checking one person or even a very few. Extremes of tolerance, reaction, etc. will be found in any group while arriving at an average or tolerable limits. The government did not pick one person to be the astronaut nor did it limit itself to seven at the start. A great number were screened. Seven was the final result. If only one had been picked and he had turned out to be the one who has recently been eliminated as to the forthcoming flight on account of a slight heart condition, our space program would now be in trouble. A group is essential. Out of twenty who took the medical checks at the Lovelace Foundation, only twelve passed. It's almost a certainty that some of these twelve would be eliminated on the next series of tests and then as a program develops still others will probably be eliminated for cause. I wish a greater number had been checked by Lovelace. Twenty I consider the minimum. There should be at least eight to ten remaining to be dealt with as a "program" after the second series of tests like the ones that were planned at Pensacola. From the standpoint of the doctors in aerospace medicine valuable information about women as potentials for space

Eisenhower Presidential Library

- 3 -

flight could be obtained from the twenty and then the twelve candidates but I'm looking beyond that to a real program. These two checks or tests do not constitute a program. They would be mostly medical research. But if handled on a group basis with satisfactory results they could lead up to a sound program. Expensive important governmental military programs are not built up around one person or two or three persons. While each active participant very understandably would like to be the first, there is glory enough for each by being part of an initial group. Within that group each should succeed on her own merits without fear or favor.

There is no real present national urgency about putting a woman into space. To attempt to do so in the near future might indeed interfere with the space program now under way which is urgent from the national standpoint. I believe you disagree with me about this and on the ground that to put a woman into space before the Russians do would be a victory - at least in the cold war. But at the same time, in arguing for speed, you said to me that there is evidence that Russian women have been training for space flights for a considerable period of time. If so they will be prepared before an American woman. And a hastily prepared flight by a less than completely trained woman could backfire. Such a flight by a woman in the near future might be regarded by a great many as somewhat in the nature of unnecessary drama. And many who would not so regard it might question why the government has gone to great expense to specially prepare the seven astronauts over a three-year period if a woman can do the same thing without long background of experience followed by careful special training.

You will properly conclude from the above that I would like to see a space program for women develop and I will certainly be helpful in this respect if the program is to be developed carefully and soundly. The first step is not to rush a woman into space at great public expense and primarily for propaganda reasons but to complete the group checks that were planned at Pensacola and which involved only a few days and practically no cost to the government. These tests had some meaning for the medicos, were, as I have just said, inexpensive, and did not crowd into an important project that it so happens - and probably for good reasons - involves only men. I don't believe there is any distinction between men and women as to ability in the air. The WASP program proved that. But it's the men who have received that long training as test pilots and therefore it's natural that such men were selected for the present phase of the astronaut program.

Women for one reason or another have always come into each phase of aviation a little behind their brothers. They should, I believe, accept this delay and not get into the hair of the public authorities

Eisenhower Presidential Library

- 4 -

about it. Their time will come and pushing too hard just now could possibly retard rather than speed that date. It's better to be sound than quick. Tests are indicated before specialized training and specialized training is needed before flight.

I compliment all of the present candidates for their ambition and their apparent willingness to go through the needed stretch of specialized training. You and I know that the second series of tests was to take not more than two weeks for the entire group. It's a pity that the candidates did not get such tests on schedule.

If you go along with the soundness of the group idea, as I hope you do, then you can be particularly helpful to a program by going out of your way to create the group image publicly.

Best wishes.

Sincerely,

JACQUELINE COCHRAN.

Eisenhower Presidential Library

BETHANY, OKLAHOMA

R

April 26, 1962

Miss Jacqueline Cochran
630 Fifth Avenue
New York 20, N.Y.

Dear Jackie:

I can't tell you how much I appreciated your detailed, interesting explanation of your feelings about a women in space program. It was certainly thoughtful of you to take the time and effort to put your views on record for me, and I do thank you.

I heartily agree with the points you made in your letter and most certainly agree that such a program should be conducted on a sound and organized basis. However, we both know that the 'race is not always to the swift, nor the battle to the strong'. There were probably many times when no one ever thought that you would be the outstanding aviatrix of this day. Many people doubted that I could ever be a pilot.

The qualification rules have been laid down for astronauts and although NASA says they have nothing against women, it just so happens that the requirements are such that no woman can meet them. I don't know much about politics but I do know that exceptions have been, and can be made without destroying the scientific basis of the program. (Glenn did not graduate from college and Slayton's heart condition had been known for three years).

No one will dispute the fact that Glenn made an exceptional astronaut. I don't think it was because of his lack of a college degree or his extensive experience as an experimental jet test pilot; but perhaps it was a result of his determination, extensive studying, application of himself totally, many years as a pilot, belief and faith in God and will to succeed. You have these qualities, and I'm sure many other pilots do who would make good astronauts. There will never be any lack of astronaut candidates and it's not important who they are but what they believe in and what they accomplish for mankind.

SUBSIDIARY OF ROCKWELL-STANDARD CORPORATION
Coraopolis, Pennsylvania

Eisenhower Presidential Library

AERO COMMANDER, INC.
SUBSIDIARY OF ROCKWELL-STANDARD CORPORATION

Page two-

The people who are saying that women should not go into space now, are the same type who were saying that man would never fly, or that women did not have the mechanical or scientific mind to pilot an airplane. All I'm saying is that I am not content to sit back and listen to their silly excuses while waiting for Russia to prove the scientific importance of putting women in space. If anyone doubts the ability of women to successfully perform the exacting duties of an astronaut in space, let them prove it by testing and training women for the job and comparing their abilities in the space flight simulators.

The 'space race' is too important and too costly ($10 million a day for the next eight years) to let anything keep us from utilizing every capability - and this includes women as well as men. We still know so very little about living and working in the space environment, but scientists agree that there are some areas in space flight where women are uniquely qualified because of basic differences that nature has bestowed upon the female race. Some hypothetical examples are: should a moon shot be delayed for months, possibly years, because of high radiation in space connected with solar flares because of the harmful effects it would have on man, whereas it would not effect women; should a space voyage to Mars or Venus not go when ready because men could not tolerate the isolation, sensory deprivation, confinement of space, boredom, monotony and tedious task during the months and years such a trip would take, if women were capable of enduring such psychological stresses?

All I'm saying is that it should be determined, through extensive testing, training, and actual space flight wherein lies the areas where nature has adapted women to operate...which would be harmful to men. There will be certain areas where men are uniquely qualified (such as in physical strength) but I believe that our space programs should go ahead as fast as possible using both men and women to the best of their unique (and sometimes totally different) abilities.

Eisenhower Presidential Library

AERO COMMANDER, INC.

SUBSIDIARY OF ROCKWELL-STANDARD CORPORATION

Page three-

I'm not much good at putting feelings into words, but hope you can make some sense out of this. I really think that we agree on the important points. I don't think it is important who is the First woman in space but I do feel that it's important to our country that she be an American. However, the prestige is secondary to the scientific aspects and I only hope and pray that the United States will use everything that God has created upon this earth, accomplishing His will. Isn't this what we try to do every day ourselves? I feel no different about space and will continue to do everything I can, with whatever tools He has given me to work with. My aim is not personal but to get a group of women included in the space program. . . on a sound, orderly scientific basis. No less can I do, with a sound conscience,

I would like very much to talk with you and hope that I'll have the opportunity in the near future. Thanks again for your thoughtful letter and may the Lord bless you, and those you love.

Sincerely,

Jerrie

Miss Jerrie Cobb, Ass't to
Vice President Marketing

JC"bd

Eisenhower Presidential Library

16 June 1972

General James H. Doolittle
702 Mutual of Omaha Building
5225 Wilshire Boulevard
Los Angeles, California 90036

Dear Jimmy:

Jackie has been in the hospital in Albuquerque for a couple of weeks but is leaving, according to plan, in her new motor home within the next couple of days to go to Sweetwater, Texas, to attend a celebration that the city is putting on honoring the WASP (Women's Air Force Service Pilots) who trained at Sweetwater during the war. Jackie, in preparation for her departure, has sent back to me a batch of mail, including your letter to her of June 1st about the Harmon Trophy Awards. She did not note on this letter that it called for any answer, but I notice that you were about to go into the hospital for a "repair job" and I just want to express the wish on behalf of Jackie and myself that everything went well, that the cutting was of a miror nature, and that you are quite your good self again.

I know what Jackie had in mind about Jerrie Cobb and she did not wish to commit it to a paper that would be generally passed around. Jackie does not know what kind of a job Miss Cobb has been doing during 1971 in the jungle. She can assume that it was a satisfactory flying job and Jackie has had some flights herself across the Brazilian jungle, as well as being down in the jungle itself. However, surely these activities on her part have not been like yourself in the same area or the activities recently of Miss Cobb.

Miss Cobb's attitude and actions several years ago in connection with her posing as a woman astronaut, accepted or about to be accepted as such, Jackie thought were thoroughly unjustified, thoroughly uncalled for and harmful in effect to women pilots as a whole. Jackie happened to know the facts in this respect. Miss Cobb on her own application went to the Lovelace Clinic and asked to be given the same medical tests that were given to the original seven astronauts. She failed this for some reason that was not too important because the tests themselves were not much

Eisenhower Presidential Library

General James H. Doolittle -2- 16 June 1972

more than the standard medical tests. She came back the second time at Randy Lovelace's suggestion, after having done some exercises, and passed the test. Randy was to give a paper in Europe some place soon thereafter on the medical aspects of space flying and, incidentally and as a matter of interest, mentioned in his paper that a woman (naming her) had also taken the same test successfully. This broke into the press and in the course of time was quite blown up to the point where a great many people thought Miss Cobb had been accepted for the astronaut program. Many of the women pilots kicked to Jackie about this and in consequence she selected twenty experienced women pilots and at her own expense had them go through the Lovelace Clinic and successfully take the same medical test that Miss Cobb had taken. Later, there was the next echelon of tests that were to be given to these twenty women plus Miss Cobb. I think these were to be given at some Naval base. Somehow Miss Cobb managed to get down there first to be tested by herself and whoever was in charge of this work in Washington called the whole thing off for all of the women in view of the poor showing that Miss Cobb made in these further tests. Jackie had no thought that any of the women would be accepted for astronaut training at any early date, but she thought that against the future and as a gesture to women as a whole these tests would have been fruitful to determine whether women were qualified, if necessary, to go into space work.

About this time there was a hearing before a committee of Congress generated by Miss Cobb and two or three other women pilots to determine whether the specifications of NASA should not be changed so as to embrace women pilots. Miss Cobb testified in favor of such change and Jackie testified against such a change. It is apparent that Jackie's action in this respect took the committee and NASA off the hook and Jackie soon thereafter was made a Special Consultant to the head of NASA. She never worked with him but has worked with the people in charge of manned space flights, etc., and is still such a consultant. She felt in her conversation with me that Miss Cobb's actions during these years were not consistent with the awarding of the Harmon Trophy to her, but she told me she would bet me two to one that the committee would do just that, and your letter indicates that this may have happened.

Best wishes to you.

Sincerely,

Floyd B. Odlum

Eisenhower Presidential Library

AUTHOR'S NOTE

In writing this book, I visited multiple archives and museums, spoke to experts, and dug up ephemera to create as complex characters as I could without taking any liberties with reality. As such, all the dialogue in this book comes from memoirs, interviews, radio transcripts, press releases, and recalled conversations. In rare instances, I modified direct quotes to avoid repetition, but I show these skips with ellipses. Only minor edits have been made to source material such as spelling out a numeral for the sake of consistent style throughout the manuscript. If I preserved a spelling error—like the loaded error of Jackie misspelling Jerrie's name as "Jerry"—I left out the traditional [sic] so it was less obtrusive to the reader. Overall, I took great pains to preserve everyone's meaning and intention. To avoid having 1,216 footnotes, all sources are listed by chapter in the bibliography.

I was lucky in writing this book that I had ample primary source material to work with, but this brought up an interesting quandary. Memories fade and one person's recollection of a conversation might differ from someone else's, so memoirs, interviews, and various other writings don't necessarily line up with a verifiable historical record. Both Jackie and Jerrie wrote multiple memoirs, which is a wonderful source for details of their personal lives, but they also introduce what I've come to call the "Jackie version" or the "Jerrie version" as the case may be, and I need to draw attention to certain moments where I know or suspect we're dealing with their "versions" of a story.

When Jackie meets Floyd on page 18, she claims she had no idea who he was, but it's quite possible she knew exactly who he was and maneuvered her way into meeting him. On page 38 when Jackie recounts the story of Cecil Allen's beheading before her first Bendix race in 1935, I know that's a Jackie version. Cecil Allen did die that night, but he crashed in a potato field in front of two witnesses, and Jackie was not one of them; she had taken off before Cecil and would only have read about the crash after her own flight. Jumping ahead to Al Shepard's 1961 *Freedom 7* mission, Jackie writes in a letter that she served as an official timer on the mission but there is no NASA source to back that up.

While Jerrie's romance with Jack Ford is incredible, she never mentions his wife, Mary Ford, in recounting the courtship in her memoirs. I haven't yet found any insight into the state of Jack and Mary's marriage or how open he was about this affair he was having with one of his employees. It's highly unlikely, though, that Jerrie didn't know her boss and lover was married. Not only was Mary a pilot and co-owner of the business, so presumably in the office from time to time, they inspired a Hollywood film: *The Lady Takes a Flyer* (1958) stars Lana Turner and Jeff Chandler as married pilots co-owning a ferrying business, and co-stars Andra Martin as a young pilot employee with whom Jeff Chandler's character has an affair.

There are inconsistencies between Jerrie's memoirs, too. On page 133 of this book, Jerrie recounts that she was released from her Ecuadorian imprisonment after twelve days. That's the figure she gives in her 1963 memoir, but writing in 1997, she says she was freed after just two days. Jerrie's first memoir is also my source for saying she spent nine hours in Dr. Shurley's isolation tank on page 216, but there's speculation she didn't actually last that long.

I chose to leave these anecdotes from both women for two simple reasons: they don't change the core story I'm telling about women in space and I felt strongly about wanting to honor how these women saw themselves. I don't care that Jackie didn't see Cecil Allen die, I care that the story became part of her personal narrative and underscores her fearlessness. We can't forget that "Jackie Cochran" was Bessie Pittman living as an ideal version of herself, so those stories say so much about the woman she chose to be and are important for understanding her character. With Jerrie, I wanted you to feel her pride in the isolation tank run, because even if she didn't last nine hours it was still a feat of mental fortitude she took great pride in. I didn't want to focus on Jack Ford's marriage because Jerrie's feeling of falling in love and loss when it ended was more important for understanding her subsequent actions. But I did want to make sure you knew about Mary Ford's existence; this detail, even with just a few mentions, is so important for understanding Jerrie's dogged adherence to her own version of events. It plants a seed of doubt; if she omitted her lover's wife in her memoirs, what other truths might she have left out from other parts of her life? My goal was for you to appreciate her very human actions and motivations, while also questioning her integrity.

I chose to put all this in the final author's note rather than footnotes in the text, which was my initial approach, because I didn't want to impact your experience meeting and reading about these women. Heroic tales are so often mixed with adrenaline, ego, heartbreak, elation, fear, and hope. This one is no exception and I didn't want to remove any of the important emotional elements. And after all, how many men can honestly say they never inflated a story to make it "better"? Our women, at the end of the day, are simply human.

ACKNOWLEDGMENTS

It's fitting that this book about two badass women wouldn't have been possible without two amazing women in my own life. First and foremost, my agent, Eve Attermann, who encouraged me to pursue this story and refused to let me send out a mediocre proposal. Second, my editor, Gretchen Young, who came in at the eleventh hour and worked to get this book done well and on time. It's very fitting that all facets of this book were done by women. Additional thanks to Jesseca Salky for navigating the scary legal end of things, Haley Weaver for making sure details didn't slip through the cracks, and Jeaneen Lund for making me look awesome.

Writing this book hinged on a lot of archival research. A world of thanks to Kevin Bailey, Kathy Struss, and the staff at the Eisenhower Presidential Library; Alexis Percle and the staff at the Johnson Presidential Library; Colin Fries and Liz Suckow in the NASA archives; Emily Mathay at the Kennedy Presidential Library; Dara Baker at the Roosevelt Presidential Library; Michael Sharaba at the International Women's Air and Space Museum; and Dallas Hanbury in the Montgomery County Probate Office.

A handful of people helped by filling in some blanks through interviews, some directly related to my research and others more informational. Thank you to Gene Nora Jessen, Jeremy Hanson, Reid Weisman, and Robert Pearlman for talking through various pieces of this narrative with me. A special thank you to Francis French for not only reading multiple drafts of my manuscript but for sharing his own archives with me, answering endless questions, and generally serving as a sounding board for ideas both good and bad. With a project like this that consumes your whole life, there are also people who don't know they played a part in helping get you through to the finish line. Justin Carter; Jason Han, Allister Buchanan, and the whole Healthfit team; Coach Joe Del Real; Geoffrey Notkin; Jeff Kluger; Matt Wood—you kept me sane, healthy, motivated, inspired, or otherwise feeling like I could keep tackling the insanity that is writing a book.

And finally, my family. Pete Conrad (the cat) for sitting on my desk nuzzling support every day. My parents for the ongoing gift of my education as well as your support and willingness to read multiple drafts as I went along. And finally, to Merrick. Putting up with someone writing a book is all the stress of being an author without the fun of living in a world. Thank you for letting me rant and vent, for encouraging my collecting of Jackie ephemera, and generally loving me through this whole messy process.

BIBLIOGRAPHY

PREFACE

"Bessie Coleman." National Aviation Hall of Fame. Accessed August 20, 2019. https://www.nationalaviation.org/our-enshrinees/coleman-bessie/.

Crouch, Tom. *A History of Aviation from Kites to the Space Age*. New York: W. W. Norton, 2004.

Spenser, Jay. *The Airplane: How Ideas Gave Us Wings*. Washington, DC: Smithsonian, 2008.

Rampton, Martha. "Four Waves of Feminism." Pacific University Oregon, accessed August 11, 2019. https://www.pacificu.edu/about/media/four-waves-feminism.

"McCarthyism and the Red Scare." UVA Miller Center, accessed August 8, 2019. https://millercenter.org/the-presidency/educational-resources/age-of-eisenhower/mcarthyism-red-scare.

Dicker, Rory. *A History of U.S. Feminisms*. Berkeley: Seal Press, 2016.

CHAPTER 1

Ocala-Star Banner, August 21, 1990, accessed August 13, 2018. https://news.google.com/newspapers?id=zOVPAAAAIBAJ&sjid=UwcEAAAAIBAJ&pg=4243%2C106832.

Ayers, Billie Pittman and Beth Dees. *Superwoman: Jacqueline Cochran, Family Memoirs about the Famous Pilot, Patriot, Wife & Business Woman*. 1st Books, 2001.

Rich, Doris L. *Jackie Cochran: Pilot in the Fastest Lane*. Gainesville: University of Florida Press, 2007.

Cochran, Jacqueline. *The Stars at Noon*. New York: Little, Brown and Company, 1954.

"Bagdad, Florida." Jacqueline Cochran: Papers, 1932–1975. Stars at Noon Series. Box 6. Eisenhower Presidential Library.

Cochran, Jacqueline and Maryann Bucknum Brinley. *Jackie Cochran: The Autobiography of the Greatest Woman Pilot in Aviation History*. New York: Bantam Books, 1987.

"Robert Harvey Cochran (1895–1958)." https://www.geni.com/people/Robert-Cochran/6000000016943174222. Accessed June 11, 2019.

Divorce Record, Montgomery County Archives 1925#3703.

"First Permanent Wave for Hair is Demonstrated 8 October 1906." History Channel

Australia. Accessed August 13, 2018. https://www.historychannel.com.au/this-day-in-history/first-permanent-wave-for-hair-is-demonstrated/.

Sheen, Maureen. "The Story of Us: Do the Wave." *American Salon*, January 20, 2016. Accessed August 13, 2018. https://www.americansalon.com/products/story-us-1910-1920-do-wave.

CHAPTER 2

"1929—Early Sound Footage of New York City." guy jones, accessed September 25, 2018. https://www.youtube.com/watch?v=yx7rrYSslYc.

"1928—More Early Sound Footage of New York City." guy jones, accessed September 25, 2018. https://www.youtube.com/watch?v=gv1vRs3l2ek.

Cochran, *The Stars at Noon*.

Rich, Doris L. *Jackie Cochran: Pilot in the Fastest Lane*. University of Florida Press, Gainsville, 2007.

Cochran and Brinley, *Jackie Cochran: The Autobiography of the Greatest Woman Pilot in Aviation History*.

Charles of the Ritz Cosmetics and Skin, accessed March 15, 2018. http://cosmeticsandskin.com/companies/charles-ritz.php.

"The Greatest Hairdresser of the 20th Century." *Tablet*, accessed March 15, 2018. http://www.tabletmag.com/jewish-arts-and-culture/190585/antoine-roman-vishniac.

Kennedy, David M. *Freedom from Fear: The American People in Depression and War, 1929–1945*. Oxford: Oxford University Press, 1999.

"Great Depression Timeline: 1929–1941." *The Balance*, accessed September 25, 2018. https://www.thebalance.com/great-depression-timeline-1929-1941-4048064.

"History of the Surf Club." Four Seasons Press Room, accessed October 8, 2018. https://press.fourseasons.com/surfside/trending-now/the-surf-club-history/.

"Miami (Molly Hemphill)." Jacqueline Cochran: Papers, 1932–1975. Stars at Noon Series. Box 6. Eisenhower Presidential Library.

"Portion of 'The Stars at Noon' Concerning Cosmetics." Jacqueline Cochran: Papers, 1932–1975. Stars at Noon Series. Box 6. Eisenhower Presidential Library.

"Atlas Corporation." Undated (1952). Odlum, Floyd Papers: 1892–1976. Box 2. Eisenhower Presidential Library.

Fisher, Kenneth L. *100 Minds that Made the Market*. Accessed March 15, 2018. https://books.google.com/books?id=hV2vXZNVhQ4C&pg=PA137&lpg=PA137&dq=Floyd+B.+Odlum&source=bl&ots=MP5oks3OoU&sig=6A-EgFE4kQhjNlosxLh3A45Fp0s&hl=en&sa=X&ved=0ahUKEwjVzaWR7O7ZAhVL7oMKHT6QCKkQ6AEIVTAI#v=onepage&q=Floyd%20B.%20Odlum&f=false.

"Business & Finance: Lady from Atlas." *Time*, October 22, 1934. Accessed August 14, 2018. http://content.time.com/time/subscriber/article/0,33009,847205,00.html.

"First Lessons in Flying." (Dictated material dated October 31, 1951.) Jacqueline Cochran: Papers, 1932–1975. Stars at Noon Series. Box 6. Eisenhower Presidential Library.

"Roosevelt Field Through the Years." *Newsday*, accessed April 2, 2018. https://www.newsday.com/long-island/nassau/roosevelt-field-through-the-years-1.10862824.

"LI Remembers Charles Lindbergh's Flight from Roosevelt Field to Paris." *Newsday*, accessed August 12, 2019. https://www.newsday.com/long-island/history/long-island-remembers-charles-lindbergh-s-flight-from-roosevelt-field-to-paris-1.13322997.

"First Lady of the Air Lanes." *New York Times*, September 25, 1938.

CHAPTER 3

Cochran and Brinley, *Jackie Cochran: The Autobiography of the Greatest Woman Pilot in Aviation History*.

Rich, *Jackie Cochran: Pilot in the Fastest Lane*.

"Easterner Here to Become Pilot." *San Diego Union*, Friday, October 7, 1932.

"Mrs. Odlum Gets Nevada Divorce." *New York Times*, October 8, 1935.

"Salton Sea State Recreation Area." California Department of Parks and Recreation. Accessed August 20, 2019. http://www.parks.ca.gov/?page_id=639.

"Portion of 'The Stars at Noon' Concerning Cosmetics." Jacqueline Cochran: Papers, 1932–1975. Stars at Noon Series. Box 6. Eisenhower Presidential Library.

"JC Book Chapter ______." Jacqueline Cochran: Papers, 1932–1975. Stars at Noon Series. Box 6. Eisenhower Presidential Library.

"Women in Aviation and Space History: Amelia Earhart." Smithsonian Air and Space Museum, accessed October 15, 2018. https://airandspace.si.edu/explore-and-learn/topics/women-in-aviation/earhart.cfm.

Thurman, Judith. "Amelia Earhart's Last Flight." *The New Yorker*, September 7, 2009. Accessed October 15, 2018. https://www.newyorker.com/magazine/2009/09/14/missing-woman.

"The Amelia I Knew." 1949. Jacqueline Cochran: Papers, 1932–1975. Articles Series. Box 1. Eisenhower Presidential Library.

"Chapter IV." Jacqueline Cochran: Papers, 1932–1975. Stars at Noon Series. Box 6. Eisenhower Presidential Library.

"Bendix, Vincent Hugo." National Aviation Hall of Fame, accessed April 4, 2018. https://www.nationalaviation.org/our-enshrinees/bendix-vincent/.

"Bendix Trophy." *Air Racing History*, accessed April 4, 2018. http://www.air-racing-history.com/Between%20the%20wars(2).htm.

"Speedy Allen Plane Plunges into Field." *New York Times*, August 30, 1935. https://timesmachine.nytimes.com/timesmachine/1935/08/31/93481313.pdf.

CHAPTER 4

Cochran, *The Stars at Noon*.

Cochran and Brinley, *Jackie Cochran: The Autobiography of the Greatest Woman Pilot in Aviation History*.

Rich, *Jackie Cochran: Pilot in the Fastest Lane*.

"Seversky Sets East-West Flight Record of 10 Hours 3 Minutes in Bendix Race Plane." *New York Times*. August 30, 1938.

"Seversky P-35." *Military Factory*, accessed August 17, 2018. https://www.militaryfactory.com/aircraft/detail.asp?aircraft_id=730.

"Hypoxia: Its Causes and Symptoms." Ross P. Cafaro, accessed August 10, 2019. https://www.ncbi.nlm.nih.gov/pmc/articles/PMC2067517/pdf/jadsa00124-0004.pdf.

"Airman Education Programs: Beware of Hypoxia." Federal Aviation Administration, accessed August 17, 2018. https://www.faa.gov/pilots/training/airman_education/topics_of_interest/hypoxia/.

Elliott, Richard G. "On a Comet Always." *New Mexico Quarterly*. Winter 1966–67. Lovelace, Dr. W. Randolph Folder. NASA Archives.

"Jacqueline Cochran (USA)." Fédération Aéronautique Internationale, accessed October 10, 2018. https://www.fai.org/record/12026.

"Merrill, Jean Batten Win Harmon Award." *New York Times*, Friday, February 18, 1938. https://timesmachine.nytimes.com/timesmachine/1938/02/18/98099198.pdf.

"TWA History." TWA Museum, accessed August 17, 2018. http://twamuseum.com/htdocs/twahistory2.htm.

"Bendix Trophy." *Air Racing History*, accessed April 4, 2018. http://www.air-racing-history.com/Between%20the%20wars(2).htm.

"Miss Cochran Wins Fast Bendix Race." *New York Times*, September 4, 1938.

"The Seversky P-35-Predecessor of the P-47 Thunderbolt." Fiddlers Green, accessed August 17, 2018. http://www.fiddlersgreen.net/models/aircraft/Seversky-P35.html.

"Miss Cochran First at Bendix Also." *New York Times*, September 4, 1938.

"Miss Cochran Wins Bendix Race; Close to Record." *New York Times*, September 4, 1938.

"The Cleveland Races: A Longer Course, Higher Danger, and Better Speeds." *Newsweek*, September 12, 1938.

"First Lady of the Air Lanes." *New York Times*, September 25, 1938.

"Chapter IV." Jacqueline Cochran: Papers, 1932–1975. Stars at Noon Series. Box 6. Eisenhower Presidential Library.

"Harper's Bazaar Insert. November 1939." Jacqueline Cochran: Papers, 1932–75. Business Series. Box 7. Eisenhower Presidential Library.

Newsweek. Volume XII, No. 11. September 12, 1938.

Cobb, Jerrie and Jane Rieker. *Woman into Space: The Jerrie Cobb Story*. Englewood Cliffs, NJ: Prentice-Hall International, Inc., 1963.[29]

CHAPTER 5

Cochran, *The Stars at Noon*.

Cochran and Brinley, *Jackie Cochran: The Autobiography of the Greatest Woman Pilot in Aviation History*.

---

29 Jerrie Cobb's memoirs, *Woman into Space* and *Solo Pilot*, were essential sources for this book in order to use her own words to give the reader a sense of who she was, both as a woman and a pilot. Her excitement, heartache, fight, and sadness could only come from her.

Rich, *Jackie Cochran: Pilot in the Fastest Lane.*

Cobb and Rieker, *Woman into Space: The Jerrie Cobb Story.*

"1028 Lectures on Aeronautics." *Flying and Popular Aviation*, September, 1940. Accessed October 16, 2018. https://books.google.com/books?id=acokk-XqWvgC&pg=PA85&lpg=PA85&dq=National+Aeronautic+Association+William+Enyart&source=bl&ots=QXc3GptUSI&sig=wEmIDUw1gaz6V_5mjEk1Wrzv-rY&hl=en&sa=X&ved=2ahUKEwjRt66OvI3eAhUMvlMKHWz3Ae4Q6AEwDXoECAAQAQ#v=onepage&q=National%20Aeronautic%20Association%20William%20Enyart&f=false.

Transcript, "Jacqueline Cochran Oral History Interview I by Joe B. Frantz." July 4, 1974. Lyndon Baines Johnson Library.

"Nomination Guidelines for the Collier Trophy." National Aeronautic Association, accessed August 20, 2018. https://naa.aero/awards/awards-and-trophies/collier-trophy/.

"Lockheed XC-35 Electra." Smithsonian National Air and Space Museum, accessed October 16, 2018. https://airandspace.si.edu/collection-objects/lockheed-xc-35-electra.

"Historical Snapshot." Boeing, accessed October 16, 2018. http://www.boeing.com/history/products/model-307-stratoliner.page.

"Douglas DC-3 Flight Notes." Krepelka, accessed October 16, 2018. http://krepelka.com/fsweb/learningcenter/aircraft/flightnotesdouglasdc3.htm.

Taves, Isabella. "Lady in a Jet." *Reader's Digest*, August 1955.

"Draft #2—JC Article for United States Medical Journal Anniversary Issue." Jacqueline Cochran: Papers, 1932–1975. Articles Series. Box 6. Eisenhower Presidential Library.

"Nomination Guidelines for the Colliers Trophy." National Aeronautic Association, accessed May 19, 2019. https://naa.aero/awards/awards-and-trophies/collier-trophy.

"Awards and Trophies, Collier (Robert J.) Trophy, 1930s." Smithsonian Air and Space Museum, accessed August 20, 2018. https://airandspace.si.edu/collection-objects/awards-and-trophies-collier-robert-j-trophy-1930s-boothby-walter-meredith.

"President to Present Collier Trophy Today." Press Release, December 17, 1940. The Franklin D. Roosevelt Presidential Library and Museum.

Dara Baker, archivist at the Franklin D. Roosevelt Presidential Library and Museum, email to author. December 21, 2018.

Woods, Randall. *LBJ: Architect of American Ambition*. New York: Free Press, 2006.

"Great Britain on the Brink." Franklin D. Roosevelt Presidential Library and Museum, accessed December 15, 2018. https://fdrlibrary.org/lend-lease.

Letter, Floyd Odlum to Morris Wilson. March 17, 1941. Jacqueline Cochran: Papers, 1932–1975. WASP Series. Box 1. Eisenhower Presidential Library.

Letter, D. D. Frye to Wm. K. Trower. May 19, 1941. Jacqueline Cochran: Papers, 1932–1975. WASP Series. Box 1. Eisenhower Presidential Library.

"Memo" (Bomber Flight). Jacqueline Cochran: Papers, 1932–1975. WASP Series. Box 1. Eisenhower Presidential Library.

Letter, from Morris Wilson to Jacqueline Cochran. March 13, 1941. Jacqueline Cochran: Papers, 1932–1975. WASP Series. Box 1. Eisenhower Presidential Library.

Letter, Hortense Wordenau to Jacqueline Cochran. June 20, 1941. Jacqueline Cochran: Papers, 1932–1975. WASP Series. Box 1. Eisenhower Presidential Library.

Note, R.A.F. Station North Weald to Jacqueline Cochran. June 28, 1941. Jacqueline Cochran: Papers, 1932–1975. WASP Series. Box 1. Eisenhower Presidential Library.

Handwritten telegram dictation, Floyd Odlum to Jacqueline Cochran. Undated (June 1941). Jacqueline Cochran: Papers, 1932–1975. WASP Series. Box 1. Eisenhower Presidential Library.

Letter, Robert Lodge to Jacqueline Cochran. June 24, 1941. Jacqueline Cochran: Papers, 1932–1975. WASP Series. Box 1. Eisenhower Presidential Library.

McDonough, Yona Zeldis. "The Women's RAF." *Smithsonian Air & Space Magazine*, May 2012. Accessed May 20, 2019. https://www.airspacemag.com/military-aviation/the-womens-raf-118165440/.

Letter, Pauline Gower to Jacqueline Cochran. September 26, 1941. Jacqueline Cochran: Papers, 1932-1975. WASP Series. Box 1. Eisenhower Presidential Library.

"Robert Olds." Veteran Tributes, accessed August 23, 2018. http://www.veterantributes.org/TributeDetail.php?recordID=1498.

"Women's Auxiliary Air Corps—Preliminary Outline of Organization." Jacqueline Cochran: Papers, 1932–1975. WASP Series. Box 1. Eisenhower Presidential Library.

Memo, Jacqueline Cochran to Colonel Robert Olds. July 5, 1941. Jacqueline Cochran: Papers, 1932–1975. WASP Series. Box 1. Eisenhower Presidential Library.

Hoyt, John R. "Women Can Fly." *Flying and Popular Aviation*, November 1941. Jacqueline Cochran: Papers, 1932–1975. WASP Series. Box 2. Eisenhower Presidential Library.

Letter, Jacqueline Cochran to Robert Olds. "Time for Starting of Ferrying by Women." July 28, 1941. Jacqueline Cochran: Papers, 1932–1975. WASP Series. Box 1. Eisenhower Presidential Library.

Letter, Jacqueline Cochran to Robert Olds. "Base Locations for Women Ferry Pilots." July 28, 1941. Jacqueline Cochran: Papers, 1932–1975. WASP Series. Box 1. Eisenhower Presidential Library.

Letter, Jacqueline Cochran to Colonel Robert Olds. July 20, 1941. Jacqueline Cochran: Papers, 1932–1975. WASP Series. Box 1. Eisenhower Presidential Library.

Memo, Jacqueline Cochran to Colonel Robert Olds. July 5, 1941. Jacqueline Cochran: Papers, 1932–1975. WASP Series. Box 1. Eisenhower Presidential Library.

"Organization of Women Pilots Division of Air Corps Ferrying Command." Jacqueline Cochran: Papers, 1932–1975. WASP Series. Box 1. Eisenhower Presidential Library.

Report, Colonel Robert Olds to Secretary of War. "Organization of a Women Pilots Division of the Air Corps Ferrying Command." Undated (1941). Jacqueline Cochran: Papers, 1932–1975. WASP Series. Box 1. Eisenhower Presidential Library.

Letter, Louis Gimbel to Lt. Col. Steadman S. Hanks. July 29, 1941. Jacqueline Cochran: Papers, 1932–1975. WASP Series. Box 1. Eisenhower Presidential Library.

Letter, Robert Olds to Mrs. B. F. Roberts. June 27, 1941. Jacqueline Cochran: Papers, 1932–1975. WASP Series. Box 1. Eisenhower Presidential Library.

Letter, Evelyn Dent to Jacqueline Cochran. July 22, 1941. Jacqueline Cochran: Papers, 1932–1975. WASP Series. Box 2. Eisenhower Presidential Library.

"Survey to all women holders of licenses." Air Corps Ferry Command. July 1941. Jacqueline Cochran: Papers, 1932–1975. WASP Series. Box 2. Eisenhower Presidential Library.

Letter, Jacqueline Cochran to Sidney Weinberg. August 7, 1941. Jacqueline Cochran: Papers, 1932–1975. WASP Series. Box 1. Eisenhower Presidential Library.

Letter, General Hap Arnold to Jacqueline Cochran. September 19, 1941. Jacqueline Cochran: Papers, 1932–1975. WASP Series. Box 1. Eisenhower Presidential Library.

Letter, Jacqueline Cochran to General Hap Arnold. October 4, 1941. Jacqueline Cochran: Papers, 1932–1975. WASP Series. Box 1. Eisenhower Presidential Library.

Arnold, Henry H. *Global Mission*. New York: Harper & Brother Publishers, 1949.

Lange, Katie. "Pearl Harbor Wasn't the Only Installation Attacked on Dec. 7." DoD Live, December 6, 2016. Accessed August 12, 2019. http://www.dodlive.mil/2016/12/06/pearl-harbor-wasnt-the-only-installation-attacked-on-dec-7/.

"FDR DECLARES WAR (12/8/41)—Franklin Delano Roosevelt, WWII, Infamy Speech." Periscope Films, accessed October 19, 2018. https://www.youtube.com/watch?v=YhtuMrMVJDk.

"December 8, 1941—Franklin Roosevelt asks Congress for a Declaration of War with Japan." The Franklin D. Roosevelt Presidential Library and Museum, accessed October 19, 2018. http://docs.fdrlibrary.marist.edu/tmirhdee.html.

CHAPTER 6

Cochran, *The Stars at Noon*.

Cochran and Brinley, *Jackie Cochran: The Autobiography of the Greatest Woman Pilot in Aviation History*.

Rich, *Jackie Cochran: Pilot in the Fastest Lane*.

"British Air Transport Auxiliary, Mary Zerbel Hooper Ford." http://www.airtransportaux.com/members/hooper-ford.html. Accessed August 9, 2020.

Arnold, Henry H. *Global Mission*. New York: Harper & Brother Publishers, 1949.

"Urgent and Important." (Copy, Original of this letter delivered by hand by Miss Cochran to General Arnold.) Undated (late 1941). Jacqueline Cochran: Papers, 1932–1975. WASP Series. Box 1. Eisenhower Presidential Library.

Letter, Glynn M. Jones to Jacqueline Cochran. "Establishment of Air Service Command Ferry Service." July 3, 1942. Jacqueline Cochran: Papers, 1932–1975. WASP Series. Box 2. Eisenhower Presidential Library.

Letter, General Arnold to Major General John F. Curry. January 17, 1942. Jacqueline Cochran: Papers, 1932–1975. WASP Series. Box 1. Eisenhower Presidential Library.

"Women Will Form a Ferry Command." *New York Times*, Friday, September 11, 1942.

"Lieutenant General Harold L. George." US Air Force, accessed August 12, 2019. http://www.af.mil/About-Us/Biographies/Display/article/107023/lieutenant-general-harold-l-george/.

Letter, Jacqueline Cochran to Hap Arnold. "Use of Women Pilots." September 11, 1942. Jacqueline Cochran: Papers, 1932–1975. WASP Series. Box 1. Eisenhower Presidential Library.

James, George. "C. R. Smith, Pioneer of Aviation as Head of American, Dies at 90."

New York Times, April 5, 1990. https://www.nytimes.com/1990/04/05/obituaries/c-r-smith-pioneer-of-aviation-as-head-of-american-dies-at-90.html.

"'WASP' is New Title for AAF Woman Pilot." August 20, 1943. Jacqueline Cochran: Papers, 1932–1975. WASP Series. Box 14. Eisenhower Presidential Library.

Memorandum for the Commanding Officer of the Army from L. S. Smith. "Women's Flying Training." September 15, 1942. Jacqueline Cochran: Papers, 1932–1975. WASP Series. Box 4. Eisenhower Presidential Library.

Memo, George E. Stratemeyer to Hap Arnold. "Flying Training for Women." October 7, 1942. Jacqueline Cochran: Papers, 1932–1975. WASP Series. Box 4. Eisenhower Presidential Library.

Memo, William Tunner to Commanding Officer, 2nd Ferrying Group. "Hiring Civilian Women Pilots." September 15, 1942. Jacqueline Cochran: Papers, 1932–1975. WASP Series. Box 4. Eisenhower Presidential Library.

"Portion of 'The Stars at Noon' Concerning Cosmetics." Jacqueline Cochran: Papers, 1932–1975. Stars at Noon Series. Box 6. Eisenhower Presidential Library.

CHAPTER 7

Cochran, *The Stars at Noon*.

Cochran and Brinley, *Jackie Cochran: The Autobiography of the Greatest Woman Pilot in Aviation History*.

Rich, *Jackie Cochran: Pilot in the Fastest Lane*.

"AAF to Deactivate WASP on December 20." Press Release. October 3, 1944. Jacqueline Cochran: Papers, 1932–1975. WASP Series. Box 12. Eisenhower Presidential Library.

Image 91-4-4752. Jacqueline Cochran Photographs. Eisenhower Presidential Library.

Leatherwood, Art. "William P. Hobby Airport." Texas State Historical Association, accessed October 22, 2018. https://tshaonline.org/handbook/online/articles/epwhe.

Merryman, Molly. *Clipped Wings: The Rise and Fall of the Women's Air Force Service Pilots (WASPs) of World War II*. New York: New York University Press, 1998.

"Lieutenant General William H. Tunner." U.S. Air Force, accessed August 12, 2019. http://www.af.mil/About-Us/Biographies/Display/Article/105384/lieutenant-general-william-h-tunner/.

Letter, Walter F. Kraus to Commanding General, Army Air Forces. "Absorption of Army Air Forces Women Pilots into Women's Army Auxiliary Corps." December 19, 1942. Jacqueline Cochran: Papers, 1932–1975. WASP Series. Box 5. Eisenhower Presidential Library.

Spring, Kelly A. "Oveta Culp Hobby." *Women's History*, 2007. Accessed August 12, 2019. https://www.womenshistory.org/education-resources/biographies/oveta-hobby.

Letter, Jacqueline Cochran to Hap Arnold. "Subject: WASP Report." June 1, 1945. Jacqueline Cochran: Papers, 1932–1975. WASP Series. Box 12. Eisenhower Presidential Library.

Transcript. "Conversation between Miss Cochran, Colonel Hobby and Colonel Carmichael." June 25, 1943. Jacqueline Cochran: Papers, 1932–1975. WASP Series. Box 5. Eisenhower Presidential Library.

Memorandum for Commanding General, AFFTC. "Subject: Women's Flying Training." January 8, 1943. Jacqueline Cochran: Papers, 1932–1975. WASP Series. Box 4. Eisenhower Presidential Library.

"Orientation by Lieutenant Fleishman to Class 43-W-4." February 16, 1943. Jacqueline Cochran: Papers, 1932–1975. WASP Series. Box 4. Eisenhower Presidential Library.

"Girl Pilots." *LIFE* magazine. July 10, 1943. Accessed August 12, 2019. https://books.google.com/books?id=MVAEAAAAMBAJ&printsec=frontcover#v=onepage&q&f=false.

Newcomb, SMSgt. Harold. "Cochran's Convent." *Airman*, Official Magazine of the US Air Force. Vol. XXI no. 5. May 20, 1977. NASA Archives.

Draft Memorandum, Hap Arnold to General Marshall. June 14, 1943. Jacqueline Cochran: Papers, 1932–1975. WASP Series. Box 5. Eisenhower Presidential Library.

Letter, Floyd Odlum to Jacqueline Cochran. May 21, 1943. Jacqueline Cochran: Papers, 1932–1975. WASP Series. Box 5. Eisenhower Presidential Library.

"'WASP' is New Title for AAF Woman Pilot." August 20, 1943. Jacqueline Cochran: Papers, 1932–1975. WASP Series. Box 14. Eisenhower Presidential Library.

Cobb and Rieker, *Woman into Space: The Jerrie Cobb Story*.

"New Uniform, Insignia Adopted for WASPs." Press Release, November 17, 1943. Jacqueline Cochran: Papers, 1932–1975. WASP Series. Box 14. Eisenhower Presidential Library.

"Fabrikoid—A Game Changer." Hagley, accessed August 31, 2018. https://www.hagley.org/about-us/news/museum-fabrikoid-game-changer.

"Harper's Bazaar Insert. November 1939." Jacqueline Cochran: Papers, 1932–75. Business Series. Box 7. Eisenhower Presidential Library.

"Training, Recruiting of WASPs Terminated Immediately." Press Release. June 26, 1944. Jacqueline Cochran: Papers, 1932–1975. WASP Series. Box 14. Eisenhower Presidential Library.

Englund, Julie I. "First-Rate, Second-Class." *Washington Post*, May 12, 2002. Cochran, Jacqueline Folder. NASA Archives.

"Hearings Before the Committee on Military Affairs House of Representatives, Seventy-Eighth Congress. H.R. 4219—Providing for the appointment of female pilots and aviation cadets of the Army Air Forces." March 22, 1944. United States Printing Office, Washington: 1944. Jacqueline Cochran: Papers, 1932–1975. WASP Series. Box 5. Eisenhower Presidential Library.

Mrs. Harvey W. Wiley to the State Chairman of Legislation, General Federation of Women's Clubs. August 8, 1944. Jacqueline Cochran: Papers, 1932–1975. WASP Series. Box 5. Eisenhower Presidential Library.

"Committee on the Civil Service House of Representatives—Investigation Concerning Inquiries Made of Certain Proposals for the Expansion and Change in Civil Service Status of the WASPS." H. Res. 16. June 5, 1944. United States Government Printing Office. Washington: 1944. Jacqueline Cochran: Papers, 1932–1975. WASP Series. Box 14. Eisenhower Presidential Library.

"Memorandum for Members of the WASP Board," from Cochran. July 8, 1944. Jacqueline Cochran: Papers, 1932–1975. WASP Series. Box 10. Eisenhower Presidential Library.

Letter, Hap Arnold to Jacqueline Cochran. "Subject: Deactivation of WASP." October 1, 1944. Jacqueline Cochran: Papers, 1932–1975. WASP Series. Box 12. Eisenhower Presidential Library.

Letter, Jacqueline Cochran to Commanding General, Army Air Force. "Inactivation of WASP." October 1, 1944. Jacqueline Cochran: Papers, 1932–1975. WASP Series. Box 12. Eisenhower Presidential Library.

Telegram, Attn Barney M. Giles Signed Arnold. Undated (1944). Jacqueline Cochran: Papers, 1932–1975. WASP Series. Box 12. Eisenhower Presidential Library.

Letter, Jacqueline Cochran (unsigned) to ALL WASP. October 2, 1944. Jacqueline Cochran: Papers, 1932–1975. WASP Series. Box 12. Eisenhower Presidential Library.

"Address by General H. H. Arnold." War Department Press Release. December 7, 1944. Jacqueline Cochran: Papers, 1932–1975. WASP Series. Box 14. Eisenhower Presidential Library.

"Statement by Miss Jacqueline Cochran on Accomplishments of WASP Program." War Department Press Release. December 19, 1944. Jacqueline Cochran: Papers, 1932–1975. WASP Series. Box 14. Eisenhower Presidential Library.

"WASP Deactivation Program to be Completed Wednesday." War Department Press Release. December 19, 1944. Jacqueline Cochran: Papers, 1932–1975. WASP Series. Box 14. Eisenhower Presidential Library.

CHAPTER 8

Cochran, *The Stars at Noon*.

Cochran and Brinley, *Jackie Cochran: The Autobiography of the Greatest Woman Pilot in Aviation History*.

Rich, *Jackie Cochran: Pilot in the Fastest Lane*.

"Miss Cochran Buys Building on 56th St." *New York Times*, September 16, 1945.

"Dr. W. Randolph Lovelace II—A Brief Biographical Sketch." Lovelace-Bataan Medical Center. 1976. Lovelace, Dr. W. Randolph Folder. NASA Archives.

Martin, Edward T. "The Hero of High-Altitude Flight." *Air Line Pilot*, February 1983. Lovelace, Dr. W. Randolph Folder. NASA Archives.

Elliott, "On a Comet Always."

Market, Howard. "Franklin D. Roosevelt's Painfully Eloquent Final Words." PBS, April 12, 2018. Accessed September 14, 2018. https://www.pbs.org/newshour/health/the-quiet-final-hours-of-franklin-d-roosevelt.

Leuchtenburg, William E. "Franklin D. Roosevelt: Impact and Legacy." UVA Miller Center, accessed August 12, 2019. https://millercenter.org/president/fdroosevelt/impact-and-legacy.

"Postwar Gender Roles and Women in American Politics." History, Art & Archives, United States House of Representatives, accessed August 12, 2019. https://history.house.gov/Exhibitions-and-Publications/WIC/Historical-Essays/Changing-Guard/Identity/.

"Daily Thought Capsules prescribed by Dr. Happiness and Family." October 1955. Jacqueline Cochran: Papers, 1932–1975. Primary Political Files. Box 10. Eisenhower Presidential Library.

"Women: Representations in Advertising." *AdAge*, accessed April 8, 2019. https://adage.com/article/adage-encyclopedia/women-representations-advertising/98938/.

Woods, *LBJ: Architect of American Ambition*.

Cobb and Rieker, *Woman into Space: The Jerrie Cobb Story*.

Wolk, Herman S. *Toward Independence: The Emergence of the U.S. Air Force 1945–1947*. Air Force History and Museum Programs, 1996. Accessed August 12, 2019. http://www.dtic.mil/dtic/tr/fulltext/u2/a433273.pdf.

"Air Force History." *Military*, accessed August 12, 2019. https://www.military.com/air-force-birthday/air-force-history.html.

Yeager, General Chuck and Leo Janos. *Yeager*. New York: Bantam Books, 1985.

Glass, Andrew. "Bernard Baruch coins term 'Cold War,' April 16, 1947." Politico, April 16, 2010. Accessed December 17, 2018. https://www.politico.com/story/2010/04/bernard-baruch-coins-term-cold-war-april-16-1947-035862.

Tolchin, Martin. "How Johnson Won Election He'd Lost." *New York Times*, February 11, 1990. https://www.nytimes.com/1990/02/11/us/how-johnson-won-election-he-d-lost.html.

CHAPTER 9

Cobb and Rieker, *Woman into Space: The Jerrie Cobb Story*.

CHAPTER 10

Cochran, *The Stars at Noon*.

Cochran and Brinley, *Jackie Cochran: The Autobiography of the Greatest Woman Pilot in Aviation History*.

Rich, *Jackie Cochran: Pilot in the Fastest Lane*.

"Air Record Disputed." *New York Times*. May 18, 1951.

"French Woman Flyer Sets New World Speed Record." *New York Times*, May 13, 1951.

Auriole, Jacqueline. *I Live to Fly*. Translated by Pamela Swinglehurst. Michael Joseph, Great Britain. 1970.

Letter, Floyd Odlum to Jacqueline Cochran. May 26, 1951. Jacqueline Cochran: Papers, 1932–1975. Trip Series. Box 3. Eisenhower Presidential Library.

Cochran, Jacqueline. "I Flew Faster than Sound." *American Weekly*, July 19, 1953. Jacqueline Cochran: Papers, 1932–1975. Scrapbook Series. Box 5. Eisenhower Presidential Library.

Letter, John Jay Hopkins to Jacqueline Cochran. March 17, 1953. Jacqueline Cochran: Papers, 1932–1975. Speed Records Series. Box 3. Eisenhower Presidential Library.

Letter, Jacqueline Cochran to Hoyt Vandenberg. "Personal and Confidential Memorandum." Unsigned and undated (1952). Jacqueline Cochran: Papers, 1932–1975. Speed Records Series. Box 3. Eisenhower Presidential Library.

Ravo, Nick. "Barry Gray, Pioneer of Talk Radio, Dies at 80." *New York Times*, December 23, 1996. Accessed November 5, 2018. https://www.nytimes.com/1996/12/23/nyregion/barry-gray-pioneer-of-talk-radio-dies-at-80.html.

Transcript, "Barry Gray Interviews Jacqueline Cochran." February 6/7, 1952. Jacqueline Cochran: Papers, 1932–1975. Eisenhower Campaign Series. Box 1. Eisenhower Presidential Library.

"U.S. Army Five-Star Generals." U.S. Army Center of Military History. Updated September 27, 2017. Accessed June 7, 2019. https://history.army.mil/html/faq/5star.html.

Smith, Jean Edward. *Eisenhower in War and Peace*. New York: Random House, 2012.
"John Hay Whitney." Britannica, accessed November 5, 2018. https://www.britannica.com/biography/John-Hay-Whitney.
"Jacqueline Cochran and the Women's Airforce Service Pilots (WASPs)." Eisenhower Presidential Library Online, accessed September 4, 2018. https://www.eisenhower.archives.gov/research/online_documents/jacqueline_cochran.html.
"Jacqueline Cochran Personal Analysis Chart." Jacqueline Cochran: Papers, 1932–1975. Business Series. Box 7. Eisenhower Presidential Library.
Cochran, Jacqueline. "The Sky Is No Limit." *Vogue*, July 1952. Jacqueline Cochran: Papers, 1932–1975. Stars at Noon Series. Box 6. Eisenhower Presidential Library.
Cochran, Jacqueline. "Women Can Fly as Well as Men." *Vogue*, August 1952. Jacqueline Cochran: Papers, 1932–1975. Stars at Noon Series. Box 6. Eisenhower Presidential Library.
"Introduction for Jacqueline Cochran." Jacqueline Cochran: Papers, 1932–1975. Eisenhower Campaign Series. Box 2. Eisenhower Presidential Library.
Letter, Jacqueline Cochran to General Eisenhower. February 15, 1952. Jacqueline Cochran Papers, Eisenhower Campaign Series. Box 2. Correspondence with General Eisenhower File. Eisenhower Presidential Library.
"News announcement regarding the New York campaign rally, February 12, 1952." Jacqueline Cochran Papers, Eisenhower Campaign Series. Box 1. Eisenhower–General File 1952. Eisenhower Presidential Library.
Letter, John Jay Hopkins to Jacqueline Cochran. March 16, 1953. Jacqueline Cochran: Papers, 1932–1975. Speed Records Series. Box 3. Eisenhower Presidential Library.
"Preliminary Flight Program for Sabre MK.3 Tests at Edwards AFB." November 20, 1952. Jacqueline Cochran: Papers, 1932–1975. Speed Records Series. Box 3. Eisenhower Presidential Library.
"Training Jackie Cochran to Become 1st Woman to Break Sound Barrier." Victoria Yeager, May 18, 2014. Accessed August 12, 2019. http://victoriayeager.com/training-jackie-cochran-to-become-1st-woman-to-break-sound-barrier/.
Letter, T. V. Maloney to Floyd Odlum. July 26, 1951. Odlum, Floyd: Papers, 1892–1976. Box 3. Eisenhower Presidential Library.
Letter, Charles Logsdon to Floyd Odlum. May 28, 1951. Odlum, Floyd: Papers, 1892–1976. Box 3. Eisenhower Presidential Library.
"Confidential for Maj. Yeager." March 23, 1953. Jacqueline Cochran: Papers, 1932–1975. Speed Records Series. Box 4. Eisenhower Presidential Library.
Memorandum to Chuck Yeager. May 26, 1953. Jacqueline Cochran: Papers, 1932–1975. Speed Records Series. Box 4. Eisenhower Presidential Library.
Note, Harry Bruno to Floyd Odlum re: phone call. March 5, 1953. Letter attached. Jacqueline Cochran: Papers, 1932–1975. Speed Records Series. Box 3. Eisenhower Presidential Library.
"Thomas K. Finletter April 24, 1950–January 20, 1953." U.S. Department of Defense. Accessed August 12, 2019. https://media.defense.gov/2016/Mar/11/2001479131/-1/-1/0/AFD-160311-466-006.PDF.
Telegram, HQ USAF Wash DC to Jacqueline Cochran. Jacqueline Cochran: Papers, 1932–1975. Speed Records Series. Box 3. Eisenhower Presidential Library.
Letter, Geoffrey Notman to Jacqueline Cochran. March 19, 1953. Jacqueline Cochran: Papers, 1932–1975. Speed Records Series. Box 3. Eisenhower Presidential Library.

"Disposition Form—Complain on Low Flying Aircraft." May 18, 1953. Jacqueline Cochran: Papers, 1932–1975. Speed Records Series. Box 3. Eisenhower Presidential Library.

Hendry, Erica R. "Today in History: Jackie Cochran Breaks the Sound Barrier." *Smithsonian,* May 28, 2010. Accessed November 6, 2018. https://www.smithsonianmag.com/smithsonian-institution/today-in-history-jackie-cochran-breaks-the-sound-barrier-130780022/.

Author interview with astronaut Reid Wiseman.

"23 May 1953." This Day in Aviation, accessed August 12, 2019. https://www.thisdayinaviation.com/23-may-1953/.

Telegram, Floyd Odlum to Jacqueline Cochran. December 1, 1952. Odlum, Floyd: Papers, 1892–1976. Box 3. Eisenhower Presidential Library.

Letter, Floyd Odlum to Charles Logsdon. December 1, 1952. Odlum, Floyd: Papers, 1892–1976. Box 3. Eisenhower Presidential Library.

Letter, Floyd Odlum to Mr. J. F. Davidson. November 15, 1952. Odlum, Floyd: Papers, 1892–1976. Box 3. Eisenhower Presidential Library.

Air Force Flight Test Center, Air Research & Development Command. AF Technical Report No. AFFTC 53-12, Establishment of a New World Speed Record with an F-86D. April 1953, Edwards Air Force Base. Jacqueline Cochran Papers, Scrapbook Series. Box 1. Eisenhower Presidential Library.

Letter, Randy Lovelace to Jacqueline Cochran. November 28, 1952. Jacqueline Cochran: Papers, 1932–1975. Speed Records Series. Box 3. Eisenhower Presidential Library.

Letter, Jacqueline Cochran to Wilf Curtis, August 4, 1953. Jacqueline Cochran: Papers, 1932–1975. Speed Records Series. Box 4. Eisenhower Presidential Library.

Place cards from Sabre (Canadair) dinner at Edwards AFB, May 1953. Jacqueline Cochran: Papers, 1932–1975. Scrapbook Series. Box 20. Eisenhower Presidential Library.

Phillips, Nicole M. "Nasogastric Tubes: An Historical Context." *MEDSURG Nursing*. April 2006, Vol. 15, No. 2.

Letter, Jacqueline Cochran to Fred Ascani, May 25, 1953. Jacqueline Cochran: Papers, 1932–1975. Speed Records Series. Box 4. Eisenhower Presidential Library.

Letter, Jacqueline Auriole to Jacqueline Cochran (original French and translation). Undated (1953). Jacqueline Cochran: Papers, 1932–1975. Speed Records Series. Box 4. Eisenhower Presidential Library.

Letter, Jacqueline Cochran to the President (Eisenhower), May 26, 1953. Jacqueline Cochran: Papers, 1932–1975. Speed Records Series. Box 4. Eisenhower Presidential Library.

CHAPTER 11

Cochran, Jacqueline. "Medical Aspects of Women as Pilots." Report for Dr. Boothby. 1953. Jacqueline Cochran: Papers, 1932–1975. Articles Series. Box 2. Eisenhower Presidential Library.

Letter, Jacqueline Cochran to Randy Lovelace. June 26, 1953. Jacqueline Cochran: Papers, 1932–1975. Articles Series. Box 2. Eisenhower Presidential Library.

Letter, Walter Boothby to Jacqueline Cochran. May 20, 1953. Jacqueline Cochran: Papers, 1932–1975. Articles Series. Box 2. Eisenhower Presidential Library.
Cochran, Jacqueline. "I Flew Faster than Sound." *American Weekly*, July 19, 1953. Jacqueline Cochran: Papers, 1932–1975. Scrapbook Series. Box 5. Eisenhower Presidential Library.
"The Strength Around Us, as told to Curtis Mitchell." Draft manuscript. 1953. Jacqueline Cochran: Papers, 1932–1975. Articles Series. Box 2. Eisenhower Presidential Library.
Cochran, *The Stars at Noon*.
Cobb and Rieker, *Woman into Space: The Jerrie Cobb Story*.
Cobb, Jerrie. *Solo Pilot*. Jerrie Cobb Foundation, Sun City Center, Florida. 1997.
"British Air Transport Auxiliary, Mary Zerbel Hooper Ford." http://www.airtransportaux.com/members/hooper-ford.html. Accessed August 9, 2020.

CHAPTER 12

Cobb, *Solo Pilot*.
"Immediate Release, James C. Hagerty, Press Secretary to the President, July 29, 1955." Eisenhower Presidential Library Online, accessed September 2, 2018. https://www.eisenhower.archives.gov/research/online_documents/igy/1955_7_29_Press_Release.pdf.
"News Report, National Academy of Sciences, National Research Council, March–April 1954." Eisenhower Presidential Library Online, accessed August 12, 2019. https://www.eisenhowerlibrary.gov/sites/default/files/research/online-documents/igy/1954-news-report.pdf.
"Korolev and Freedom of Space: February 14, 1955–October 4, 1957." NASA History Online, accessed November 8, 2018. https://history.nasa.gov/monograph10/korspace.html.

CHAPTER 13

"Speech Before Palm Springs Republican Assembly. January 27, 1956." Jacqueline Cochran: Papers, 1932–1975. Primary Political Files. Box 3. Eisenhower Presidential Library.
"JC Book Politics, Retyped March 1970." Red, handwritten. Jacqueline Cochran: Papers, 1932–1975. Stars at Noon Series. Box 6. Eisenhower Presidential Library.
"Brief Summary of Enclosed Platform of Jacqueline Cochran Odlum" incl. "Things I Believe," "Why I Am a Candidate," and "Platform and Beliefs," undated (1955–56). Jacqueline Cochran: Papers, 1932–1975. Primary Political Files. Box 18. Eisenhower Presidential Library.
Rich, *Jackie Cochran: Pilot in the Fastest Lane*.
"Cochran's Airborne Campaign." *LIFE*, May 7, 1956. Jacqueline Cochran: Papers, 1932–1975. Primary Political Files. Box 18. Eisenhower Presidential Library.
"Desert Dwellers for 20 Years." *Palm Springs Villager*, May 1956. Jacqueline Cochran: Papers, 1932–1975. Primary Political Files. Box 18. Eisenhower Presidential Library.
"Jacqueline Cochran Odlum Leads Race for Congress." *Riverside Imperial Bulletin*,

May–June 1956. Jacqueline Cochran: Papers, 1932–1975. Primary Political Files. Box 18. Eisenhower Presidential Library.

"Salute to Jackie" ticket, May 27, 1956. Jacqueline Cochran: Papers, 1932–1975. Primary Political Files. Box 5. Eisenhower Presidential Library.

"7000 Undeterred by Dust, Attend Salute for Jackie." *Indio News*, May 29, 1956. Jacqueline Cochran: Papers, 1932–1975. Primary Political Files. Box 5. Eisenhower Presidential Library.

Cobb, *Solo Pilot*.

"SAUND, Dalip Singh (Judge)." History, Art & Archives, United States House of Representatives, accessed September 4, 2018. http://history.house.gov/People/detail/21228.

Letter, W. Sterling Cole(?) to Jacqueline Cochran. April 19, 1956. Jacqueline Cochran: Papers, 1932–1975. Primary Political Files. Box 8. Eisenhower Presidential Library.

Letter, Jimmy Doolittle to Jacqueline Cochran. April 16, 1956 (Florence Walsh to Mr. Ainsworth. Undated. Attached). Jacqueline Cochran: Papers, 1932–1975. Primary Political Files. Box 8. Eisenhower Presidential Library.

Letter, E. N. Eisenhower to Jacqueline Cochran. May 7, 1956. Jacqueline Cochran: Papers, 1932–1975. Primary Political Files. Box 8. Eisenhower Presidential Library.

Letter, George C. Kenney to Jacqueline Cochran. May 7, 1956. Jacqueline Cochran: Papers, 1932–1975. Primary Political Files. Box 8. Eisenhower Presidential Library.

Letter, Jacqueline Cochran to George C. Kenney. May 18, 1956. Jacqueline Cochran: Papers, 1932–1975. Primary Political Files. Box 8. Eisenhower Presidential Library.

Letter, John Reed Kilpatrick to Jacqueline Cochran. August 15, 1956. Jacqueline Cochran: Papers, 1932–1975. Primary Political Files. Box 8. Eisenhower Presidential Library.

"Excerpts from My Mother India by Dalip Singh Saund." Undated (1956) Jacqueline Cochran: Papers, 1932–1975. Primary Political Files. Box 18. Eisenhower Presidential Library.

Friedrich, Pieter. "First Asian in U.S. Congress Was a Sikh Inspired by Civil Rights Principles." *India West*, August 5, 2012. Accessed August 12, 2019. https://www.indiawest.com/blogs/first-asian-in-u-s-congress-was-a-sikh-inspired/article_1f76fde2-3b71-11e5-ac41-1bafc2ce7063.html.

"Rank, Odious, Prejudice..." *Beaumont Gazette*, October 25, 1956. Odlum, Floyd Papers: 1892–1976. Box 115. Eisenhower Presidential Library.

"To the Many Volunteer Workers in the Jacqueline Cochran Odlum for Congress Committee." Undated (1956). Odlum, Floyd Papers: 1892–1976. Box 115. Eisenhower Presidential Library.

Letter, Admiral USN (Retired) to Editor of *Press-Enterprise*, Riverside. October 29, 1956. Odlum, Floyd Papers: 1892–1976. Box 115. Eisenhower Presidential Library.

Letter, Unknown (former WASP) to Editor, *Press-Enterprise*, Riverside. Draft. October 25, 1956. Odlum, Floyd Papers: 1892–1976. Box 115. Eisenhower Presidential Library.

Telegram, Jacqueline Cochran to D. S. Saund. November 7, 1956. Odlum, Floyd Papers: 1892–1976. Box 115. Eisenhower Presidential Library.

Letter, Dwight Eisenhower to Jacqueline Cochran. November 7, 1956. Odlum, Floyd Papers: 1892–1976. Box 115. Eisenhower Presidential Library.

"Salute to Jackie Riverside County Fair Grounds." Stage Show Agenda. May 27, 1956. Jacqueline Cochran: Papers, 1932–1975. Primary Political Files. Box 5. Eisenhower Presidential Library.

"Invitation to the Ranch 3:30 Sunday, May 27." 1956. Jacqueline Cochran: Papers, 1932–1975. Primary Political Files. Box 5. Eisenhower Presidential Library.

"Note on People Speaking for JCO at Rally in Indio, Sunday, May 27." 1956. Jacqueline Cochran: Papers, 1932–1975. Primary Political Files. Box 5. Eisenhower Presidential Library.

"Coachella Valley Committee for Jacqueline Cochran Odlum for Congress." May 27, 1956. Jacqueline Cochran: Papers, 1932–1975. Primary Political Files. Box 5. Eisenhower Presidential Library.

Letter, Rosalind Russel to Jacqueline Cochran. Undated. Jacqueline Cochran: Papers, 1932–1975. Primary Political Files. Box 5. Eisenhower Presidential Library.

"Release by Jacqueline Cochran Odlum." Undated. Jacqueline Cochran Papers, 1932–75. Primary Political Files. Eisenhower Presidential Library.

CHAPTER 14

Cobb and Rieker, *Woman into Space: The Jerrie Cobb Story*.

"Jerrie Cobb (USA)." Fédération Aéronautique International (FAI), accessed August 12, 2019. https://www.fai.org/record/8111.

Ferrell, Don. "State's Flying Ambassador Ends Record Hop Still Fresh." *Daily Oklahoman*, May 26, 1957. Accessed August 10, 2019. https://www.newspapers.com/image/450218692/?terms=Jerrie%2BCobb.

"Jerrie Reaches Altitude Record." *Daily Oklahoman*, June 14, 1957. Accessed August 17, 2019. https://www.newspapers.com/image/450227914/?terms=Jerrie%2BCobb.

"Jerrie Cobb (USA)." Fédération Aéronautique International (FAI), accessed June 11, 2018. https://www.fai.org/record/jerrie-cobb-usa-10283.

Glenn, John with Nick Taylor. *John Glenn: A Memoir*. Bantam Books, New York. 1999.

"Important for Women to Know Their Skin Type." August 7, 1957. Jacqueline Cochran: Papers, 1932–1975. Business Series. Box 7. Eisenhower Presidential Library.

"On Television." *New York Times*, Tuesday October 1, 1957. Accessed August 18, 2019. https://timesmachine.nytimes.com/timesmachine/1957/10/01/91165013.pdf.

"*Name That Tune* Episode dated 4 October 1957." IMDb online, accessed June 11, 2018. https://www.imdb.com/title/tt5734410/.

Name That Tune Wiki, accessed June 11, 2018. http://name-that-tune.wikia.com/wiki/Name_That_Tune_(1953-1959).

Dickson, Paul. *Sputnik: The Shock of the Century*. New York: Berkley Books, 2007.

Woods, *LBJ: Architect of American Ambition*.

Mieczkowski, Yanek. *Eisenhower's Sputnik Moment: The Race for Space and World Prestige*. Ithaca, New York: Cornell University Press, 2013.

AFTER SPUTNIK, OCTOBER 1957: A SPACE INTERLUDE

"Round the World in 96 Minutes." *New York Times*, Sunday October 6, 1957. https://timesmachine.nytimes.com/timesmachine/1957/10/06/91166843.pdf.

Blakeslee, Alton L. "Visual Spotting of Soviet's Baby Moon Hard, Depends on Radio Tracking." *Daily Oklahoman*, Sunday October 6, 1957. Accessed August 19, 2019. https://www.newspapers.com/image/450186647/?terms=sputnik.

"Dwight D. Eisenhower Executive Office Building, Washington, DC." U.S. General Services Administration. Accessed November 14, 2018. https://www.gsa.gov/historic-buildings/dwight-d-eisenhower-executive-office-building-washington-dc.

Mieczkowski, *Eisenhower's Sputnik Moment: The Race for Space and World Prestige*.

"Official White House Transcript of President Eisenhower's Press and Radio Conference #123." Eisenhower Presidential Library Online, accessed June 12, 2018. https://www.eisenhower.archives.gov/research/online_documents/sputnik/10_9_57.pdf.

Reedy, George. *Lyndon B. Johnson: A Memoir*. New York: Andrews and McMeel, Inc., 1982.

"Remarks by the Vice President on Space and Space Exploration in 1957 and 1958," with memo, GER to LBJ attached, February 22, 1962. Vice Presidential Aide's Files of George Reedy. Box 12. The Lyndon Baines Johnson Library.

"Record of the Vice President on Space Activities." 1957–1962. Vice Presidential Aide's Files of George Reedy. Box 14. The Lyndon Baines Johnson Library.

Wang, Zuoyue. *In Sputnik's Shadow: The President's Science Advisory Committee and Cold War America*. New Brunswick, NJ: Rutgers University Press, 2008.

Dickson, Paul. *Sputnik: The Shock of the Century*. New York: Berkley Books, 2007.

S. Res. 256. "Resolution Establishing a Special Committee on Astronautics and Space Exploration." February 5, 1958. United States Senate, 1949–1961. Committee on Armed Services, Special Cmte. on Space and Astronautics. Box 357. The Lyndon Baines Johnson Library.

Woods, *LBJ: Architect of American Ambition*.

Glenn, *John Glenn: A Memoir*.

Letter, Lyndon Johnson to Dwight Eisenhower. March 25, 1958. United States Senate, 1949–1961. Subject Files, 1958. Box 630. The Lyndon Baines Johnson Library.

S. Res. 3609, Bill Submitted by LBJ. April 14, 1958. United States Senate, 1949–1961. Committee on Armed Services Special Cmte. on Space and Astronautics. Box 357. The Lyndon Baines Johnson Library.

Opening Statement by Chairman Lyndon B. Johnson Before the Senate Special Committee on Space and Astronautics. May 6, 1958. United States Senate, 1949–1961. Committee on Armed Services Special Cmte. on Space and Astronautics. Box 357. The Lyndon Baines Johnson Library.

Memo from MacIntyre. May 12, 1958. United States Senate, 1949–1961. Committee on Armed Services Special Cmte. on Space and Astronautics. Box 357. The Lyndon Baines Johnson Library.

"Previous FAI General Conferences." Fédération Aéronautique Internationale

(FAI), accessed November 14, 2018. https://www.fai.org/previous-fai-general-conferences.
Elliott, "On a Comet Always."
"Minutes of the Meeting of the Working Group on Human Factors and Training." April 24–25,1958. Lovelace, Dr. W. Randolph Folder. NASA Archives.
"Immediate Release, James C. Hagerty, Press Secretary to the President, July 29, 1957." Eisenhower Presidential Library Online, accessed November 14, 2018. https://www.eisenhower.archives.gov/research/online_documents/nasa/Binder17.pdf.
Swenson, Loyd, et. al. "Beginnings of Space Medicine," in *This New Ocean: A History of Project Mercury*, NASA. Accessed August 12, 2019. https://history.nasa.gov/SP-4201/ch2-2.htm.
"Invitation to Apply for Position of Research Astronaut-Candidate." NASA Project A Announcement No. 1. December 22, 1958. Accessed August 18, 2019. https://history.nasa.gov/40thmerc7/invite.pdf.
Swenson, Loyd, et. al. "Project Astronaut," in *This New Ocean: A History of Project Mercury*, NASA. Accessed August 12, 2019. https://history.nasa.gov/SP-4201/ch5-8.htm.
Swenson, Loyd S., Jr., James M. Grimwood, and Charles C. Alexander. *This New Ocean: A History of Project Mercury*. NASA Historical Series SP-4201. Washington. 1966.
"Telephone Conversation with Dr. T. Keith Glennan," from Lee D. Saegesser. January 24, 1995. Courtesy of Francis French.
"Memorandum of Conference with the President." A. J. Goodpaster. December 22, 1958. Courtesy of Francis French.
Burgess, Colin. *Selecting the Mercury Seven: The Search for America's First Astronauts*. Chichester, UK: Springer-Praxis, 2011.

CHAPTER 15

Cobb and Rieker, *Woman into Space: The Jerrie Cobb Story*.
"Mrs. Odlum Elected Head of Flier Group." *Los Angeles Times*, April 17, 1958. Accessed August 10, 2019. https://www.newspapers.com/image/381007740/?terms=Jacqueline%2BCochran%2C%2BFAI.
"International Aviation Education Commission of FAI." March 16, 1959. Jacqueline Cochran Papers, 1932–75. Federation Aeronautique Internationale Series (FAI). Box 11. Eisenhower Presidential Library.
"Astronautics Sub-Committee of International Sporting Aviation Commission of FAI." March 16, 1959. Jacqueline Cochran Papers, 1932–75. Federation Aeronautique Internationale Series (FAI). Box 11. Eisenhower Presidential Library.
"Astronautics Sub-Committee of International Sporting Aviation Commission of FAI." March 16, 1959. Jacqueline Cochran: Papers, 1932–1975. Federation Aeronautique Internationale Series (FAI). Box 11. Eisenhower Presidential Library.
Letter, Jacqueline Cochran to the President, Real Aero Club d'Espagna. February 1, 1959. Jacqueline Cochran Papers, 1932–75. Federation Aeronautique Internationale Series (FAI). Box 11. Eisenhower Presidential Library.

Letter, Jacqueline Cochran to James Walsh. February 23, 1959. Jacqueline Cochran: Papers, 1932–1975. Federation Aeronautique Internationale Series (FAI). Box 11. Eisenhower Presidential Library.

"Memorandum from Miss Jacqueline Cochran." January 21, 1959. Jacqueline Cochran: Papers, 1932–1975. Federation Aeronautique Internationale Series (FAI). Box 11. Eisenhower Presidential Library.

"150 Delegates to FAI Plan Visit Here." *Desert Sun*, April 16, 1958. Accessed August 18, 2019. https://www.newspapers.com/image/238623577/?terms=Jacqueline%2BCochran%2C%2BFAI.

"Hotel Accommodations for Members of Education Committee & Sub-Committee of CASA—Las Vegas." Undated. Jacqueline Cochran Papers, 1932–75. Federation Aeronautique Internationale Series (FAI). Box 11. Eisenhower Presidential Library.

"40th Anniversary of the Mercury 7." NASA, accessed June 12, 2018. https://history.nasa.gov/40thmerc7/documents.htm.

"Press Conference, Mercury Astronaut Team, Thursday 9 April 1959." Transcript. NASA, accessed June 12, 2018. https://history.nasa.gov/40thmerc7/presscon.pdf.

"Press Conference Introducing 7 Mercury Astronauts 1959 Part 1/3." Bazi, accessed November 15, 2018. https://www.youtube.com/watch?v=FXj5lc_QUOM.

Glenn, *John Glenn: A Memoir*.

Coffey, Ivy. "Sooner Pilot Hopes She Will Be Flying a Record-Breaker." *Daily Oklahoman*. April 12, 1959.

"Skyways Aircraft Ferrying Service, Inc. v. Stanton." Justia, accessed November 16, 2018. https://law.justia.com/cases/california/court-of-appeal/2d/242/272.html.

Author interview with Francis French. July 31, 2019.

Letter, Don Flickinger to Jerrie Cobb. December 7, 1959. Courtesy of Francis French.

Letter, Don Flickinger to Randy Lovelace, December 20, 1959. Courtesy of Francis French.

Caruthers, Osgood. "Back of Moon Seen First Time; Photo by Soviet Rocket Shows Fewer Craters than Face." *New York Times*, October 27, 1959. Accessed November 17, 2018. https://timesmachine.nytimes.com/timesmachine/1959/10/27/83442730.pdf.

Memorandum, Glen Wilson to Ken Belieu. "Significance of Soviet Moon Photographs." October 28, 1959. United States Senate, 1949–1961. Subject Files, 1959. Box 730. The Lyndon Baines Johnson Library.

Mieczkowski, *Eisenhower's Sputnik Moment: The Race for Space and World Prestige*.

Letter, Jacqueline Cochran to Lyndon Johnson. November 21, 1959. LBJA Famous Names. Box 2. The Lyndon Baines Johnson Library.

Letter, Lyndon Johnson to Jacqueline Cochran. December 2, 1959. United States Senate, 1949–1961. "Master File" Index. Nor–Oa-Oe. Box 138. The Lyndon Baines Johnson Library.

Memorandum to Senator Johnson. November 25, 1959. United States Senate, 1949–1961. Subject Files, 1959. Box 730. The Lyndon Baines Johnson Library.

Woods, *LBJ: Architect of American Ambition*.

Look magazine. Vol. 24 No. 3, February 2, 1959.

CHAPTER 16

Cobb and Rieker, *Woman into Space: The Jerrie Cobb Story.*

"Wally Funk Schedule," courtesy of Francis French.

Steadman, Bernice Trimble. *Tethered Mercury: A Pilot's Memoir the Right Stuff...but the Wrong Sex*. Traverse City, Michigan: Aviation Press, 2001.

Woods, *LBJ: Architect of American Ambition.*

Caro, Robert A. *The Years of Lyndon B. Johnson: Master of the Senate*. New York: Vintage Books, 2003.

Caro, Robert A. *The Years of Lyndon Johnson: Means of Ascent*. New York: Vintage Books, 1990.

Caro, Robert A. *The Years of Lyndon Johnson: The Path to Power.* New York: Vintage Books, 1990.

Caro, Robert A. *Lyndon Johnson: The Passage of Power.* New York: Vintage Books, 2012.

Letter, Lyndon Johnson to Floyd Odlum. November 18, 1960. United States Senate, 1949–1961. "Master File" Index. Nor–Oa-Oe. Box 138. The Lyndon Baines Johnson Library.

CHAPTER 17

Cobb and Rieker, *Woman into Space: The Jerrie Cobb Story.*

"A Woman Passes Test Given to 7 Astronauts." *New York Times*, August 19, 1960.

"A Lady Proves She's Fit for Space Flight." *LIFE*, August 29, 1960.

"Spacewoman Ready for Flights with Men." *Washington Star*, August 24, 1960.

"Girls Ride to Space Still Long Way Off." *Dayton Daily News*, September 29, 1960.

Letter, Clark T. Randt to Randy Lovelace. September 23, 1960. Women/Mercury Astronauts Folder. NASA Archives.

Ackmann, Martha. *Mercury 13: The True Story of Thirteen Women and the Dream of Space Flight*. New York: Random House Trade Paperbacks, 2003.

Steadman, *Tethered Mercury: A Pilot's Memoir the Right Stuff...but the Wrong Sex.*

Dietrich, Marion. "First Woman into Space." *McCall's*, September 1961.

Letter, Jeanne Williams to Jacqueline Cochran. November 30, 1960. Jacqueline Cochran: Papers, 1932–1975. General Files Series. Box 139. Eisenhower Presidential Library.

Marion Dietrich to Jan Dietrich. January 13, 1961. Courtesy of International Women's Air and Space Museum.

Letter, Wally Funk to Jay Shurley. October 25, 1960. Courtesy of Francis French.

Letter, Jerrie Cobb to Wally Funk. November 30, 1960. Courtesy of Francis French.

Funk, Wally. "NASA Oral History," interview by Carol Butler. NASA, July 18, 1999. Accessed August 19, 2019. https://historycollection.jsc.nasa.gov/JSCHistoryPortal/history/oral_histories/NASA_HQ/Aviatrix/FunkW/FunkW_7-18-99.htm.

Letter, Jay Shurley to Wally Funk. November 2, 1960. Courtesy of Francis French.

Letter, Wally Funk to Randy Lovelace. November 5, 1960. Courtesy of Francis French.

Letter, Randy Lovelace to Wally Funk. November 11, 1960. Courtesy of Francis French.

Letter, Wally Funk to Randy Lovelace. November 21, 1960. Courtesy of Francis French.

Letter, Jerrie Cobb to Wally Funk. November 30, 1960. Courtesy of Francis French.

Woods, *LBJ: Architect of American Ambition*.

Weitekamp, Margaret A. *Right Stuff Wrong Sex: America's First Woman in Space Program*. Baltimore: Johns Hopkins University Press, 2004.

Letter, Jacqueline Cochran to Randy Lovelace. November 28, 1960. Jacqueline Cochran: Papers, 1932–1975. General Files Series. Box 139. Eisenhower Presidential Library.

Interoffice Communication. "Information from Mrs. Quarles re Women Pilots." December 6, 1960. Jacqueline Cochran: Papers, 1932–1975. General Files Series. Box 138. Eisenhower Presidential Library.

Letter, Jacqueline Cochran to Randy Lovelace. November 28, 1960. Jacqueline Cochran: Papers, 1932–1975. General Files Series. Box 139. Eisenhower Presidential Library.

Letter, Jacqueline Cochran to Randy Lovelace. December 27, 1960. Jacqueline Cochran: Papers, 1932–1975. General Files Series. Box 139. Eisenhower Presidential Library.

"Jacqueline Cochran Stresses Top Role for Women in Space." *Dallas Morning News*, Wednesday, January 18, 1961.

CHAPTER 18

Letter, Randy Lovelace to Floyd Odlum. June 8, 1961. Jacqueline Cochran: Papers, 1932–1975. General Files Series. Box 139. Eisenhower Presidential Library.

Letter, Jacqueline Cochran to Randy Lovelace. January 14, 1961. Jacqueline Cochran: Papers, 1932–1975. General Files Series. Box 139. Eisenhower Presidential Library.

Ackmann, *Mercury 13: The True Story of Thirteen Women and the Dream of Space Flight*.

Letter, Jan Dietrich to Floyd Odlum. January 23, 1961. Jacqueline Cochran: Papers, 1932–1975. General Files Series. Box 139. Eisenhower Presidential Library.

Letter, Jan Dietrich to Jacqueline Cochran. January 23, 1961. Jacqueline Cochran: Papers, 1932–1975. General Files Series. Box 139. Eisenhower Presidential Library.

Letter, Floyd Odlum to Randy Lovelace. February 6, 1961. Jacqueline Cochran: Papers, 1932–1975. General Files Series. Box 139. Eisenhower Presidential Library.

Letter, Randy Lovelace to Wally Funk. February 2, 1961. Courtesy of Francis French.

Webb, James. "NASM Oral History Project." Interview by David DeVorkin and Joe Tatarewicz. Washington, DC. February 22, 1985. National Air and Space Museum. Accessed August 19, 2019. https://airandspace.si.edu/research/projects/oral-histories/TRANSCPT/WEBB1.HTM.

Bizony, Piers. *The Man Who Ran the Moon: James E. Webb, NASA, and the Secret History of Project Apollo*. New York: Thunder's Mouth Press, 2006.

Transcript, James E. Webb Oral History Interview I, 4/29/69, by T. H. Baker, Internet

Copy, LBJ Library. Accessed August 19, 2019. http://www.lbjlibrary.net/assets/documents/archives/oral_histories/webb-j/webb.pdf.

Woods, *LBJ: Architect of American Ambition.*

Dryden, Hugh L. "Oral History Interview." Interviewed by Walter D. Sohier, Arnold W. Frutkin, and Eugene M. Emme. Washington, DC., March 26, 1964. John F. Kennedy Library. Accessed August 19, 2019. https://www.jfklibrary.org/sites/default/files/archives/JFKOH/Dryden%2C%20Hugh%20L/JFKOH-HLD-01/JFKOH-HLD-01-TR.pdf.

Interoffice Memorandum to Jackie from Floyd. March 6, 1961. Jacqueline Cochran: Papers, 1932–1975. General Files Series. Box 139. Eisenhower Presidential Library.

Dietrich, Marion. "First Woman into Space." *McCall's*, September 1961.

Glenn, *John Glenn: A Memoir.*

Wainwright, Loudon S. "The Chosen Three For First Space Ride." *LIFE*, March 3, 1961.

Steadman, *Tethered Mercury: A Pilot's Memoir the Right Stuff... but the Wrong Sex.*

Weitekamp, *Right Stuff Wrong Sex: America's First Woman in Space Program.*

Doran, Jamie and Piers Bizony. *Starman: The Truth Behind the Legend of Yuri Gagarin.* London: Bloomsbury, 1998.

"Reds Orbit and Return Spaceman." *Philadelphia Daily News*, April 12, 1961.

Brooks, Courtney G., James M. Grimwood, and Loyd S. Swenson, Jr. *Chariots for Apollo: The NASA History of Manned Lunar Spacecraft to 1969.* Mineola, New York: Dover Publications, 2009.

"Soviet Orbits Man and Recovers Him; Space Pioneer Reports: 'I Feel Well'; Sent Messages While Circling Earth." *New York Times*, Wednesday, April 12, 1961.

"News Conference 9, April 12, 1961." John F. Kennedy Library and Museum online, accessed September 17, 2018. https://www.jfklibrary.org/Research/Research-Aids/Ready-Reference/Press-Conferences/News-Conference-9.aspx.

Hearing Before the Committee on Science and Astronautics, US House of Representatives. (H.R 6169.) April 12, 1961. Vice President, 1961–1963. Box 116. The Lyndon Baines Johnson Library.

"Draft of Letter from the President to the Vice President." April 1961. Vice President, 1961–1963. Box 117. The Lyndon Baines Johnson Library.

"Record of the vice President on Space Activities during 1957–1962" prepared by Eileen Galloway. March 9, 1962. Vice Presidential Aide's Files of George Reedy. Box 12. The Lyndon Baines Johnson Library.

Woods, *LBJ: Architect of American Ambition.*

"The Bay of Pigs." JFK Presidential Library and Museum Online, accessed November 19, 2018. https://www.jfklibrary.org/learn/about-jfk/jfk-in-history/the-bay-of-pigs.

Jessen, Gene Nora. "New Horizons—Irene Leverton (7/23/2017)." Ninety-Nines online, accessed November 18, 2018. https://www.ninety-nines.org/NH-Irene_Leverton_8.htm.

"Irene H. Leverton." Smithsonian National Air and Space Museum Online. https://airandspace.si.edu/support/wall-of-honor-irene-h-leverton. Accessed October 22, 2019.

Memorandum, John F. Kennedy to Lyndon Johnson. April 20, 1961. White House Famous Names, Service Set Box 6. The Lyndon Baines Johnson Library.

"Memorandum for the President." April 28, 1961. White House Famous Names, Service Set Box 6. The Lyndon B. Johnson Library.

Cochran, Jacqueline. "Women in Space." *Parade*, April 30 , 1961. Jacqueline Cochran: Papers, 1932–1975. Articles Series. Box 7. Eisenhower Presidential Library.

"Parade Article," April 18, 1961. Jacqueline Cochran: Papers, 1932–1975. Articles Series. Box 7. Eisenhower Presidential Library.

Letter, Myrtle Thompson Cagle to Jacqueline Cagle. May 1, 1961. Jacqueline Cochran: Papers, 1932–1975. General Files Series. Box 139. Eisenhower Presidential Library.

Letter, Carra Elaine Harrison to Jacqueline Cochran. June 3, 1961. Jacqueline Cochran: Papers, 1932–1975. General Files Series. Box 139. Eisenhower Presidential Library.

Letter, Bobby Galbreath to Jacqueline Cochran. May 12, 1961. Jacqueline Cochran: Papers, 1932–1975. General Files Series. Box 139. Eisenhower Presidential Library.

Letter, Myrtle Thompson Cagle to Jacqueline Cochran. June 20, 1961. Jacqueline Cochran: Papers, 1932–1975. General Files Series. Box 139. Eisenhower Presidential Library.

Cobb, Jerrie. "Woman's Participation in Space" speech at Aviation/Space Writers Association 23rd Annual Meeting and News Conference. May 1, 1961. NASA Archives.

"Last-Minute Qualms" in Swenson et. al., *This New Ocean*. Accessed August 19, 2019. https://history.nasa.gov/SP-4201/ch11-3.htm.

Letter, Jacqueline Cochran to Judith Ostrozny. June 6, 1961. Jacqueline Cochran: Papers, 1932–1975. General Files Series. Box 139. Eisenhower Presidential Library.

Letter, Lyndon Johnson to James Webb. May 5, 1961. 1962 Subject File. Science—Space and Aeronautics. [Women in Space] The Lyndon Baines Johnson Library.

Letter, Lyndon Johnson to James Webb. May 5, 1961. Vice President, 1961–1963. Box 116. The Lyndon Baines Johnson Library.

French, Francis and Colin Burgess. *Into that Silent Sea: Trailblazers of the Space Era, 1961–1965*. University of Nebraska Press, Lincoln. 2007.

Cobb and Rieker, *Woman into Space: The Jerrie Cobb Story*.

Letter, Randy Lovelace to Wally Funk. May 17, 1961. Courtesy of Francis French.

Letter, Pat Jetton to Jacqueline Cochran. March 17, 1961. Jacqueline Cochran: Papers, 1932–1975. General Files Series. Box 139. Eisenhower Presidential Library.

Letter, Wally Funk to Jacqueline Cochran. March 17, 1961. Jacqueline Cochran: Papers, 1932–1975. General Files Series. Box 139. Eisenhower Presidential Library.

Letter, Frances Bera to Jacqueline Cochran. March 29, 1961. Jacqueline Cochran: Papers, 1932–1975. General Files Series. Box 139. Eisenhower Presidential Library.

Letter, Rhea Hurrle to Jacqueline Cochran. May 20, 1961. Jacqueline Cochran: Papers, 1932–1975. General Files Series. Box 139. Eisenhower Presidential Library.

Letter, Bernice Steadman to Jacqueline Cochran. May 6, 1961. Jacqueline Cochran: Papers, 1932–1975. General Files Series. Box 139. Eisenhower Presidential Library.

Letter, Irene Leverton to Jacqueline Cochran. June 23, 1961. Jacqueline Cochran: Papers, 1932–1975. General Files Series. Box 139. Eisenhower Presidential Library.

Letter, Jerry Sloan to Jacqueline Cochran. May 21, 1961. Jacqueline Cochran: Papers, 1932–1975. General Files Series. Box 139. Eisenhower Presidential Library.

"Address to Joint Session of Congress May 25, 1961." John F. Kennedy Presidential Library and Museum online, accessed September 17, 1961. https://www.jfklibrary.org/Asset-Viewer/xzw1gaeeTES6khED14P1Iw.aspx.

"President Kennedy's Special Message to the Congress on Urgent National Needs, May 25, 1961." John F. Kennedy Presidential Library and Museum online, accessed August 12, 2019. https://www.jfklibrary.org/archives/other-resources/john-f-kennedy-speeches/united-states-congress-special-message-19610525.

Swanson, Glen E. ed. *Before this Decade Is Out: Personal Reflections on the Apollo Program*. Mineola, New York: Dover Publications, 2012.

Levine, Arnold S. *Managing NASA in the Apollo Era*. NASA SP-4102. The National Aeronautics and Space Administration, Washington, DC, 1982. Accessed August 19, 2019. https://history.nasa.gov/SP-4102.pdf.

Remarks by James E. Webb, Administrator. First National Conference on the Peaceful Uses of Space. Tulsa, Oklahoma. May 26, 1961. Cobb, Jerrie Folders. NASA Archives.

CHAPTER 19

Cobb, Jerrie. Untitled recommendations to NASA. June 15, 1961. Cobb, Jerrie #2. NASA Archives.

Ackmann, *Mercury 13: The True Story of Thirteen Women and the Dream of Space Flight*.

Letter, Floyd Odlum to Randy Lovelace. May 31, 1961. Jacqueline Cochran: Papers, 1932–1975. General Files Series. Box 139. Eisenhower Presidential Library.

Letter, Jerrie Cobb to Jacqueline Cochran (handwritten, Hotel to Crillon). Undated (May 29, 1961). Jacqueline Cochran: Papers, 1932–1975. General Files Series. Box 185. Eisenhower Presidential Library.

Letter, Secretary (Jacqueline Cochran's) to Jerrie Cobb. May 31, 1961. Jacqueline Cochran: Papers, 1932–1975. General Files Series. Box 185. Eisenhower Presidential Library.

Letter, Jerrie Cobb to Jacqueline Cochran (handwritten, Hotel to Crillon). Undated (June 1, 1961). Jacqueline Cochran: Papers, 1932–1975. General Files Series. Box 185. Eisenhower Presidential Library.

Letter, Randy Lovelace to Floyd Odlum. June 8, 1961. Jacqueline Cochran: Papers, 1932–1975. General Files Series. Box 139. Eisenhower Presidential Library.

Beeson, Lovelace, et al. to Abraham Hyatt. "Life Sciences Program Leading to Manned Landing on the Moon." May 31, 1961. Lovelace, Dr. W. Randolph Folder. NASA Archives.

Letter to Randy Lovelace. June 16, 1961. Jacqueline Cochran: Papers, 1932–1975. General Files Series. Box 139. Eisenhower Presidential Library.

Weitekamp, *Right Stuff Wrong Sex: America's First Woman in Space Program*.

Letter to Randy Lovelace. June 16, 1961. Jacqueline Cochran: Papers, 1932–1975. General Files Series. Box 139. Eisenhower Presidential Library.

Author correspondence with Francis French. August 10, 2019.

Author interview with Gene Nora Jessen née Stumbough and Sarah Ratley. June 2017.

Letter, Randy Lovelace to Wally Funk. July 8, 1961. Courtesy of Francis French.

Letter, Randy Lovelace to Jacqueline Cochran. July 12, 1961. Jacqueline Cochran: Papers, 1932–1975. General Files Series. Box 140. Eisenhower Presidential Library.

Letter from Randy Lovelace (form letter). July 8, 1961. Jacqueline Cochran: Papers, 1932–1975. General Files Series. Box 140. Eisenhower Presidential Library.

"Letter Randy is sending out to the women." Undated. Jacqueline Cochran: Papers, 1932–1975. General Files Series. Box 140. Eisenhower Presidential Library.

Letter, Randy Lovelace to Wally Funk. July 12, 1961. Courtesy of Francis French.

Letter, Jacqueline Cochran to Wally Funk. July 12, 1961. Courtesy of Francis French.

Letter, Jacqueline Cochran to Myrtle Thompson Cagle. July 12, 1961. Jacqueline Cochran: Papers, 1932–1975. General Files Series. Box 140. Eisenhower Presidential Library.

Letter, Floyd Odlum to Randy Lovelace. July 19, 1961. Jacqueline Cochran: Papers, 1932–1975. General Files Series. Box 140. Eisenhower Presidential Library.

Letter, Jacqueline Cochran to Jerrie Cobb. July 10, 1961. Jacqueline Cochran: Papers, 1932–1975. General Files Series. Box 140. Eisenhower Presidential Library.

Letter, Jacqueline Cochran to Marion Dietrich. July 12, 1961. Jacqueline Cochran: Papers, 1932–1975. General Files Series. Box 140. Eisenhower Presidential Library.

Jessen, Gene Nora. *The Fabulous Flight of the Three Musketeers*. 1st Books. 2009.

Letter, Wally Funk to Jacqueline Cochran. July 19, 1961. Jacqueline Cochran: Papers, 1932–1975. General Files Series. Box 140. Eisenhower Presidential Library.

Letter, Sarah Lee Gorelick to Jacqueline Cochran. July 20, 1961. Jacqueline Cochran: Papers, 1932–1975. General Files Series. Box 140. Eisenhower Presidential Library.

Steadman, *Tethered Mercury: A Pilot's Memoir the Right Stuff...but the Wrong Sex*.

Author email correspondence with Al Hallonquist.

"Mercury-Redstone 4." NASA, accessed August 12, 2019. https://www.nasa.gov/mission_pages/mercury/missions/libertybell7.html.

Author interview with Gene Nora Jessen. March 28, 2019.

Author correspondence with Francis French. August 10, 2019.

Swanson, ed. *Before this Decade Is Out: Personal Reflections on the Apollo Program*.

Brooks, Grimwood, and Swenson. *Chariots for Apollo: The NASA History of Manned Lunar Spacecraft to 1969*. Mineola, New York: Dover Publications, 2009.

Letter, "Personal and Confidential Memorandum." Jacqueline Cochran to Robert Pirie. August 1, 1961. Jacqueline Cochran: Papers, 1932–1975. General Files Series. Box 138. Eisenhower Presidential Library.

Cobb and Rieker, *Woman into Space: The Jerrie Cobb Story*.

Letter, Jerrie Cobb to Wally Funk. July 24, 1961. Courtesy of Francis French.

Shurley, Jay. "Assessment of Women Astronaut's Performance in Sensory Deprivation." August 8, 1961. Courtesy of Francis French.

Letter, Hugh Dryden to Robert Pirie. October 2, 1961. Jacqueline Cochran: Papers, 1932–1975. General Files Series. Box 138. Eisenhower Presidential Library.

Letter, James Webb to Lyndon Johnson. August 23, 1961. Vice President, 1961–1963. Box 116. The Lyndon Baines Johnson Library.

Memorandum to the Vice President from E. C. Welsh. "Summary Report on Space Council Activities." July 28, 1961. Vice President, 1961–1963. Box 117. The Lyndon Baines Johnson Library.

Letter, Glenn Dawn to Lyndon Johnson. July 23, 1961. Vice President, 1961–1963. Box 116. The Lyndon Baines Johnson Library.

Letter, Ruben R. Nuñez to Lyndon Johnson. September 5, 1961. Vice President, 1961–1963. Box 116. The Lyndon Baines Johnson Library.

Letter, Marjory Hopkins to LBJ. August 27, 1961. Vice President, 1961–1963. Box 116. The Lyndon Baines Johnson Library.

Letter, Wally Funk to Randy Lovelace. May 31, 1961. Courtesy of Francis French.

Memo to Phillip Slood(?) re: Summer Employees (Jerrie Cobb). June 30, 1962. Cobb, Jerrie #1. NASA Archives.

Letter, Jacqueline Cochran to Randy Lovelace. September 1, 1961. Jacqueline Cochran: Papers, 1932–1975. General Files Series. Box 140. Eisenhower Presidential Library.

Letter, Randy Lovelace to Jacqueline Cochran. September 11, 1961. Jacqueline Cochran: Papers, 1932–1975. General Files Series. Box 140. Eisenhower Presidential Library.

Letter, Randy Lovelace to Jacqueline Cochran. August 23, 1961. Jacqueline Cochran: Papers, 1932–1975. General Files Series. Box 140. Eisenhower Presidential Library.

Letter, Randy Lovelace to Myrtle Thompson Cagle. August 21, 1961. Jacqueline Cochran: Papers, 1932–1975. General Files Series. Box 140. Eisenhower Presidential Library.

Letter, Randy Lovelace to Jan Dietrich. August 21, 1961. Jacqueline Cochran: Papers, 1932–1975. General Files Series. Box 140. Eisenhower Presidential Library.

Letter, Randy Lovelace to Marion Dietrich. August 21, 1961. Jacqueline Cochran: Papers, 1932–1975. General Files Series. Box 140. Eisenhower Presidential Library.

Letter, Randy Lovelace to Wally Funk. August 21, 1961. Jacqueline Cochran: Papers, 1932–1975. General Files Series. Box 140. Eisenhower Presidential Library.

Letter, Randy Lovelace to Sarah Lee Gorelick. August 21, 1961. Jacqueline Cochran: Papers, 1932–1975. General Files Series. Box 140. Eisenhower Presidential Library.

Letter, Randy Lovelace to Janey Hart. August 21, 1961. Jacqueline Cochran: Papers, 1932–1975. General Files Series. Box 140. Eisenhower Presidential Library.

Letter, Randy Lovelace to Jean Hixson. August 21, 1961. Jacqueline Cochran: Papers, 1932–1975. General Files Series. Box 140. Eisenhower Presidential Library.

Letter, Randy Lovelace to Rhea Hurrle. August 21, 1961. Jacqueline Cochran: Papers, 1932–1975. General Files Series. Box 140. Eisenhower Presidential Library.

Letter, Randy Lovelace to Irene Leverton. August 23, 1961. Jacqueline Cochran: Papers, 1932–1975. General Files Series. Box 140. Eisenhower Presidential Library.

Letter, Randy Lovelace to Bernice Steadman. August 21, 1961. Jacqueline Cochran: Papers, 1932–1975. General Files Series. Box 140. Eisenhower Presidential Library.

Letter, Randy Lovelace to Gene Nora Steadman. August 21, 1961. Jacqueline Cochran: Papers, 1932–1975. General Files Series. Box 140. Eisenhower Presidential Library.

Glenn, *John Glenn: A Memoir*.

Dietrich, Marion. "First Woman into Space." *McCall's*, September 1961.

Cox, Donald. "Women Astronauts." *Space World*, vol 1 no 10. September 1961. Cobb, Jerrie #2. NASA Archives.

CHAPTER 20

Telegram, Randy Lovelace to Wally Funk. September 12, 1961. Courtesy of Francis French.

Telegram, Jeanne Williams to Jaqueline Cochran. September 12, 1961. Jacqueline Cochran: Papers, 1932–1975. General Files Series. Box 140. Eisenhower Presidential Library.

Cobb and Rieker, *Woman into Space: The Jerrie Cobb Story*.

Letter, Hugh Dryden to Robert Pirie. October 2, 1961. Jacqueline Cochran: Papers, 1932–1975. General Files Series. Box 138. Eisenhower Presidential Library.

Letter, Robert Pirie to Jacqueline Cochran. October 9, 1961. Jacqueline Cochran: Papers, 1932–1975. General Files Series. Box 138. Eisenhower Presidential Library.

Letter, Jacqueline Cochran to Patricia Meholic. February 15, 1962. Jacqueline Cochran: Papers, 1932–1975. General Files Series. Box 139. Eisenhower Presidential Library.

Letter, Patricia Meholic to Jacqueline Cochran. November 7, 1961. Jacqueline Cochran: Papers, 1932–1975. General Files Series. Box 139. Eisenhower Presidential Library.

Letter, Stanley Marshall to Jacqueline Cochran. October 25, 1961. Jacqueline Cochran: Papers, 1932–1975. General Files Series. Box 139. Eisenhower Presidential Library.

Letter, Jacqueline Cochran to June Moyer. August 10, 1961. Jacqueline Cochran: Papers, 1932–1975. General Files Series. Box 139. Eisenhower Presidential Library.

Letter, June Moyer to Jacqueline Cochran. July 25, 1961. Jacqueline Cochran: Papers, 1932–1975. General Files Series. Box 139. Eisenhower Presidential Library.

Letter, Jacqueline Cochran to Paula Sue Campbell. June 6, 1961. Jacqueline Cochran: Papers, 1932–1975. General Files Series. Box 139. Eisenhower Presidential Library.

Letter, Paula Sue Campbell to Jacqueline Cochran. May 2, 1961. Jacqueline Cochran: Papers, 1932–1975. General Files Series. Box 139. Eisenhower Presidential Library.

Letter, Jacqueline Cochran to Sandra Massaro. June 6, 1961. Jacqueline Cochran: Papers, 1932–1975. General Files Series. Box 139. Eisenhower Presidential Library.

Letter, Sandra Massaro to Jacqueline Cochran. May 10, 1961. Jacqueline Cochran: Papers, 1932–1975. General Files Series. Box 139. Eisenhower Presidential Library.

Letter, Jacqueline Cochran to Alicia McFadden. February 13, 1962. Jacqueline Cochran: Papers, 1932–1975. General Files Series. Box 139. Eisenhower Presidential Library.

Letter, Alicia McFadden to Jacqueline Cochran. December 30, 1961. Jacqueline Cochran: Papers, 1932–1975. General Files Series. Box 139. Eisenhower Presidential Library.

Letter, Jacqueline Cochran to Alicia McFadden. June 6, 1961. Jacqueline Cochran:

Papers, 1932–1975. General Files Series. Box 139. Eisenhower Presidential Library.

Letter, Alicia McFadden to Jacqueline Cochran. May 2, 1961. Jacqueline Cochran: Papers, 1932–1975. General Files Series. Box 139. Eisenhower Presidential Library.

Letter, Jacqueline Cochran to Sue Udell. June 1, 1961. Jacqueline Cochran: Papers, 1932–1975. General Files Series. Box 139. Eisenhower Presidential Library.

Letter, Sue Udell to Jacqueline Cochran. April 30, 1960. Jacqueline Cochran: Papers, 1932–1975. General Files Series. Box 139. Eisenhower Presidential Library.

Letter, Jacqueline Cochran to Sophia M. Payton. October 9, 1961. Jacqueline Cochran: Papers, 1932–1975. General Files Series. Box 139. Eisenhower Presidential Library.

Letter, Sophia M. Payton to Jacqueline Cochran. incl. Aviation Background. September 24, 1961. Jacqueline Cochran: Papers, 1932–1975. General Files Series. Box 139. Eisenhower Presidential Library.

Letter, Floyd Odlum to Randy Lovelace. November 17, 1961. Jacqueline Cochran: Papers, 1932–1975. General Files Series. Box 139. Eisenhower Presidential Library.

Letter, Randy Lovelace to James Webb. September 29, 1961. Jacqueline Cochran: Papers, 1932–1975. General Files Series. Box 140. Eisenhower Presidential Library.

Bizony, *The Man Who Ran the Moon: James E. Webb, NASA, and the Secret History of Project Apollo.*

Glenn, *John Glenn: A Memoir.*

Memo, John F. Kennedy to Lyndon Johnson. February 21, 1962. White House Famous Names, Service Set Box 6. The Lyndon B. Johnson Library.

Memorandum for the Administrator, Deputy Administrator, Associate Administrator. Subject: MA-6 Contingencies from O. B. Lloyd, Jr. January 16, 1962. Vice Presidential Aide's Files of George Reedy. Box 12. The Lyndon B. Johnson Library.

Ackmann, *Mercury 13: The True Story of Thirteen Women and the Dream of Space Flight.*

Letter, Jerrie Cobb to Jacqueline Cochran. March 8, 1962. Jacqueline Cochran: Papers, 1932–1975. General Files Series. Box 139. Eisenhower Presidential Library.

Letter, Jerrie Cobb to Floyd Odlum. March 8, 1962. Jacqueline Cochran: Papers, 1932–1975. General Files Series. Box 139. Eisenhower Presidential Library.

Letter, Floyd Odlum to Randy Lovelace. March 29, 1962. Jacqueline Cochran: Papers, 1932–1975. General Files Series. Box 139. Eisenhower Presidential Library.

"The Friendship 7 Mission: A Major Achievement and a Sign of More to Come." NASA, accessed September 20, 2018. https://history.nasa.gov/friendship7/.

Memorandum, George Reedy to the Vice President. February 21, 1962. Vice Presidential Aide's Files of George Reedy. Box 12. The Lyndon Baines Johnson Library.

"Space for Women?" presented by Jerrie Cobb, First Women's Space Symposium. February 22, 1962. Los Angeles, CA. Vice President, 1961–1963. Box 183. The Lyndon Baines Johnson Library.

"Address by Dr. Edward C. Welsh, Prepared for Delivery Before the Women's Space Symposium, Los Angeles, California, Thursday, February 22, 1962." February 23, 1962. Vice President, 1961–1963. Box 183. The Lyndon Baines Johnson Library.

"Could Put Women in Space Now." UPI, Los Angeles. February 25, 1962. NASA Archives.

"Transcript of Senator Keating's Interview of Mr. Webb for Senator's Radio and TV

Program." Undated (March 6, 1962). LBJ Library. VP Aide's Files of George Reedy. Space [March 9, 1962] [Statements on the Space Program] [1 of 2] The Lyndon Baines Johnson Library.

"NASA to Select Additional Astronauts." March 21, 1962. VP Aide's Files of George Reedy. Space [March 9, 1962] [Statements on the Space Program] [1 of 2] The Lyndon Baines Johnson Library.

Memorandum to the Vice President. Subject: Space Council Meeting. July 5, 1962. Vice President, 1961–1963. Box 183. The Lyndon Baines Johnson Library.

"Memorandum for the Chairman of National Aeronautics and Space Council." March 13, 1962. VP Aide's Files of George Reedy. Space [March 9, 1962] [Statements on the Space Program] [1 of 2] The Lyndon Baines Johnson Library.

Letter, Lyndon Johnson to John F. Kennedy. March 23, 1962. VP Aide's Files of George Reedy. Space [March 9, 1962] [Statements on the Space Program] [1 of 2] The Lyndon Baines Johnson Library.

Memo from Welsh to Lyndon Johnson. March 14, 1962. Papers of Lyndon B. Johnson. Vice President, 1961–1963. Box 183. The Lyndon Baines Johnson Library.

"Request for Highest National Priority for the Apollo Program." March 13, 1962. Vice President, 1961–1963. Box 183. The Lyndon Baines Johnson Library.

National Security Action Memorandum No. 144. April 11, 1962. VP Aide's Files of George Reedy. Space [March 9, 1962] [Statements on the Space Program] [1 of 2] The Lyndon Baines Johnson Library.

Letter, James Webb to John F. Kennedy. March 13, 1962. Vice President, 1961–1963. Box 183. The Lyndon Baines Johnson Library.

Memorandum for the Honorable Robert S. McNamara, Secretary of Defense. March 23, 1962. Vice President, 1961–1963. Box 183. The Lyndon Baines Johnson Library.

Memorandum, E. C. Welsh to Mr. Hale, Subject: US—Soviet Cooperation in Space. March 3, 1962. Vice Presidential Aide's Files of George Reedy. Box 12. The Lyndon Baines Johnson Library.

Letter, John F. Kennedy to Sergei Khrushchev. March 7, 1962. VP Aide's Files of George Reedy. Space [March 9, 1962] [Statements on the Space Program] [1 of 2] The Lyndon Baines Johnson Library.

Memorandum, E. C. Welsh to Vice President (LBJ). March 19, 1962. VP Aide's Files of George Reedy. Space [March 9, 1962] [Statements on the Space Program] [1 of 2] The Lyndon Baines Johnson Library.

"Clinical Record of Donald K. Slayton," from Lawrence E. Lamb, MD. Undated. Vice President, 1961–1963. Box 183. The Lyndon Baines Johnson Library.

"Carpenter Replaces Slayton as MA 7 Pilot." March 21, 1962. VP Aide's Files of George Reedy. Space [March 9, 1962] [Statements on the Space Program] [1 of 2] The Lyndon Baines Johnson Library.

Letter, Lawrence E. Lamb to Brigadier General Charles Roadman. March 5, 1962. Vice Presidential Aide's Files of George Reedy. Box 12. The Lyndon Baines Johnson Library.

Steadman, *Tethered Mercury: A Pilot's Memoir the Right Stuff...but the Wrong Sex.*

Invitation to Senator Hart reception, circa 1958. LBJA Congressional File. Box 45. The Lyndon Baines Johnson Library.

Telegram, Lyndon Johnson to Senator Hart. November 5, 1958. LBJA Congressional File. Box 45. The Lyndon Baines Johnson Library.

Letter, Lyndon Johnson to Governor Hart. November 14, 1958. LBJA Congressional File. Box 45. The Lyndon Baines Johnson Library.

Woods, Randall. *LBJ: Architect of American Ambition*. Free Press, New York, 2006.

Letter, Philip Hart to Lyndon Johnson. November 6, 1958. LBJA Congressional File. Box 45. The Lyndon Baines Johnson Library.

Letter, Lyndon Johnson to George Reedy. March 8, 1962. 1962 Subject File. Science—Space and Aeronautics. [Women in Space] The Lyndon Baines Johnson Library.

"Memo on requested meeting with the Vice President." March 7, 1962. Papers of Lyndon B. Johnson. Vice President, 1961–1963. Box 183. The Lyndon Baines Johnson Library.

Letter, Jerrie Cobb to Jacqueline Cochran. March 8, 1962. Jacqueline Cochran: Papers, 1932–1975. General Files Series. Box 139. Eisenhower Presidential Library.

Letter, Jerrie Cobb to Floyd Odlum. March 8, 1962. Jacqueline Cochran: Papers, 1932–1975. General Files Series. Box 139. Eisenhower Presidential Library.

"Senate Wife Could Be First Woman in Space." *Sunday Star*, March 11, 1962. 1962 Subject File. Science—Space and Aeronautics. [Women in Space] The Lyndon Baines Johnson Library.

"Senator's Wife Wants Distaff Elbow Room Asks Space for Women in Space." *Washington Post*, March 14, 1962.

Letter, Catherine Smith to Lyndon Johnson. March 15, 1962. Vice President, 1961–1963. Box 183. The Lyndon Baines Johnson Library.

Letter, Mrs. George B. Ward to Lyndon Johnson. March 19, 1962. 1962 Subject File. Science—Space and Aeronautics. [Women in Space] The Lyndon Baines Johnson Library.

Letter, Sue Ann Winkelman to Lyndon Johnson. March 21, 1962. Vice President, 1961–1963. Box 183. The Lyndon Baines Johnson Library.

Letter, Rachel D. Jones to Lyndon Johnson. March 13, 1962. Vice President, 1961–1963. Box 183. The Lyndon Baines Johnson Library.

Letter, Howard F. White to Lyndon Johnson. March 13, 1962. Vice President, 1961–1963. Box 183. The Lyndon Baines Johnson Library.

Letter, Fritzi Mann to Lyndon Johnson. March 22, 1962. Vice President, 1961–1963. Box 183. The Lyndon Baines Johnson Library.

Letter, Carole Glad to Lyndon Johnson. August 15, 1962. Vice President, 1961–1963. Box 183. The Lyndon Baines Johnson Library.

Letter, Lyndon Johnson to Carole Glad. August 21, 1962. Vice President, 1961–1963. Box 183. The Lyndon Baines Johnson Library.

Letter, Lyndon Johnson to Catherine Smith. March 24, 1962. 1962 Subject File. Science—Space and Aeronautics. [Women in Space] The Lyndon Baines Johnson Library.

Letter, A. Bachelor to Lyndon Johnson. March 15, 1962. Papers of Lyndon B. Johnson. Vice President, 1961–1963. Box 183. The Lyndon Baines Johnson Library.

"Background for your conference at 11am." March 14, 1962. Papers of Lyndon B. Johnson. Vice President, 1961–1963. Box 183. The Lyndon Baines Johnson Library.

Letter, Lyndon Johnson to James Webb. March 15, 1962. Vice President, 1961–1963. Box 183. The Lyndon Baines Johnson Library.

CHAPTER 21

"The Lyndon Baines Johnson Room." Office of the Senate Curator, accessed August 19, 2019. https://www.senate.gov/artandhistory/art/resources/pdf/Lyndon_B._Johnson_Room.pdf.

Ackmann, *Mercury 13: The True Story of Thirteen Women and the Dream of Space Flight*.

Letter, Lyndon Johnson to Jacqueline Cochran. March 5, 1958. United States Senate, 1949–1961. "Master File" Index. Cla-Cle–Coe-Cok. Box 37. The Lyndon Baines Johnson Library.

Letter, Lyndon Johnson to Jacqueline Cochran. April 13, 1962. 1962 Subject File. Science—Space and Aeronautics. [Women in Space] The Lyndon Baines Johnson Library.

Letter, Jerrie Cobb to James Webb. March 30, 1962. Courtesy of Francis French.

"Moon Flight and Women." Jacqueline Cochran: Papers, 1932–1975. Articles Series. Box 8. Eisenhower Presidential Library.

News clipping enclosed in Letter, Roy Marshall to Lyndon Johnson. November 21, 1965. White House Central File Name File. Box 22. Folder: Odlum, A–K. The Lyndon Baines Johnson Library.

Letter, Jacqueline Cochran to Jerrie Cobb. March 23, 1962. Jacqueline Cochran: Papers, 1932–1975. General Files Series. Box 139. Eisenhower Presidential Library.

Letter, Jacqueline Cochran to Myrtle Thompson Cagle. March 29, 1962. Jacqueline Cochran: Papers, 1932–1975. General Files Series. Box 139. Eisenhower Presidential Library.

Letter, Jacqueline Cochran to Jan Dietrich. March 29, 1962. Jacqueline Cochran: Papers, 1932–1975. General Files Series. Box 139. Eisenhower Presidential Library.

Letter, Jacqueline Cochran to Bernice Steadman. April 12, 1962. Jacqueline Cochran: Papers, 1932–1975. General Files Series. Box 139. Eisenhower Presidential Library.

Letter, Jacqueline Cochran to Marion Dietrich. March 29, 1962. Jacqueline Cochran: Papers, 1932–1975. General Files Series. Box 139. Eisenhower Presidential Library.

Letter, Jacqueline Cochran to Mary Wallace Funk. April 12, 1962. Jacqueline Cochran: Papers, 1932–1975. General Files Series. Box 139. Eisenhower Presidential Library.

Letter, Jacqueline Cochran to Sara Lee Gorelick. April 12, 1962. Jacqueline Cochran: Papers, 1932–1975. General Files Series. Box 139. Eisenhower Presidential Library.

Letter, Jacqueline Cochran to Jane B. Hart. April 12, 1962. Jacqueline Cochran: Papers, 1932–1975. General Files Series. Box 139. Eisenhower Presidential Library.

Letter, Jacqueline Cochran to Jean Hixson. April 12, 1962. Jacqueline Cochran: Papers, 1932–1975. General Files Series. Box 139. Eisenhower Presidential Library.

Letter, Jacqueline Cochran to Rhea Hurrle. April 12, 1962. Jacqueline Cochran: Papers, 1932–1975. General Files Series. Box 139. Eisenhower Presidential Library.

Letter, Jacqueline Cochran to Irene Leverton. April 12, 1962. Jacqueline Cochran: Papers, 1932–1975. General Files Series. Box 139. Eisenhower Presidential Library.

Letter, Jacqueline Cochran to Bernice Steadman. April 12, 1962. Jacqueline Cochran: Papers, 1932–1975. General Files Series. Box 139. Eisenhower Presidential Library.

Letter, Floyd Odlum to Randy Lovelace. March 29, 1962. Jacqueline Cochran: Papers, 1932–1975. General Files Series. Box 139. Eisenhower Presidential Library.

Letter, Jacqueline Cochran to Lyndon Johnson. March 26, 1962. 1962 Subject File. Science—Space and Aeronautics. [Women in Space] The Lyndon Baines Johnson Library.

Letter, Jacqueline Cochran to Olin Teague. April 12, 1962. Jacqueline Cochran: Papers, 1932–1975. General Files Series. Box 139. Eisenhower Presidential Library.

Letter, Jacqueline Cochran to Jane Rieker. April 12, 1962. (Note: sub letter to Jane Nash attached). Jacqueline Cochran: Papers, 1932–1975. General Files Series. Box 139. Eisenhower Presidential Library.

Letter, Jacqueline Cochran to Jane Rieker. June 6, 1962. Jacqueline Cochran: Papers, 1932–1975. General Files Series. Box 138. Eisenhower Presidential Library.

Letter, Jacqueline Cochran to Jane Nash. April 2, 1962. Jacqueline Cochran: Papers, 1932–1975. General Files Series. Box 139. Eisenhower Presidential Library.

Letter, Olin Teague to Jacqueline Cochran. April 26, 1962. Jacqueline Cochran: Papers, 1932–1975. General Files Series. Box 139. Eisenhower Presidential Library.

Letter, Robert Pirie to Jacqueline Cochran. April 30, 1962. Jacqueline Cochran: Papers, 1932–1975. General Files Series. Box 139. Eisenhower Presidential Library.

Letter, Hugh Dryden to Jacqueline Cochran. April 26, 1962. Jacqueline Cochran: Papers, 1932–1975. General Files Series. Box 139. Eisenhower Presidential Library.

Letter, Robert Gilruth to Jacqueline Cochran. April 27, 1962. Jacqueline Cochran: Papers, 1932–1975. General Files Series. Box 139. Eisenhower Presidential Library.

Letter, Walt Willams to Jacqueline Cochran. April 27, 1962. Jacqueline Cochran: Papers, 1932–1975. General Files Series. Box 139. Eisenhower Presidential Library.

Letter, Randy Lovelace to Floyd Odlum. April 24, 1962. Jacqueline Cochran: Papers, 1932–1975. General Files Series. Box 139. Eisenhower Presidential Library.

Jessen, *The Fabulous Flight of the Three Musketeers.*

Cobb and Rieker, *Woman into Space: The Jerrie Cobb Story.*

Letter, Gene Nora Stumbough to Jacqueline Cochran. June 11, 1962. Jacqueline Cochran: Papers, 1932–1975. General Files Series. Box 139. Eisenhower Presidential Library.

Letter, Jerrie Cobb to James Webb, March 30, 1962. Courtesy of Francis French.

Letter, Jerrie Cobb to Lyndon Johnson. April 17, 1962. Vice President, 1961–1963. Box 183. The Lyndon Baines Johnson Library.

Letter, Lyndon Johnson to Jerrie Cobb. April 23, 1962. Vice President, 1961–1963. Box 183. The Lyndon Baines Johnson Library.

Letter, Jerrie Cobb to Lyndon Johnson. July 5, 1962. Vice President, 1961–1963. Box 183. The Lyndon Baines Johnson Library.

Letter, Jerrie Cobb to Jacqueline Cochran. April 26, 1962. Jacqueline Cochran: Papers, 1932–1975. General Files Series. Box 139. Eisenhower Presidential Library.

Space Council Meetings. April 24, 1962. Vice President, 1961–1963. Box 183. The Lyndon Baines Johnson Library.

Interoffice Communication from Miss Walsh to Miss Cochran "Women Astronauts" March 20, 1962. Jacqueline Cochran: Papers, 1932–1975. General Files Series. Box 140. Eisenhower Presidential Library.

"Women Astronauts," from NASA Office of Legislative Affairs. March 20, 1962. Cobb, Jerrie #1. NASA Archives.

Inter-Office Communication. "Eloise Engle." Undated (1961/1962) Jacqueline Cochran: Papers, 1932–1975. General Files Series. Box 138. Eisenhower Presidential Library.

Letter, Eloise Engle to Jacqueline Cochran. May 29, 1962. Jacqueline Cochran: Papers, 1932–1975. General Files Series. Box 138. Eisenhower Presidential Library.

Letter, Jacqueline Cochran to Eloise Engle. May 28, 1962. Jacqueline Cochran: Papers, 1932–1975. General Files Series. Box 138. Eisenhower Presidential Library.

Engle, Eloise. "Brainy Gals Help Push America's Race into Space." *Dodge News Magazine*, April 1962. Jacqueline Cochran: Papers, 1932–1975. General Files Series. Box 138. Eisenhower Presidential Library.

Letter, Randy Lovelace to Eloise Engle. December 21, 1961. Jacqueline Cochran: Papers, 1932–1975. General Files Series. Box 138. Eisenhower Presidential Library.

Letter, Jacqueline Cochran to Eloise Engle, December 13, 1961. Jacqueline Cochran: Papers, 1932–1975. General Files Series. Box 138. Eisenhower Presidential Library.

Letter, Jacqueline Cochran to Eloise Engle, December 9, 1961. Jacqueline Cochran: Papers, 1932–1975. General Files Series. Box 138. Eisenhower Presidential Library.

Letter, Hugh L. Dryden to Jacqueline Cochran. June 18, 1962. Cochran, Jacqueline folder. NASA Archives.

"Geraldyn M. Cobb" in *Current Biography*, February, 1961. Cobb, Jerrie #1. NASA Archives.

Undated memo "Written by Brownstein." Cobb, Jerrie #1. NASA Archives.

"Records (Jacqueline Cochran)." Fédération Aéronautique Internationale, accessed September 17, 2018. https://www.fai.org/records?record=Jacqueline+Cochran.

"Aviatrix Has 2 Big Ambitions." Lawrence, Mass. April 3, 1962. Jacqueline Cochran: Papers, 1932–1975. Scrapbook Series. Box 20. Eisenhower Presidential Library.

"49 Records in U.S.-Germany Hop Claimed by Jacqueline Cochran." April 23, 1962. Jacqueline Cochran: Papers, 1932–1975. Scrapbook Series. Box 20. Eisenhower Presidential Library.

"Jacqueline Cochran ('By Golly') Arrives Here." *Seattle Post-Intelligencer*, May 8, 1962. Jacqueline Cochran: Papers, 1932–1975. Scrapbook Series. Box 20. Eisenhower Presidential Library.

Letter, Jacqueline Cochran to James Webb. May 15, 1962. Courtesy of Francis French.

Note, Elvis J. Stahr to James Webb, attached to Jacqueline Cochran to Jerrie Cobb, March 23, 1962. Undated. Cochran, Jacqueline folder. NASA Archives.

Letter, James Webb to Elvis J. Stahr. May 18, 1962. Cochran, Jacqueline folder. NASA Archives.

Letter, James Webb to Jacqueline Cochran. May 24, 1962. Courtesy of Francis French.

Letter, Jacqueline Cochran to James Webb. June 14, 1962. Cochran, Jacqueline folder. NASA Archives.

Memo to Phillip Slood(?) re: Summer Employees (Jerrie Cobb). June 30, 1962. Cobb, Jerrie #1. NASA Archives.

Hearings Before the Special Subcommittee on the Selection of Astronauts, July 17 and 18, 1962. US Government Printing Office. Washington: 1962.

"Miller, George Paul." Biographical Directory of the United States Congress, accessed December 1, 2018. http://bioguide.congress.gov/scripts/biodisplay.pl?index=M000727.

Letter, Jacqueline Cochran to James Webb. June 15, 1962. Jacqueline Cochran: Papers, 1932–1975. General Files Series. Box 138. Eisenhower Presidential Library.

Steadman, *Tethered Mercury: A Pilot's Memoir the Right Stuff...but the Wrong Sex*.

"Discrimination on Women in Space Charged." *Los Angeles Times*, June 15, 1962.

Letter, Lyndon Johnson to Victor Anfuso. April 21, 1961. Johnson Vice President, 1961–1963. Box 2. The Lyndon Baines Johnson Library.

Letter, Lyndon Johnson to Antionio Ambrosini. April 6, 1962. Vice President, 1961–1963. Box 181. The Lyndon Baines Johnson Library.

Letter, Lyndon Johnson to Victor Anfuso. February 28, 1962. Vice President, 1961–1963. Box 2. The Lyndon Baines Johnson Library.

Letter, Lyndon Johnson to Victor Anfuso. January 10, 1962. Vice President, 1961–1963. Box 2. The Lyndon Baines Johnson Library.

Letter, Jacqueline Cochran to Janey Hart. June 15, 1962. Jacqueline Cochran: Papers, 1932–1975. General Files Series. Box 138. Eisenhower Presidential Library.

Letter, Jacqueline Cochran to James Webb. June 15, 1962. Jacqueline Cochran: Papers, 1932–1975. General Files Series. Box 138. Eisenhower Presidential Library.

Letter, Jacqueline Cochran to Hugh Dryden. June 18, 1962. Cochran, Jacqueline folder. NASA Archives.

Letter, Jacqueline Cochran to Curtis Le May. June 16, 1962. Jacqueline Cochran: Papers, 1932–1975. General Files Series. Box 138. Eisenhower Presidential Library.

Letter, Jacqueline Cochran to Robert Pirie. July 11, 1962. Jacqueline Cochran: Papers, 1932–1975. General Files Series. Box 138. Eisenhower Presidential Library.

"Copies of Statement No. 1 Went to." undated (July 1962) Jacqueline Cochran: Papers, 1932–1975. General Files Series. Box 138. Eisenhower Presidential Library.

Letter, Hugh Dryden to Jacqueline Cochran. June 26, 1962. Jacqueline Cochran: Papers, 1932–1975. General Files Series. Box 138. Eisenhower Presidential Library.

Letter, James Webb to Jacqueline Cochran. July 12, 1962. Jacqueline Cochran: Papers, 1932–1975. General Files Series. Box 138. Eisenhower Presidential Library.

Letter, Robert Gilruth to Jacqueline Cochran. June 26, 1962. Jacqueline Cochran: Papers, 1932–1975. General Files Series. Box 138. Eisenhower Presidential Library.

Letter, Randy Lovelace to Jacqueline Cochran. June 29, 1962. Jacqueline Cochran: Papers, 1932–1975. General Files Series. Box 138. Eisenhower Presidential Library.

Letter, Janey Hart to Jacqueline Cochran. June 22, 1962. Jacqueline Cochran: Papers, 1932–1975. General Files Series. Box 138. Eisenhower Presidential Library.

Letter, Richard Hines to Jacqueline Cochran. July 9, 1962. Jacqueline Cochran: Papers, 1932–1975. General Files Series. Box 138. Eisenhower Presidential Library.

Letter, Jerrie Cobb to Myrtle Thompson Cagle. Undated. Courtesy of the International Women's Air and Space Museum.

Letter, Jacqueline Cochran to Victor Anfuso. July 11, 1962. Jacqueline Cochran: Papers, 1932–1975. General Files Series. Box 138. Eisenhower Presidential Library.

"Lunar Orbit Rendezvous News Conference on Apollo Plans at the NASA Headquarters on July 11, 1962." Jacqueline Cochran: Papers, 1932–1975. Scrapbook Series. Box 4. Eisenhower Presidential Library.

Ertel, Ivan D. and Mary Louise Morse. *The Apollo Spacecraft, Volume I Through November 7, 1962.* The NASA Historical Series. Office of Technology Utilization, NASA. Washington, DC, 1969.

"NASA Outlines Apollo Plans." Transcript. NASA News Release. July 11, 1962 NASA MSC, Houston, Texas.

Letter, Jacqueline Cochran to Richard Hines. July 24, 1962. Jacqueline Cochran: Papers, 1932–1975. General Files Series. Box 139. Eisenhower Presidential Library.

Interoffice Communication from Miss Schroeder to Miss Cochran. July 12, 1962. Jacqueline Cochran: Papers, 1932–1975. General Files Series. Box 138. Eisenhower Presidential Library.

Letter, Myrtle Thompson Cagle to Jacqueline Cochran. August 8, 1962. Jacqueline Cochran: Papers, 1932–1975. General Files Series. Box 140. Eisenhower Presidential Library.

Letter, Marion Dietrich to Jacqueline Cochran. August 7, 1962. Jacqueline Cochran: Papers, 1932–1975. General Files Series. Box 140. Eisenhower Presidential Library.

"Women in Space Washington Trip." July 1962. Jacqueline Cochran: Papers, 1932–1975. General Files Series. Box 138. Eisenhower Presidential Library.

CHAPTER 22

"Women in Space Washington Trip." July 1962. Jacqueline Cochran: Papers, 1932–1975. General Files Series. Box 138. Eisenhower Presidential Library.

"Women Pilots Make Bid For a Chunk of Space." *New York Herald Tribune*, July 18, 1962.

Ackmann, *Mercury 13: The True Story of Thirteen Women and the Dream of Space Flight*.

Hearings before the Special Subcommittee on the Selection of Astronauts, July 17 and 18, 1962. US Government Printing Office. Washington: 1962.

Cobb and Rieker, *Woman into Space: The Jerrie Cobb Story*.

Letter, Lyndon Johnson to Senator James G. Fulton. November 13, 1961. Vice President, 1961–1963. Box 116. The Lyndon Baines Johnson Library.

CHAPTER 23

Telegram, Jerrie Cobb to Myrtle Thompson Cagle, July 18, 1962. Courtesy of the International Women's Air and Space Museum.

"Lady Fliers Ask to be Astronauts, too; Plead for Co-Eds in Space." *Wall Street Journal*, Wednesday, July 18, 1962.

"Give Us Space Role, Women Pilots Urge." *Chicago Tribune*, July 18, 1962.

"Women Pilots Make Bid For a Chunk of Space." *New York Herald Tribune*, July 18, 1962.

"Women Fliers Try to Crack Barriers on Space Travel." *Washington Post*, July 18, 1962.

Stanford, Neal. "High G." *The Christian Science Monitor*, Friday, October 3, 1962.

"Let the Girls, Bless 'Em, Loose in Space!" *Washington Daily News*, August 8, 1962.

"Why Not Put a Lady in Space?" *New York World-Telegram & Sun*, August 7, 1962. Jacqueline Cochran: Papers, 1932–1975. Scrapbook Series. Box 20. Eisenhower Presidential Library.

"Space Suffragette." *Daily Oklahoman*, July 27, 1962.

"Woman Flier Claims She 'Won't Give Up.'" *Los Angeles Times*, July 19, 1962.

"Of Sex & Spacenicks." *New York Daily News*, July 18, 1962.

"Lady Fliers Ask to be Astronauts, too; Plead for Co-Eds in Space." *Wall Street Journal*, Wednesday, July 18, 1962.

Letter, Vivienne Mudd to Jacqueline Cochran. July 19, 1962. Jacqueline Cochran: Papers, 1932–1975. General Files Series. Box 140. Eisenhower Presidential Library.

"Why Not Women in Outer Space?" *S.F. News Call Bulletin*, July 24, 1962.

"2 Astronauts 'Scrub' Bid of Women Pilots." *Chicago Tribune*, July 19 1962.

"Glenn Would Yield Space in Space." *Washington Post*, July 19, 1962.

"Now Women Try to Invade Last Male Frontier—Space." *Miami Herald*, Cobb, Jerrie folders. NASA Archives.

"The Question Should There Be Women Astronauts? Where Asked Boston Airport." *Boston Herald*, Monday, August 27, 1962.

Moore, Bill. "Still Struggling to Get a Chance at Space." *Kansas City Star*, September 28, 1962.

Telegram, Jerrie Cobb to the White House. July 20, 1962. John F. Kennedy Library.

Letter, James Webb to Jerrie Cobb. August 3, 1962. John F. Kennedy Library.

Letter, Jerrie Cobb to James Webb. August 7, 1962. Cobb, Jerrie folder. NASA Archives.

Letter, Jacqueline Cochran to Robert Ruark. August 24, 1962. Jacqueline Cochran: Papers, 1932–1975. General Files Series. Box 140. Eisenhower Presidential Library.

Letter, Jacqueline Cochran to Phyllis Battelle. July 27, 1962. Jacqueline Cochran: Papers, 1932–1975. General Files Series. Box 140. Eisenhower Presidential Library.

"Interesting Questions for some one to ask Miss Cobb" (FBO). Undated. Jacqueline Cochran: Papers, 1932–1975. General Files Series. Box 138. Eisenhower Presidential Library.

Letter, Jacqueline Cochran to Randy Lovelace. July 26, 1962. Jacqueline Cochran: Papers, 1932–1975. General Files Series. Box 140. Eisenhower Presidential Library.

Letter, Randy Lovelace to Jacqueline Cochran. July 16, 1962. Jacqueline Cochran: Papers, 1932–1975. General Files Series. Box 140. Eisenhower Presidential Library.

Letter, Jacqueline Cochran to Bernice Steadman. August 1, 1962. Jacqueline Cochran:

Papers, 1932–1975. General Files Series. Box 140. Eisenhower Presidential Library.

Steadman, *Tethered Mercury: A Pilot's Memoir the Right Stuff... but the Wrong Sex.*

Letter, Bernice Steadman to Jacqueline Cochran. August 12, 1962. Jacqueline Cochran: Papers, 1932–1975. General Files Series. Box 140. Eisenhower Presidential Library.

Letter, Myrtle Thompson Cagle to Jacqueline Cochran. August 8, 1962. Jacqueline Cochran: Papers, 1932–1975. General Files Series. Box 140. Eisenhower Presidential Library.

Letter, Jan Dietrich to Jacqueline Cochran. August 12, 1962. Jacqueline Cochran: Papers, 1932–1975. General Files Series. Box 140. Eisenhower Presidential Library.

Letter, Marion Dietrich to Jacqueline Cochran. August 7, 1962. Jacqueline Cochran: Papers, 1932–1975. General Files Series. Box 140. Eisenhower Presidential Library.

Letter, Irene Leverton to Jacqueline Cochran. August 8, 1962. Jacqueline Cochran: Papers, 1932–1975. General Files Series. Box 140. Eisenhower Presidential Library.

Letter, Jean Hixson to Jacqueline Cochran. October 25, 1962. Jacqueline Cochran: Papers, 1932–1975. General Files Series. Box 140. Eisenhower Presidential Library.

Letter, Rhea Hurrle to Jacqueline Cochran. October 26, 1962. Jacqueline Cochran: Papers, 1932–1975. General Files Series. Box 140. Eisenhower Presidential Library.

Letter, Jerrie Cobb to FLATS. August 22, 1962. Courtesy of the International Women's Air and Space Museum.

Telegram, Jerrie Cobb to John F. Kennedy. August 13, 1962. John F. Kennedy Library.

Letter, James Webb to Jerrie Cobb. August 20, 1962. Cobb, Jerrie #2. NASA Archives.

Letter, Jerrie Cobb to James Webb. October 25, 1962. Cobb, Jerrie #2. NASA Archives.

"9 New Astronauts Named to Train for Moon Flights." *New York Times*, September 18, 1962.

"Elliott McKay See, Jr." Arlington Cemetery, accessed August 12, 2019. http://www.arlingtoncemetery.net/emsee.htm.

"Biography of Neil Armstrong." NASA, accessed August 12, 2019. https://www.nasa.gov/centers/glenn/about/bios/neilabio.html.

Klesius, Mike. "Neil Armstrong's X-15 flight over Pasadena." *Air & Space Magazine*, May 20, 2009. Accessed August 12, 2019. https://www.airspacemag.com/daily-planet/neil-armstrongs-x-15-flight-over-pasadena-59458462/.

Cobb and Rieker, *Woman into Space: The Jerrie Cobb Story*.

"Women in Space." Speech given by Miss Jerrie Cobb. Zonta Club of Cleveland. November 28, 1962. Jacqueline Cochran Papers, Speech Series. Box 9. Eisenhower Library.

"Expenses of Medical Checks at Lovelace Clinic." Undated (1961). Jacqueline Cochran: Papers, 1932–1975. General Files Series. Box 140. Eisenhower Presidential Library.

"Speech Given by Miss Jacqueline Cochran; The Job to Be Done." November 28, 1962. Jacqueline Cochran: Papers, 1932–1975. Speech Series. Box 9. Eisenhower Presidential Library.

Letter, Jacqueline Cochran to James Webb. January 14, 1963. Jacqueline Cochran:

Papers, 1932–1975. General Files Series. Box 140. Eisenhower Presidential Library.

Letter, James Webb to Jacqueline Cochran. February 13, 1963. Cochran, Jacqueline folder. NASA Archives.

Letter, Hugh Dryden to Jacqueline Cochran. January 22, 1963. Cochran, Jacqueline folder. NASA Archives.

Letter, James Webb to George Miller. March 4, 1963. Cochran, Jacqueline folder. NASA Archives.

"The Job to be Done, Extension of Remarks of Hon. George P. Miller." April 4, 1963. Congressional Record A2058. Cochran, Jacqueline folder. NASA Archives.

Letter, Jerrie Cobb to John F. Kennedy, March 13, 1963. John F. Kennedy Library.

Letter, James Webb to Jerrie Cobb. April 5, 1963. Cobb, Jerrie #1. NASA Archives.

"Appointment Affidavits." Jacqueline Cochran as NASA Consultant. June 11, 1963. Jacqueline Cochran: Papers, 1932–1975. General Files Series. Box 227. Eisenhower Presidential Library.

Note to Mr. Webb from R. P. Young. June 10, 1963. Cochran, Jacqueline folder. NASA Archives.

Burgess, Colin and Rex Hall. *The First Soviet Cosmonaut Team*. UK: Springer-Praxis, 2009.

"Why Valentina and Not Our Gal?" *Berkshire Eagle*, June 21, 1963.

"Women Pilots Angry at Webb." *Washington Post*, June 20, 1963.

"Was Spacewoman's Flight Necessary?" *Richmond Time-Dispatch*, June 18, 1963.

Letter, Jacqueline Cochran to James Webb. July 2, 1963. Jacqueline Cochran: Papers, 1932–1975. General Files Series. Box 140. Eisenhower Presidential Library.

"Myrtle Thompson Cagle: It Should Have Been Me." *Raleigh NC News & Observer*, July 9, 1963.

Memorandum, E. C. Welsh to the Vice President Subject: Military vs. Non-Military Space Activities. January 19, 1963. White House Famous Names, Service Set Box 6. The Lyndon Baines Johnson Library.

Memorandum for the Vice President. July 29, 1963. White House Famous Names, Service Set Box 6. The Lyndon Baines Johnson Library.

Memorandum for the Vice President. April 9, 1963. White House Famous Names, Service Set Box 6. The Lyndon Baines Johnson Library.

"Address at 18th U.N. General Assembly, 20 September 1963." John F. Kennedy Presidential Library and Museum online, accessed September 14, 2018. https://www.jfklibrary.org/Asset-Viewer/Archives/JFKPOF-046-041.aspx.

Woods, *LBJ: Architect of American Ambition*.

Selverstone, Marc J. "John F. Kennedy: The American Franchise." UVA Miller Center, accessed August 12, 2019. https://millercenter.org/president/kennedy/the-american-franchise.

Tenenhaus, Sam. "Kennedy's Death, a Turning Point for a Nation Already Torn," *New York Times*, November 21, 2013, accessed August 19, 2019. https://www.nytimes.com/2013/11/22/us/in-kennedys-death-a-turning-point-for-a-nation-already-torn.html.

Letter, Jerrie Cobb to Lyndon Johnson. February 10, 1964. Cobb, Jerrie #1. NASA Archives.

Letter, Lyndon Johnson to Jerrie Cobb. March 17, 1964. Cobb, Jerrie #1. NASA Archives.

Cobb, *Solo Pilot*.

Cochran and Brinley, *Jacqueline Cochran: The Autobiography of the Greatest Woman Pilot in Aviation History*.

Letter, Hugh Dryden to Clinton P. Anderson. Mar 2, 1964. Cobb. Jerrie #1. NASA Archives.

"Newsbureau, Lockheed-California Company. New World Speed Record Set By Jacqueline Cochran." May 18, 1964. Jacqueline Cochran: Papers, 1932–1975. Speed Records Series. Box 10. Eisenhower Presidential Library.

"Newsbureau, Lockheed-California Company." Undated. Jacqueline Cochran: Papers, 1932–1975. Speed Records Series. Box 10. Eisenhower Presidential Library.

"Newsbureau, Lockheed-California Company." May 2, 1963. Jacqueline Cochran: Papers, 1932–1975. Speed Records Series. Box 9. Eisenhower Presidential Library.

"Jacqueline Cochran (USA)." Fédération Aéronautique Internationale, accessed August 12, 2019. https://www.fai.org/record/jacqueline-cochran-usa-13041.

"1 June 1964." This Day in Aviation, accessed August 12, 2019. https://www.thisdayinaviation.com/1-june-1964/.

EPILOGUE

Koepcke, Juliane. *When I Fell From the Sky*. Accessed August 12, 2019. https://books.google.com/books?id=JV4zDAAAQBAJ&printsec=frontcover&dq=Koepcke,+Juliane+(2011).+When+I+Fell+From+the+Sky&hl=en&sa=X&ved=0ahUKEwiD65O9vNXjAhUEQKwKHRmgBFIQ6AEIKjAA#v=onepage&q&f=false.

Letter, Floyd Odlum to Jimmy Dolittle. June 16, 1972. Jacqueline Cochran: Papers, 1932–1975. General Files Series. Box 219. Eisenhower Presidential Library.

Letter, Jimmy Doolittle to Jacqueline Cochran. June 1, 1972. Jacqueline Cochran: Papers, 1932–1975. General Files Series. Box 219. Eisenhower Presidential Library.

"Sooner Aviatrix Honored by Nixon." *Daily Oklahoman*, Friday, September 21, 1973.

Haitch, Richard. "Follow-Up on the News; Foiled Astronaut." *New York Times*, June 26, 1983. Accessed, August 12, 2019. https://www.nytimes.com/1983/06/26/nyregion/follow-up-on-the-news-foiled-astronaut.html.

"Biographical sketch." (June 1972) Jacqueline Cochran: Papers, 1932–1975. General Files Series. Box 219. Eisenhower Presidential Library.

"Cobb, Geraldyn 'Jerrie' M." National Aviation Hall of Fame, accessed August 12, 2019. https://www.nationalaviation.org/our-enshrinees/cobb-geraldyn-jerrie-m/.

Jessen, *NBC Dateline* (Jim Cross Productions). Women/Mercury Astronauts. NASA Archives.

Dunn, Marcia. "NASA Pioneer Asks for Her Shot at Space." *Washington Post*, July 13, 1998. Accessed August 12, 2019. https://www.washingtonpost.com/archive/politics/1998/07/13/nasa-pioneer-asks-for-her-shot-at-space/fc9f402e-eee9-44fa-adc2-fb4e179256e2/?noredirect=on&utm_term=.6baef3af2463.

Cochran and Brinley, *Jacqueline Cochran: The Autobiography of the Greatest Woman Pilot in Aviation History*.

Ayers and Dees, *Superwoman: Jacqueline Cochran, Family Memoirs about the Famous Pilot, Patriot, Wife & Business Woman*.

Clarke, David. *Wall Street Gothic: The Unlikely Rise and Tragic Fall of Financier Floyd B. Odlum*. 2018.

Wright, Robert A. "Floyd Odlum and the Work Ethic." *New York Times*, January 28, 1973. Accessed August 19, 2019. https://www.nytimes.com/1973/01/28/archives/floyd-odlum-and-the-work-ethic-floyd-odlum-and-the-work-ethic.html.

McQuiston, John T. "Floyd B. Odlum, Financier, 84, Dies." *New York Times*, June 18, 1976. https://www.nytimes.com/1976/06/18/archives/floyd-b-odlum-financier-84-dies.html.

YOUR BOOK CLUB RESOURCE

READING GROUP GUIDE

Q&A WITH AMY SHIRA TEITEL

1. *Fighting for Space* is a truly riveting, epic read. Who knew how much LBJ's state of mind played a part in America's space program. And the complexity of the cast of characters—from the passionate pilots like Jerrie and Jackie to scientists like Randy—who played such important roles in space history.

How did you first become interested in their story? And what drew you to choose to share this story in particular?

This story is one that's been around in various incarnations for years, and on the surface it's pretty simple: a group of young women led by the intrepid Jerrie Cobb are denied their chance to break a glass ceiling. It's sexism, plain and simple. These retellings always feature a Disneyesque villain named Jackie Cochran who, much like Maleficent, swoops down to thwart the younger rival before retreating to her castle in the woods to confer with her pet raven.

The thing is, when you start really digging into the story, you find those retellings are wrong. It's not a feminist epic. If anything, it's a realistic feminist story with flaws and infighting on both sides. People have this idea that a feminist story has to be all women against all men, but this shows the reality is human opinion vs. human opinion.

I went through an evolution with this story. I started, like so many people reading about it for the first time, enthralled by Jerrie's fight. But when I started reading about Jackie, she morphed from villain to voice of reason in her approach to women in space in the early 1960s. My feelings very quickly changed. Jackie, as we know, was the only one with relevant experience when it came to women in space—she was both the leader of the WASPs and a female test pilot. It didn't make sense to me that she was always presented as this one-dimensional villain when she's about the most complex and wonderful figure I'd ever come across.

In a sea of wrong versions of this story, I finally realized that a dual biography approach highlighting Jackie and Jerrie would be both interesting and necessary. I think it's hugely important we take these stories as human stories with all the messy opinions and perspectives thrown in. So few things are ever cut-and-dry!

2. You're a spaceflight historian—what exactly does that entail? And when did you become interested in learning about space?

Space is a childhood fascination for me. When I was seven, I did a second-grade project on Venus and I thought it was the coolest thing ever. Venus is close to Earth in size, but it rotates backward and is as hot as an oven on the surface (which to my childhood mind meant it was Earth inside out)...but you can see it every night without a telescope! I mean, that's about the coolest thing possible.

I started collecting space books, and one of them, which I still have, had a two-page spread on the Moon featuring a drawing of two cartoon astronauts in front of a lunar module. I was floored. I couldn't believe that people had actually *walked* on the Moon! And that I had never heard of it! I started learning everything I could about the Apollo program, and the more I learned the more questions I had. That's the thing with space: the more you know the bigger the questions and answers become.

For me, my career has basically been satisfying my childhood curiosity and trying to impart that curiosity to others. Even people who wouldn't call themselves "space people" are a little fascinated by what's beyond our planet. Whether it's life on distant moons or seeing Pluto for the first time, there's something for everyone. I love giving people a moment of awe about something they didn't even know they were interested in.

3. You've actually worked with NASA yourself, right? In what capacity? Did it feel like a childhood dream come true?

It was! I was an embedded journalist with New Horizons mission to Pluto team in 2015, so basically I helped translate the amazing things the science team uncovered into plain English so the world could be as excited as we were. My main contribution was a nearly daily video series for the month of encounter called "Pluto in a Minute." It was exactly what it sounds like: one thing I thought everyone needed to know about Pluto presented in about a minute.

I effectively let my curiosity guide me. For example, we issued so many press releases saying it took nine years to get to Pluto, but I wanted to know what path we took to get there. I got to sit down with Yanping Guo, who actually designed the trajectory, and learn all about it, then distilled it down so nonscientists could understand how amazing it was that this little spacecraft actually got all the way to Pluto.

Hands down the best moment was being in the team meeting when we got the first close-up image of Pluto. It was about five-thirty in the morning and some people had been up all night...and this huge, clear shot of the planet came up on the screens. It was amazing for me. I could only imagine how it would feel for the scientists whose whole careers have been spent on this tiny, distant world! It was an incredible moment.

4. What's been your experience writing as a woman in space and science more generally? How was this book different?

Being a woman in science can be challenging, and being a young woman in the already small world of space history even more so. Science is male-dominated, as is history. Much in the same way the women in this book have to go the extra mile to prove themselves capable of doing the same things men do, women, not just myself, are always having to prove that we deserve to be on the panels next to men and have earned those keynote speaker spots. It's like there's this expectation that women can't do the same work as men, or that young women are somehow less capable than older men simply because they haven't been in a field as long. I've had men go out of their way to prove me wrong, only to end up putting their feet in their own mouths.

Then there's the aspect of "look." Every woman has a unique style, and how a woman feels like her best and most powerful self is extremely personal. Personally, I color my hair, wear makeup, and always give talks in heels so I come across taller than my five-foot-nothing frame. But just like when Barry Gray tells Jackie she doesn't look like a pilot, I have people telling me that I don't look like a historian. Or, better yet, that they would take me more seriously if I didn't dress up or wear makeup. Knowing that you're being discounted before you even talk about the work you've devoted years to is both heartbreaking and maddening.

It took me a while to stop dressing how I thought I should look, but I finally ditched the dress pants and button-down shirts for dresses and fun prints, and it makes a huge difference. I'm so much more confident giving a talk or appearing at an event when I feel like my outside matches my inside. It's a minor detail, but when you know you're the underdog going into an event, every detail matters.

In a lot of ways this book was cathartic because I got to dig into that duality of women needing to balance professionalism with style... okay, let's just admit that I'm hugely on team Jackie. I love that she had this cosmetics empire while also breaking records left and right. I totally love that not only did she refuse to compromise her femininity and love of style, she used it as a power move. The gall to make people wait while you touch up your lipstick after winning a cross-country air race? I love how she takes what would be a moment of superficiality and turns it on its head. She's such an inspiring example of not giving a damn what people say and making them pay attention to who you are and what you do, not what you look like.

5. **So much has changed since the 1962 hearing—being a female astronaut is no longer unheard of, for starters. What, if any, steps do you think we need to make to forge forward for gender equality in science?**

Representation matters, and we're still seeing a dearth of women in science of all backgrounds and ethnicities. It's not that they don't exist, it's that it's so hard for women to really gain a platform. Both TV and digital networks prefer male hosts. If the choice is between a male generalist and a female expert, they'll still pick the man. And the issue becomes generational. Little girls grow up without science role models and their interest drops off. There's plenty of data to show that while boys and girls love science in grade school, by middle school, girls' interest drops off largely due to social pressures of being "cool." I've heard the same thing anecdotally from friends of mine who are teachers or parents, and they've also told me about the drop-off of teenage girls liking science because they want to date. Somewhere along the way boys need to learn not to be intimidated by smart girls.

I wish I had a solution! A lot needs to change. The people who make science TV need to broaden their thinking on what a "scientist" looks like, but also women in science need to put themselves out there. Many are, and it's so wonderful to see. Be a woman in science and also have pink hair and tattoos, and love music, and have style! Show little girls that you don't have to compromise who you are to do what you love, and that badasses in science are still badasses!

6. **Throughout the book, Jackie and Jerrie never get on the same page, even though they both want the same thing, a woman in space program. If you could sit Jackie and Jerrie down to have a clarifying conversation, what would you wish they would say to each other?**

I wish I could have been at that dinner in Cocoa Beach in 1962! Honestly, I wish Jerrie had listened to Jackie. I mean, she did things her way in that she badgered and nagged decision-makers in an attempt to forcibly change their minds, and what did it get her? Nothing. Jackie saw things clearly. She knew that the medical tests were only a small part of the qualifications to become an astronaut, and also knew that forcing an issue rarely gets things done.

Let's play alternate history for a second. If Jerrie had listened to Jackie and they'd started a larger research program, would Jerrie have flown in space? Probably not, because she didn't have what NASA needed in an astronaut at the time. Not to mention she wasn't exactly the team player or well-rounded person NASA wanted to fly in space. All those screenings John Glenn went through were to select for personality and attitude. I don't know that Jerrie could have passed those tests; remember that even Jay Shurley had some misgivings about her psychological performance. There's also the issue of her poor performance in Pensacola that Floyd mentions, not to mention the circulatory issues the Lovelace tests uncovered. But she might then have been part of a pioneering program that, like the WASPs, stands as a model of women proving their mettle in a male-dominated world.

Jerrie never understood that piloting skill and specifically jet-flying experience were more important than medical fitness. Understanding that might have given her a chance to occupy a better spot in history. Maybe. As it stands, I don't really know what to make of her legacy.

7. ***Fighting for Space* is so thoroughly researched, it feels like we're *there* in that moment of history, experiencing it alongside Jackie, Jerrie, and everyone else. What was the research process like for this book? How long did it take you? What was the most difficult chapter to write?**

Hands down the most difficult part of the story to write was the WASP arc. It's so important to understand how this first big women's flying program developed and the issues it faced, but it's such a huge story! It's a book in itself—and in fact there are plenty of them out there. Distilling that down while retaining the drama and personality I wanted to make Jackie come to life without overcomplicating it to the reader involved a lot of flash cards to keep my acronyms, dates, people, and places straight.

But on the whole, the writing process was awesome, and admittedly that's my writer-self talking! I love this story, and getting to live in it for years was so much fun, not to mention seeing the pieces falling into place and realizing that I had something great happening.

Even the research was fun; again, that's my inner archivist-nerd talking. Jackie, thankfully, was an epic pack rat and her whole collection is at the Eisenhower Presidential Library in Abilene, Kansas. I loved going to that tiny town and digging through her boxes of letters. It's amazing to hold a note that LBJ wrote her, to leaf through all her pictures and see her story come to life. That's where I found all the letters from the women, too, the original handwritten pages. Holding those letters is holding history.

It was the same at the LBJ Presidential Library in Austin and the NASA Archives in DC. Every page is a piece of history, but there are pages missing. It became this game of tracking down a letter or a detail, calling archivists all over the country to find the missing puzzle piece. I wasn't always successful—some things have disappeared over the years or just aren't accessible—but some finds were total gems. I mean, I got Bessie Cochran's divorce filing record! How cool is that!

Tracking down details led to some strange things, too. I don't live too far from Indio so I went to see the former Cochran-Odlum Ranch as well as Jackie's very modest grave. I hunted down all kinds of Jackie ephemera on eBay and ended up with a few compacts, a Perk Up Stick, some luggage, and even the copy of *The Fun of It* that Amelia inscribed to Jackie.

All told, from first deciding this was the book I needed to write to getting it done, there were about three solid years of digging, writing, and polishing.

8. On a similar note, were there any details of their story that didn't make it into the book? Of these, which do you most wish you'd been able to share?

There were a lot of little details and vignettes that I couldn't fit in, either because they felt like they were derailing the narrative or didn't fit the time frame. Jackie wrote a lot more about her friendship with Amelia Earhart that I would have loved to include, and details about her cosmetics empire. There were also a lot of elements about her relationship with Floyd that are so endlessly heartwarming—Floyd himself is a fascinating person—but felt too shoehorned in when all was said and done.

The other thing I would have loved to include would have been more letters. The amount of correspondence I have is staggering—letters between the Lovelace women, internal NASA letters, and even just letters from citizens about women in space. I couldn't reference every letter—it wouldn't have added much to the story—but they make for some really interesting reading!

9. *Fighting for Space* isn't your first book; you also wrote *Breaking the Chains of Gravity*, the fascinating story of spaceflight before NASA. How did the writing process for this book compare?

It was a very different beast. *Breaking the Chains* was a story I wanted to tell to demystify NASA's origins. People are always saying "If we could put a man on the Moon in nine years why can't we XYZ?" The hope with that book was to show the rich backstory underscoring this government agency everyone recognizes but no one knows about. That said, it was a much less personal undertaking.

Fighting for Space was a much harder story to write. This one wasn't about streamlining to draw out a narrative thread; this book was about setting the record straight on a story that I don't think has ever been told right. To that end, this book took a lot more digging into archives and more figuring out how to take a letter or other detail I found.

Stylistically, this was also a very different book. *Breaking the Chains* is meant to inform. *Fighting for Space* is meant to engage. I wanted to write something that feels like a novel and really draws you in. I read a lot of Sophie Kinsella and Mary Kay Andrews (aka Kathy Hogan Trocheck) while writing this book to help me think through writing a fun page-turner, though, of course, I had to work within the confines of what really happened. Throughout the whole writing process I was trying to find moments that aligned to create parallels between Jackie and Jerrie, or those overlaps that allowed their narratives to cross in perfect ways. And I found some great ones! It was so perfect that both women experienced personal tragedy in 1956—Jerrie ending her relationship with Jack Ford and Jackie losing the congressional election. Reading fiction helped me identify those moments as devices that could shape the reader's experience.

10. What's your favorite space history story? Is there another story, like that of the Lovelace graduates, that you feel has been misrepresented?

Favorite story, if we're taking this one out of contention, would have to be Apollo 13. It's such a mind-boggling story—an oxygen tank ruptures midway to the Moon and yet NASA and its contractors pull together to save the crew. The ingenuity of going to the Moon is incredible in itself. Add the problem-solving that got the crew home safely and it's a fascinating story that is so much more human than technical. On a personal note, the book cowritten by the mission's commander, Jim Lovell, and Jeff Kluger was the book that made me want to write; it's so brilliantly exciting but also teaches you so much about how Apollo worked...and how insane it was that people went to the Moon! That emotional element probably explains why I love the story so much.

But I would argue most of space history is misrepresented. We have a tendency to blindly celebrate Apollo without taking into consideration the messy political landscape that created it or the flawed legacy it left behind. Apollo was amazing, but trying to recreate a similar program to return to the Moon or go to Mars is setting us up for failure. It's a prime example of why we need to study and understand history to learn from it, not repeat it. I firmly believe that the circumstances that enabled the Moon landing in 1969 will never happen again; it was a crash solution to a short-term goal and yielded purpose-built technology that couldn't do much else besides go to the Moon. We need to keep the inspiration of Apollo but change the way we think about space exploration and start laying a real foundation for a long-term presence in space.

11. If you could travel back in time, which part of history would you want to witness? And which part would you want to be a part of making?

This one is such a toss between 1930s aviation and the Moon landing, but I'm probably going to have to say the latter. The whole Apollo era is so endlessly fascinating to me, just the amount of innovation and technology that developed in such a short amount of time was so awe-inspiring. I would have loved to see it unfold! Of course, we now get into the issue that women couldn't fly in space at that exciting time... Being involved in any part of Apollo would be amazing to me, because there were women involved!

12. Who is your personal space hero?

Pete Conrad was the first astronaut I identified as a favorite and he always will be. He's even the namesake of my cat!

Pete actually gives us a great example to understand NASA's changing astronaut qualifications in this book. Pete made it to the final round of tests in 1959; he got to the Lovelace Clinic along with John Glenn for the medical tests. But he wasn't picked to be an astronaut because NASA didn't think he had the right personality to fly in space. He was, however, chosen as part of the second astronaut class along with Neil Armstrong and Elliot See and flew four awesome and long-duration missions—Gemini 5; Gemini 11; Apollo 12, wherein he became the third man to walk on the Moon; and Skylab 1. So this goofy guy that NASA didn't think had the right mental stability and personality for spaceflight ended up one of the most flown astronauts of the Apollo era! Personality mattered.

As a kid, though, I loved Pete because he was human. If you read the Apollo 12 transcript or watch the Apollo 12 episode of *From the Earth to the Moon* you can see he was fun and funny, and when I was little, he brought out the humanity of the Apollo program. Reading about Pete made me realize that going to the Moon was ultimately a human endeavor, and that storytelling through a human lens is a powerful approach to science and history.

13. What's next on the horizon?

For another big project, I'm not sure yet. I'm excited to venture into whatever new world my next book takes me to, but before I do anything else I'm going to take time off to read some books that have nothing at all to do with pilots!

YOUR
BOOK
CLUB
RESOURCE

VISIT
GCPClubCar.com

to sign up for the **GCP Club Car** newsletter, featuring exclusive promotions, info on other **Club Car** titles, and more.

@grandcentralpub

@grandcentralpub

@grandcentralpub